# Contents

# Preface

*The Australian Biology Dictionary* has been primarily written for use by students studying Biology at the Higher School Certificate level. It incorporates a number of improvements over dictionaries presently being used for this purpose. This can best be seen when comparing entries for the same word.

The style of the following entry is common to many biology dictionaries:

**abdomen (1)** Vertebrate body region containing viscera (e.g. intestine, liver, kidneys) other than heart and lungs; bounded anteriorly in mammals but not other classes by a diaphragm. **(2)** Posterior arthropod trunk, exhibiting TAGMOSIS in insects, but not in crustaceans.

This is how the same entry appears in this book:

**abdomen 1** In *vertebrates, the internal part of the body that contains organs other than heart and lungs. See Fig. D2. **2** In *arthropods, the rearmost part of the body containing the reproductive organs and part of the digestive system. Compare thorax. See Fig. I5.

The improvements found in this dictionary include:

1 A language level more appropriate to senior secondary school students;
2 Up-to-date entries reflecting the recent changes in Biology. This second edition has required more than 300 new or changed entries;
3 The inclusion of many non-biological words where they are important to the understanding of Biology. This will help those not studying Physics or Chemistry, as well as students for whom English is their second language;
4 Australian examples used to illustrate the entries wherever possible;
5 Alternative words, spellings, etc. clearly given;
6 Scientific names in italics;
7 Many medical words of particular interest to teenagers;
8 A comprehensive classification of living things.

We are sure *The Australian Biology Dictionary* will be of great assistance to students of Biology in Australian schools.

David Heffernan and Ruth Miller

# Using this dictionary

Like all dictionaries, a number of symbols and conventions have been used in *The Australian Biology Dictionary*.

1 Most words have only a relatively short entry. However, where the entry is much longer, the first sentence or two will give you the basic meaning of the word. Second and subsequent paragraphs can be read if you require further information.
2 The symbol * indicates that the word marked is very important in the explanation, and its meaning can be found elsewhere in the dictionary.
3 The term See also... indicates where further explanation can be found.
4 The term Compare... indicates that there is a contrasting term which you can look up if it will help in your understanding.
5 Generally, a definition is under its more commonly used abbreviation, e.g. DNA rather than deoxyribonucleic acid.

Rather than repeat details of the classification of living things, and the geological time scale, they have been tabulated in Appendices 1 and 2 respectively.

Happy searching!

# Acknowledgements

The publishers are grateful for permission to reproduce the following photographs:

- Eva Boogaard, Lochman Transparencies, for the cover photo: detail of Vermilion Biscuit Star, *Pentagonaster dubeni,* shot under water with a Nikon 801, Marmion Marine Park, WA, magnification 1:1. Enlarged (X11) and digitally manipulated.
- Dr Stuart Craig, CSIRO Division of Plant Industry, Canberra, for the scanning electron micrograph on the inside front cover: *Arabidopsis* pollen (X146).
- Eric Hines, CSIRO Division of Entomology, Canberra, for the scanning electron micrograph on the inside back cover: the Australian dung beetle *Amphistomus cunninghamensis* (X53).


# A personal request

The preparation of the *The Australian Biology Dictionary* has been a very demanding task. Apart from the standard types of errors that creep into such a book, important areas can be accidentally omitted. Hence we would very much welcome constructive comments from users of this dictionary as to how it can be improved and made more useful for future students.

# Aa

***A. tumefaciens*** *See Agrobacterium tumefaciens.*

**abaxial** Of a leaf surface, facing away from the stem or axis (or turned towards the base). Compare adaxial.

**abdomen** 1 In *vertebrates, the internal part of the body that contains organs other than heart and lungs. See Fig. D2. 2 In *arthropods, the rear-most part of the body containing the reproductive organs and part of the digestive system. Compare thorax. See Fig. I5.

**abductor** A muscle that moves one part away from another. e.g. *Tricep. See also adductor. See Fig. B4.

**abiogenesis** See spontaneous generation.

**abiotic** Non-living. Compare biotic.

**aboral** On the opposite side of the body from the mouth.

**aborigine** Used to refer to the people living in a country from the earliest times. The *indigenous peoples. e.g. Australian Aborigines.

**abscess** A collection of *pus in a cavity, usually accompanied by swelling and *inflammation. Bacteria are usually the cause.

**abscisic acid** Called **ABA** for short. A plant *growth substance that is involved in *abscission (leaf fall, fruit drop), regulation of stomata, and in inhibiting the flowering of *long-day plants when under short day conditions. It acts against the effects of growth promoting substances. See Fig. A1.

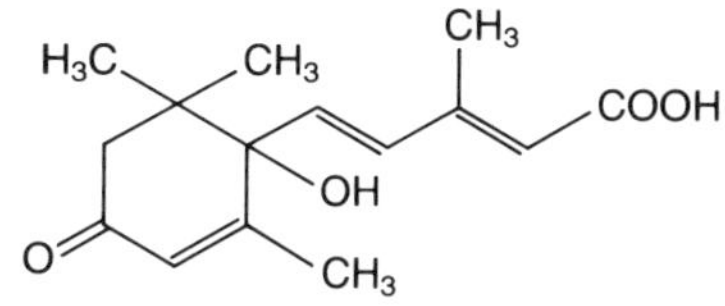


**Fig. A1 Abscisic acid**

**abscission layer** Used to refer to the special weak layer of cells at the base of a stalk that allows the dropping of leaves, flowers, fruits, or stems at the end of the growing season. See abscisic acid, deciduous.

**absorption** The movement of water and dissolved substances into a cell, organ or organism. e.g. Absorption of food into the bloodstream from the *alimentary canal. The up-take of water and dissolved minerals by plant roots, mainly in the *root hair zone. Compare adsorption.

**absorption spectrum** That part of the *electromagnetic spectrum that, when viewed through a *spectroscope, appears black because it has passed through a material and the radiation has been absorbed by the material. Each element or molecule absorbs certain parts of the spectrum, and their presence in a material can be determined by examining its absorption spectrum. See also light. See Fig. A3.

**abundance** The numbers of organisms per unit area which live in a particular part of their *distribution.

**abyssal** Relating to the lowest

parts of the ocean beyond the ★continental shelf, where at depths of 2000 metres or more, conditions are cold (near 0 °C) and dark. See Fig. O1.

***Acacia*** A large genus of dicots, most of which are trees (★wattles), and are especially common in Australia. The floral emblem of Australia is *Acacia pyenantha* (Golden wattle). See Appendix 1.

**acarpous** Not producing fruit.

**acclimation** The process of becoming used to or able to survive a new and different climate. Compare acclimatisation, which is sometimes used with this meaning.

**acclimatisation** The process of becoming used to a new and different climate because of the activities of humans. e.g. Animals in zoos. Compare acclimation.

**accommodation** Changing the focus of the eye. Many mammals, birds and reptiles do so by changing the curvature of the ★lens. Many fish and amphibians move the lens backwards and forwards compared to the retina. See eye, and Fig. E9.

**acellular** Not divided into cells; a mass of protoplasm with many nuclei e.g. ★Striated muscle fibres in animals have many nuclei, as does a stage in the life-cycle of certain ★slime moulds.

**acetic acid** A colourless liquid, formula $CH_3COOH$, found in vinegar. It is also the product of bacterial digestion of plants in the stomach of ★ruminants (e.g. cows), which is then absorbed into the bloodstream.

**acetyl CoA** Short for **acetyl coenzyme A**. A molecule that can give its (2 carbon) ★acetyl group to help synthesise ★fatty acids, or to the ★Krebs' (TCA) cycle where it is broken down to yield energy (in the form of ATP). See also co-enzyme A.

**acetyl group** The organic group $CH_3CO$–.

**acetylcholine** One of the chemicals (★neurotransmitters) that carry the nerve impulse across the gap in the ★synapse. See Fig. A2.

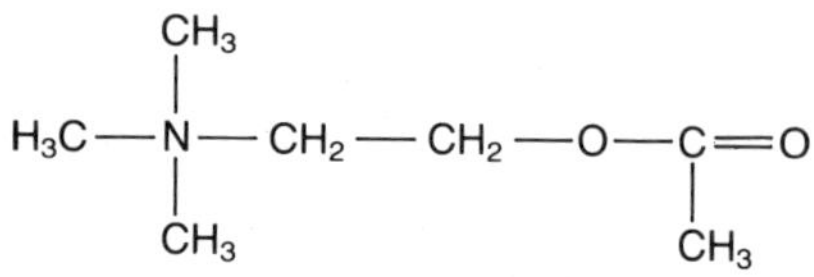

**Fig. A2 Acetylcholine**

**achene** A simple dry ★indehiscent (not opening) fruit with one seed that has developed from a single ★superior ★carpel. e.g. *Clematis*. See fruit (Fig. F4).

**Acheulian** A cultural period in the early stone age (Lower Palaeolithic), during which ★*Homo erectus* flourished with characteristic stone tools much more refined than those of the previous Oldowan culture. See culture.

**Achilles tendon** Also called the **hamstring**. The ★tendon attaching the calf muscle at the rear of the leg, to the heelbone.

**achromatic** Used to describe lenses of ★microscopes; the lenses are constructed to remove coloured rings from around the images that are produced by the microscope.

**acid** 1 A substance that releases hydrogen ions ($H^+$) when mixed with water. 2 A substance that accepts electrons or donates protons. 3 A substance with a ★pH less than 7. e.g. ★Acetic acid; ★amino acid. Compare alkali.

**acid rain** Rain made acidic by dissolving air polluting gases such as sulfur dioxide and nitrogen oxides.

Has severe effects on various plants (e.g. conifers), animals such as fish in lakes, as well as the outer surface of buildings. See also pollution.

**acne** An inflammation of the ⋆sebaceous (oil) glands in the skin that often results in the formation of ⋆pimples. See also skin.

**acorn-worm** A marine worm-like animal with a ⋆notochord. See Hemichordata and Appendix 1.

**acquired characteristic** A trait or character that is not inherited, but is gained during an organism's lifetime sometimes by use or disuse. e.g. Tail-less merino sheep; a person with large biceps. See Lamarckism.

**acquired immune deficiency syndrome** See AIDS.

**acquired immunity** The type of ⋆immunity an organism develops against certain microbes and other foreign substances during its lifetime. It is further divided into: **a Passive immunity** Immunity comes from ⋆anti-bodies obtained from outside that body. They could arrive across the ⋆placenta from the mother, or from a ⋆serum (e.g. diphtheria and tetanus) administered by injection. This type of immunity only lasts as long as the antibody is present. **b Active immunity** The body produces its own antibodies and ⋆lymphocytes against an invading ⋆antigen. The antigen could be an infection by bacteria, virus, etc., or a ⋆toxin due to the bite of a snake, spider, etc. The antigen can also be administered in a very weakened form as a ⋆vaccine. Either way, the immunity is much longer lasting due to the ⋆immunological memory.

**acrosome** The front part of the head of the ⋆sperm cell that contains the enzymes needed to break through the membrane of the ⋆ovum. See Fig. S6.

**ACTH** Short for **adrenocorticotrophic hormone**. Also called **corticotrophin**. A hormone produced by the front lobes (anterior) of the pituitary that helps control the release of hormones from the ⋆adrenal cortex, especially during times of stress. It also helps control the production of the pigment ⋆melanin. The release of ACTH is itself controlled by hormones released from the ⋆hypothalamus.

**actin** One of the two major proteins of muscle. See also myosin, myofibril and muscle fibre.

**actinomycin D** An ⋆antibiotic, first obtained from soil bacteria, that blocks DNA and RNA synthesis in certain bacteria.

**action potential** A temporary reversal of voltage across a membrane due to movement of potassium and sodium ions. In nerve cells it is caused by the passing of a nerve ⋆impulse. In muscle cells, it results in contraction. See Fig. I3. Compare refractory period, all-or-nothing rule.

**action spectrum** The wavelengths of light that are best at stimulating or inhibiting photochemical reactions. e.g. ⋆Photosynthesis (Fig. A3).

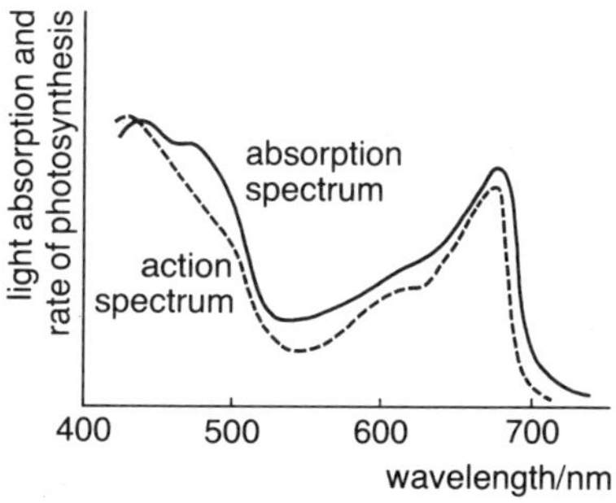


Comparing the light absorption spectrum and action spectrum of photosynthesis for a green plant

**Fig. A3 Action and absorption spectrum**

**activated sludge** A substance made up largely of bacteria and protozoans that has been used in sewage treatment to remove organic material.

**activation energy** The minimum amount of energy that must be supplied from an outside source before a chemical reaction will proceed. ⋆Enzymes reduce the activation energy needed.

**active immunity** See acquired immunity.

**active site** That part of an enzyme molecule into which the reacting molecule (⋆substrate) fits during the reaction catalysed by the enzyme. See Fig. A4.

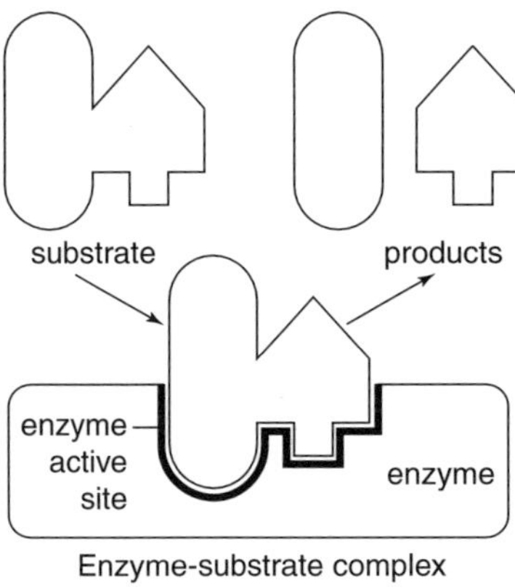


**Fig. A4 Active site**

**active transport** The process in which energy (in the form of ⋆ATP) is used to move a substance across a membrane from a region of low to one of high concentration. e.g. Move dissolved minerals from the soil into the ⋆roots (see active uptake below); or absorbing the products of digestion into cells lining the ⋆small intestine. Compare passive transport.

**active uptake** Also called **active absorption**. The process by which energy is used to absorb minerals into plant roots from the soil. See active transport. Compare passive transport.

**acuminate** A term used to describe the shape of a leaf where it gradually narrows to a point. See Figure L1.

**Adam's apple** A lump on the throat due to cartilage associated with the ⋆epiglottis inside. It is larger and more visible in the male. See Fig. D2.

**adaptation** 1 Any evolutionary change in the characteristics of an organism that makes it better able to survive in its environment. These can be either ⋆physiological, ⋆structural or ⋆behavioural. 2 The process by which a sense organ ceases to respond to a constant ⋆stimulus. e.g. Strong odours become less noticeable with time; the eye becomes more sensitive to light when in the dark for some time.

**adaptive radiation** The development or evolution from one ancestral species of a number of new species each adapted to new ⋆habitats and ways of life. e.g. Various types of kangaroo that now live on the open plains, in swamps, on rocky outcrops, and in trees, are believed to have evolved from one ancestral species.

**adaxial** Of a leaf surface, facing towards the stem or axis (or turned away from the base). Compare abaxial.

**addiction** A state of psychological and/or physical ⋆dependence on a drug.

**adductor** A muscle that moves one part towards another. e.g. ⋆Bicep. Compare abductor. See Fig. B4.

**adenine** A nitrogen-containing organic base that is one of the 'letters' of the genetic code of DNA and RNA. Chemically it is a ⋆purine. (See Fig. B3.)

**adenoids** Also called **pharyngeal tonsils**. An enlarged mass of

★lymphatic tissue in the upper part of the back of the nasal cavity. See tonsils.

**adenosine** A ★nucleoside where ribose sugar and adenine are chemically linked together. (See Fig. B3.)

**adenosine diphosphate** See ADP and ATP.

**adenosine monophosphate** See AMP and ATP.

**adenosine triphosphate** See ATP.

**adenovirus** A ★virus containing a small double strand of DNA within a protein coat shaped to have 20 faces. Responsible for acute respiratory disease in humans.

**ADH** Short for **antidiuretic hormone**. Also called ★**vasopressin**. This hormone can reduce the volume of urine produced by the ★kidneys by increasing reabsorption in the tubules. (See Fig. K1.)

**adhesion** A sticking together of two unlike objects or substances. Compare cohesion.

**adipose** Relating to ★fat. e.g. ★Adipose tissue.

**adipose tissue** Tissues that contain ★fat.

**adjuvant** A substance that increases the production and lifetime of ★antibodies after a body is injected with an ★antigen. e.g. Antigens may be absorbed onto a ★gel such as aluminium hydroxide.

**adnate** When unlike organs are attached along the whole or most of their length. e.g. Stamens attached to flower petals.

**ADP** Short for **adenosine diphosphate**. The compound formed by the hydrolysis (which here removes a phosphate) of an ATP molecule. See ATP and Fig. A10.

**adrenal cortex** See adrenal gland.

**adrenal gland** An ★endocrine gland located on the top of each kidney. The cortex (outer surface) is the source of various ★steroid hormones whose release is controlled by ★ACTH (e.g. ★androgens, ★oestrogens, ★corticosteroids). The medulla (inner core) secretes ★adrenaline (epinephrine) and ★noradrenaline (norepinephrine).

**adrenal medulla** See adrenal glands.

**adrenaline** Also called **epinephrine**. A hormone produced in the ★adrenal medulla. Adrenaline increases the concentration of sugar in the blood, raises blood pressure and rate of heartbeat, and increases muscular power and resistance to fatigue. Thus adrenaline helps prepare the body for stress etc. ('fear, flight and fright'). Also acts as a chemical transmitter of impulses across the junction of the ★synapse. See also noradrenaline and Figs A5 and E8.

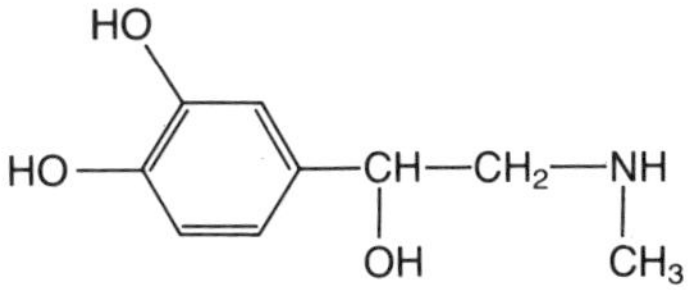


**Fig. A5 Adrenaline**

**adrenocorticotropic hormone** See ACTH.

**adsorption** The attracting of a layer of gases, liquids, or dissolved solids onto a surface. e.g. Charcoal adsorbs gases in gas-masks. Compare absorption.

**adventitious roots** Roots that grow from unexpected positions e.g. Prop-roots in ★mangroves and figs. Roots that grow from stem ★cuttings.

**aerial** In the air. e.g. Aerial roots.

**aerobic** Any biological process that needs the presence of oxygen, such as aerobic ⋆respiration. Compare anaerobic.

**aerosol** A ⋆colloid where microscopic solid or liquid particles tend to remain dispersed in a gas rather than to settle. e.g. Dust, smoke, fog and mist.

**aestivation** Spending the summer (very hot conditions) in a torpid or inactive condition. e.g. Some Queensland lungfish bury themselves in mud as the streams dry out. The Bogong moth migrates to cooler high country in summer and aestivates in caves. Compare hibernation, dormancy.

**aetiology** The study of causes, often used in reference to disease.

**afferent** Travelling or bringing inwards towards a central point. Often applied to nerves and blood vessels. e.g. Afferent nerves are the sensory nerves carrying impulses to the central nervous system. See also efferent and Fig. R2.

**aflatoxin** A toxin produced by fungi in the genus *Aspergillus*, which sometimes grow on stored nuts and cereals in humid climates, and which can cause liver disease, including cancers.

**afterbirth** The ⋆**placenta**. Called the afterbirth as it is expelled soon after a baby is born. See Fig. E4.

**agar** A jelly-like material (⋆gelatinous) obtained from the cell walls of certain red seaweeds. It is used for growing micro-organisms in laboratory ⋆cultures.

**age** **1** The time from birth of an organism to the present. **2** A particular period of life or development. e.g. Stone age; age of dinosaurs.

**agglutination** The clumping of cells due to the reaction of molecules on their surfaces with ⋆antibodies in the surrounding medium. Used to help identify blood groups, bacteria, etc. See also blood groups.

**aggression** In animal behaviour, actions such as threat or attack, usually associated with mating and territories. See also agonistic.

**Agnatha** Also called the **jawless fish**. A class of fish-like vertebrates without fins, and having a skeleton of ⋆cartilage. e.g. Hagfish and lampreys. See Appendix 1.

**agonistic** In animal behaviour, actions associated with ⋆aggression, including postures, appeasements, and flight (departing). Many ⋆displays are agonistic, keeping actual fighting to a minimum.

***Agrobacterium tumefaciens*** A bacterium that contains a ⋆plasmid often used as a ⋆vector to insert genes into the cells of plants. See genetic engineering.

**AIDS** Short for **acquired immune deficiency syndrome**. A ⋆sexually transmissible disease (STD) that is due to an initial infection by a ⋆virus (⋆HIV), which results in the lowering of the victim's immunity, and their death by a second unrelated disease (often a skin cancer called Kaposi's sarcoma, or a rare type of pneumonia caused by a protozoan). The virus appears to reduce the number of ⋆T-cells of the ⋆immune system by invading them, destroying some of them and rendering others ineffective by altering the relative numbers of killer T-cells, helper T-Cells and suppressor T-cells.

**air bladder** Air-filled sacs used for buoyancy. e.g. ⋆Swim bladders in fish. Air bladders allow some seaweeds to float. See also lungfish.

**alanine (ala)** One of the 20 common ⋆amino acids found in proteins. It is also used in the making of ⋆pantothenic acid (a B group vitamin). See Fig. A6.

**albedo** The fraction of light reflected from a surface compared to that falling onto the surface.

**albinism** The state of being an albino, an organism with pale skin, light hair and pink eyes due to a hereditary lack of the pigment ⋆melanin.

**albumen** **1** The white of an egg containing several types of protein. **2** The ⋆endosperm in plant seeds.

**albumin** A group of water-soluble proteins found in blood ⋆plasma, which are important in maintaining correct concentrations of blood relative to cells.

**alcohol** **1** A class of chemical compounds with formula R–OH, where R– is an alkyl group such as $CH_3$–, $CH_3CH_2$– (or $C_2H_5$–), $CH_3CH_2CH_2$–, etc. **2** The colourless intoxicating liquid called ⋆ethanol or ethyl alcohol ($C_2H_5OH$). See also fermentation.

**algae** Singular is **alga**. Photosynthetic plants belonging to any of several Phyla (Divisions) of the plant Kingdom, and some photosynthetic single celled forms often regarded as ⋆protists. The plant body does not have true stems or leaves. Algae are mainly aquatic (marine and freshwater), and include the ⋆seaweeds and the ⋆phytoplankton. See also Appendix 1.

**alimentary canal** An animal's ⋆digestive passage, from mouth to ⋆anus. Commonly called the gut. See digestive system (Fig. D2).

**alkali** **1** A substance that releases hydroxide ions ($OH^-$) when mixed with water. **2** A proton acceptor or electron donor. **3** A substance with ⋆pH greater than 7. e.g. ⋆Bicarbonate ion. Compare acid.

**alkaloid** Any of a group of nitrogen-containing organic compounds obtained from plants and having useful medicinal and similar properties. Some are ⋆analgesics (pain relievers) or ⋆anaesthetics. e.g. ⋆Morphine, ⋆cocaine, quinine, and ⋆caffeine. Others are poisonous e.g. Strychnine, ⋆nicotine.

**all-or-none rule** A situation where a stimulus must reach a ⋆threshold before there is any response. e.g. Nerve ⋆impulse, contraction of muscle fibres.

**allantois** A membrane found associated with the embryos of reptiles, birds, and mammals. **1** In reptiles and birds the allantois extends from the inside the hindgut out to near the shell. Inside this sac it stores wastes. The outer surface of the allantois is covered with blood vessels through which the embryo obtains oxygen. **2** In mammals, the allantois eventually forms part of the umbilical cord and placenta. See embryo (Fig. E4).

**allele** Alternative forms of a gene. They occur at the same position (⋆locus) on the paired (⋆homologous) chromosomes. e.g. Human blood type has alleles A, B and O. Although a gene can have many alleles, there is usually only two forms in each cell. See also dominance, recessive, and sex linked.

**allelomorph** An allele. See above.

**Allen's rule** Within a species, appendages such as ears, noses and limbs tend to be shorter in colder climates. Compare Bergmann's rule.

**allergy** An exaggerated response by the ⋆immune system. e.g. Hay fever, asthma and hives. See also histamine and antihistamine.

**alliance** A series of ecosystems where the ⋆associations are similar, with related dominant (canopy) species, and possibly the same or related understorey species.

**alligator** The broad snouted member of the order of reptiles ⋆Crocodilia, being found in south east America (*Alligator mississipiensis*). No teeth are visible along the side of the closed jaw.

**allopatric** Two or more populations of the same species that are prevented from breeding due to geographical separation. Compare sympatric.

**alluvium** Sediments such as sand, mud, etc., recently deposited from water, and that have not yet formed a soil.

**alternate** Concerning leaves, arranged singly on alternate sides along a stem. Compare opposite. (See Fig. L1.)

**alternation of generations** 1 A life cycle where a generation using ⋆sexual reproduction will alternate with a generation using ⋆asexual reproduction. e.g. Jellyfish, ferns. The sexual and asexually reproducing forms are often very different from each other. See also gametophyte, sporophyte, life cycles. 2 A life cycle where a ⋆haploid generation alternates with a ⋆diploid generation. e.g. In ⋆moss, the haploid plant is leafy and easy to see while the diploid plant is the stalk and capsule which grows out of the top. In ⋆ferns, the diploid plant is the most obvious whilst the haploid stage is the small heart-shaped gamete plant.

**altricial** Birds that are confined to their nesting place for a period after the eggs hatch. e.g. Tree-nesting birds like parrots, eagles. Compare precocial.

**altruism** Self-sacrifice for the benefit of others. See sociobiology.

**alveolus** Plural is **alveoli**. A tiny air sac, covered with blood ⋆capillaries, at the end of the passageway into the ⋆lungs. Oxygen passes from the air in the alveolus into the blood of the capillaries, while carbon dioxide passes from the blood to the air. Although tiny, the 300 million or so alveoli in human lungs increase the surface area to volume ratio so that diffusion is more efficient.

**Alzheimer's disease** A progressive degenerative disease of the brain now considered a leading cause of ⋆dementia among the old. After death, Alzheimer's patients show nerve cell loss, tangled fibres, particles of abnormal protein and reduced amounts of the brain's ⋆neurotransmitters such as ⋆acetylcholine.

**Amaryllidaceae** The amaryllis family of monocots including the genera *Narcissus* (Daffodils and Jonquils), and *Agave* (century plant). See Appendix 1.

**ambient** Completely surrounded. e.g. Ambient temperature is the temperature of the air surrounding us.

**ambulatory** Capable of walking or moving from place to place.

**amines** A group of organic compounds with general formula $R-NH_2$, where R is a wide variety of ⋆functional groups. e.g. Methylamine $CH_3NH_2$; glutamine which is an amino acid.

**amino acids** Organic molecules containing nitrogen in the form of $-NH_2$, and an acidic $-COOH$ group, both attached to the same carbon atom. The more than 20 different naturally occurring amino acids are the 'building blocks' (monomers) of ⋆protein molecules. ⋆Essential amino acids cannot be made by our bodies, and thus must be present in our diets. See also peptides, polypeptides. (See Fig. A6.)

alanine
(Ala)

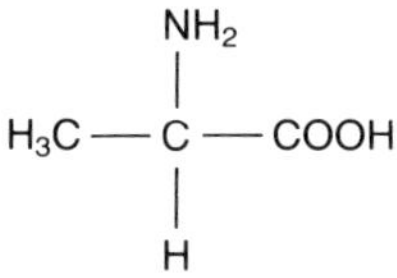

arginine
(Arg)

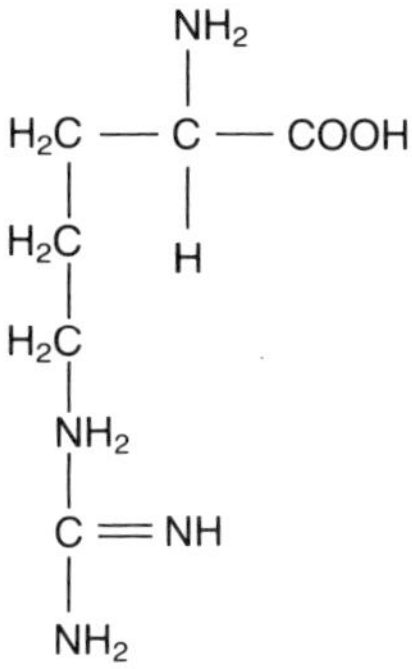

asparagine
(Asn)

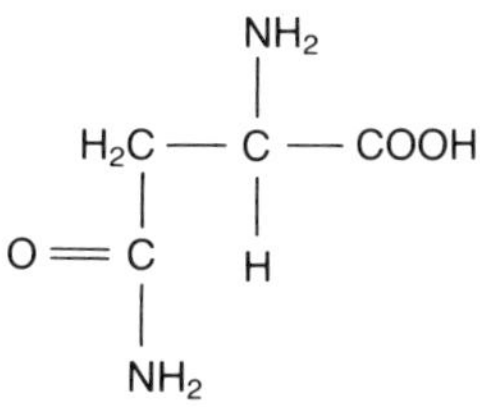

aspartic acid
(Asp)

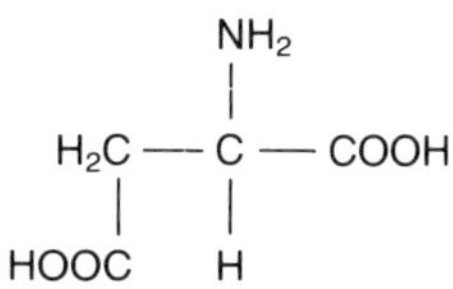

cysteine
(Cys)

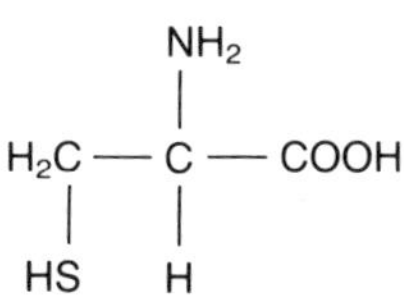

glutamic acid
(Glu)

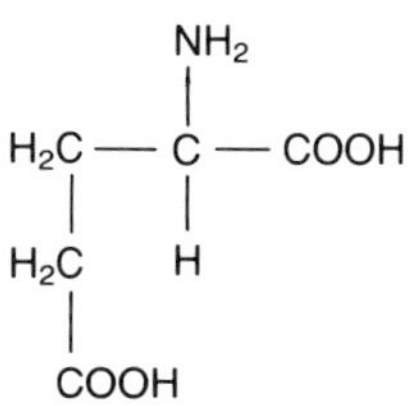

glutamine
(Gln)

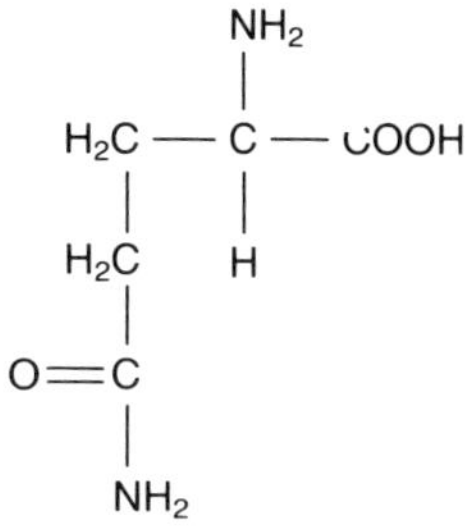

glycine
(Gly)

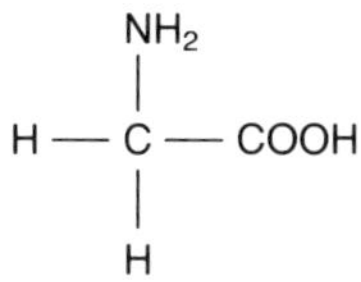

histidine
(His)

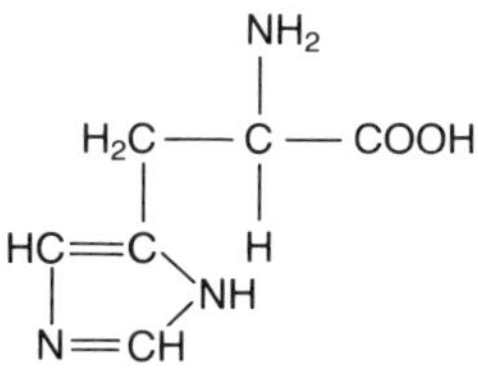

isoleucine
(Ile)

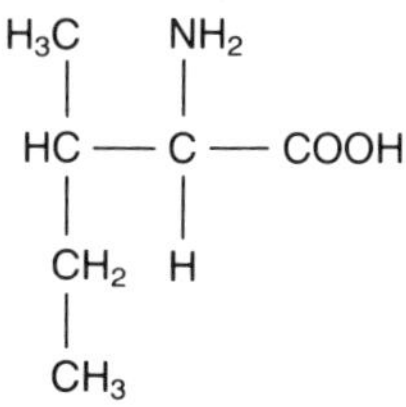

leucine
(Leu)

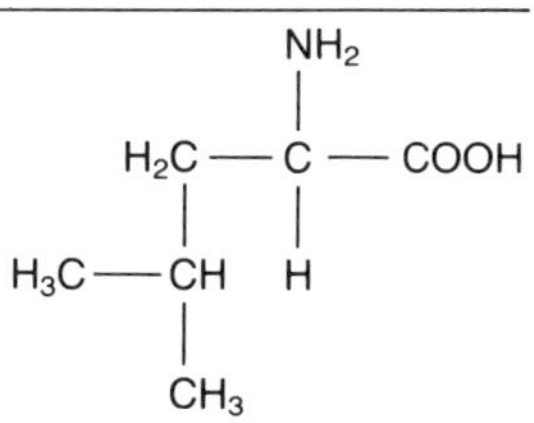

lysine
(Lys)

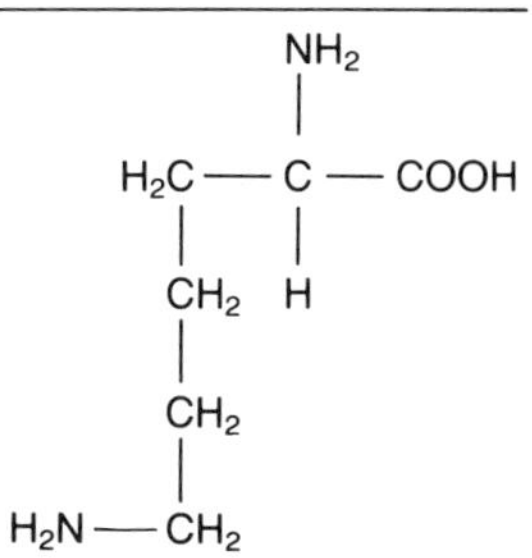

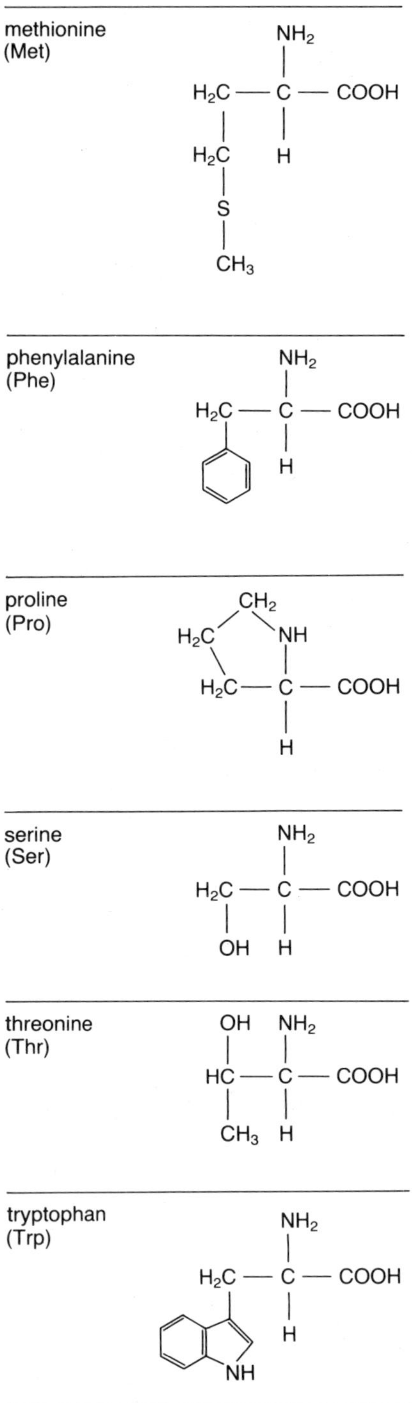


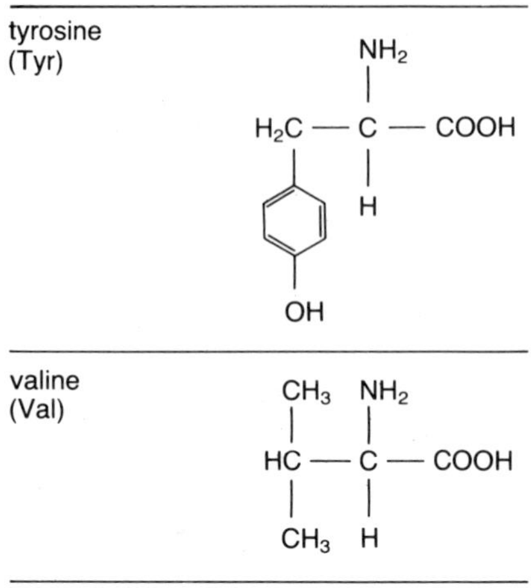


**Fig. A6 Amino acids**

**ammonia** The $NH_3$ molecule, forming a colourless strong smelling gas. Biologically it is produced by certain bacteria as they breakdown amino acids.

**amniocentesis** The removal of some ⋆amniotic fluid for medical examination.

**amnion** The fluid-filled sac that surrounds the ⋆embryo in reptiles, birds and mammals. See amniotic fluid below. (See Fig. E4.)

**amniotic fluid** The watery fluid in the ⋆amnion that helps cushion the embryo. (See Fig. E4.)

***Amoeba*** A genus of single celled protozoans that moves by means of ⋆pseudopodia. Most are free-living in the soil, mud and fresh water, while a few are ⋆parasitic. (See Fig. E6.)

**amoeboid** Moving like an amoeba, using ⋆pseudopodia. e.g. Phagocytes such as amoeba and white blood cells engulf foreign particles in an amoeboid manner. (See Fig. E6).

**AMP** Short for **adenosine monophosphate**. The compound formed by the ⋆hydrolysis (which here removes a phosphate) of an ADP molecule. See ATP for details. See also cyclic AMP.

**amphetamine** A stimulating drug that is similar in its action and molecular structure to ⋆adrenaline and ⋆noradrenaline.

**Amphibia** A class of vertebrates which live part or most of their time on land, but must return to water to reproduce, as fertilisation is external and the eggs have only a gelatinous covering. The eggs hatch to produce a larval (⋆tadpole) stage which usually undergoes ⋆metamorphosis to form the adult. Their skin is moist, and they have a three chambered heart. e.g. Frogs, toads, newts, salamanders. See Appendix 1.

**Amphineura** A class of marine molluscs which mainly live on the rocky shore, and whose shell is made up of 8 overlapping plates. e.g. ⋆Chitons. See also Appendix 1.

**amphipod** Any of the small 'jumping' crustaceans, order Amphipoda. e.g. Sand-hoppers, beach 'fleas', etc. Compare Isopoda. See Appendix 1.

**amplexus** A kind of embrace during mating when fertilisation is external. e.g. Frogs where the male climbs on the back of the female. Compare coitus.

**amplification** See biological magnification.

**ampulla** A small bladder-like structure. e.g. Attached to the leaves of some aquatic plants to help them float.

**amylase** Also called **diastase**. Any member of a group of enzymes that gradually break up (⋆hydrolyse) starch or glycogen molecules into separate glucose molecules. e.g. ⋆Ptyalin in human saliva starts the process in the mouth. It is completed in the small intestine by ⋆pancreatic amylase.

**anabolic steroid** Any ⋆steroid compound that helps increase tissue growth, especially of muscles. e.g. Male sex hormones called ⋆androgens. There are also synthetic examples which can cause liver damage.

**anabolism** The ⋆synthesis of complex molecules from simpler ones. Compare catabolism.

**anabranch** A branch of a stream that leaves the main channel. It may or may not rejoin again further downstream.

**anaemia** Too little ⋆haemoglobin, or too few red ⋆blood cells, resulting in insufficient oxygen reaching the tissues of the body.

**anaerobic** In the absence of oxygen. e.g. Anaerobic ⋆respiration, as occurs during ⋆fermentation. See also lactic acid. Compare aerobic.

**anaesthetic** A substance that produces a loss of sensation such as pain. e.g. Ether, chloroform.

**analgesic** Something that reduces or removes pain. e.g. Aspirin.

**analogous** Parts of different species that have similar function, but a different structure and embryonic development. e.g. Wings of birds and insects. Compare with homologous and convergent evolution.

**anaphase** A stage in both ⋆mitosis and ⋆meiosis where the ⋆chromatids of each chromosome separate and move to opposite ends of the cell (⋆poles).

**anatomy** **1** The structure of an animal or plant, or of its parts. **2** The study of the structure of plants or animals.

**androecium** Collective name for the male parts (⋆stamens) of a flower. (See Fig. F2.) Compare gynoecium.

**androgen** Any of a group of male sex hormones, produced in the ★testes and ★adrenal cortex, that help control the development of ★secondary sexual characteristics. e.g. ★Testosterone (Fig. T1), androsterone. See also anabolic steroid.

**anemone** 1 Marine cnidarian (coelenterate) animals of class Anthozoa, with cylindrical bag-like bodies topped with a ring of tentacles. Common on rock platforms and coral reefs. See Fig. P8. 2 Any plant of the genus *Anemone*.

**anemophily** Plant ★pollination brought about by the wind.

**aneurism** A 'ballooning' of the walls of arteries, veins or heart ventricles due to weakening.

**Angiospermae** The **flowering plants**. A major Phylum (Division) of vascular plants which have their reproductive parts in ★flowers, and whose ★seeds are contained in a ★fruit (matured ovary). See also monocotyledons, dicotyledons and Appendix 1.

***Angophora*** Also called '**apples**'. Any species of tree genus *Angophora*, with features very similar to ★eucalypts, but whose mature leaves are usually ★opposite. e.g. *A. costata* or Smooth-barked 'apple' found near Sydney.

**angstrom** Symbol Å. A unit of length used for very small measurements. One angstrom is $10^{-10}$ metres.

**animal** Any living organism that can be told apart from plants in the following ways: Cannot make its own food (★heterotrophic), and thus must feed on other animals or plants; more able to move around and respond to ★stimuli; a more compact and definite body shape; body cells lack a cell wall and do not contain chloroplasts. Compare plants, fungi, protists, Monera.

**animal rights** The fair and humane treatment of animals by society. See also antivivisection and bioethics.

**animal welfare** See animal rights.

**Animalia** The animal kingdom that includes the ★animals. See Appendix 1.

**anion** A negatively charged ion. e.g. Chloride $Cl^-$, hydroxide $OH^-$. Compare cation.

**Annelida** A phylum of soft, round, worms with ringed or segmented bodies, and relatively complex digestive, circulatory and nervous systems. e.g. Earthworms, leeches, beachworms. See Appendix 1.

**annual** A flowering plant that completes its life cycle (from seed germination to seed production) and dies within a single growing season. Compare biennial, ephemeral, perennial.

**annual ring** See growth rings.

**annular thickening** A ring-like pattern of secondary wall thickening in xylem vessels and tracheids. See xylem and Fig. X1.

**Anoplura** An order of insects which includes the blood sucking ★lice, parasitic on birds and mammals. See Appendix 1.

**anorexia** Lack of appetite. e.g. Anorexia nervosa, a mental disorder causing aversion to food which may lead to malnutrition.

**ant** Small, thin-waisted insects of order Hymenoptera, family Formicidae, usually having some form of social organisation with several ★castes. Wings are lost soon after dispersal to establish a new nest. See Appendix 1.

**anteater** In Australia, either the monotreme ★echidnas or the marsupial ★numbat. In other parts of the world placentals such as pangolins and aardvarks.

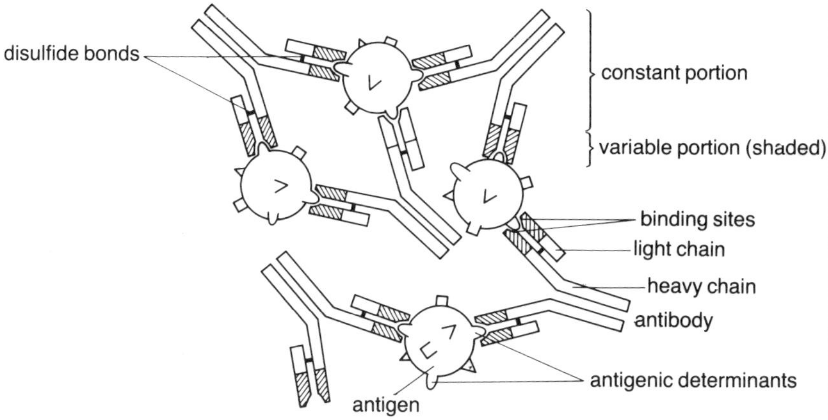


**Fig. A7 Antibody**

***Antechinus*** A genus of small carnivorous marsupial mammals made famous by *A. stuartii*, the males of which die from exhaustion after the mating season. See marsupial mouse and Appendix 1.

**antenatal** Before birth or during pregnancy.

**antenna** Plural is **antennae**. Long paired sensory appendages on the head of many arthropods (not spiders).

**anterior** The front end of an organism. With animals that walk upright (⋆bipedal) like humans, the surface furthest from the backbone. Compare posterior. Fig. S11.

**anther** The ⋆pollen containing part at the top of the ⋆stamens of flowers. Compare stigma. Fig. F2.

**antheridium** The sex organ that produces male gametes in lower plants such as algae, fungi, bryophytes (mosses, liverworts, etc.), and pteridophytes (ferns, etc.). Compare archegonium.

**anthesis** The period during which a flower is open and can be pollinated.

**Anthocerotopsida** A class of bryophytes similar in appearance to ⋆liverworts. e.g. Hornworts. See Appendix 1.

**Anthozoa** A class of cnidarians (coelenterates) that includes the ⋆sea-anemones and ⋆corals. See Appendix 1.

**Anthropoidea** The **anthropoid apes**. A suborder of primates that includes monkeys, ⋆apes and humans. Compare prosimian. See Appendix 1.

**anthropology** The study of the origins, variety and culture of humans.

**anthropomorphism** Attributing human characteristics, abilities or feelings to non-human organisms.

**antibiotic** A substance, originally obtained from moulds (fungi) and bacteria, that inhibits the growth of, or destroys, other microorganisms, especially bacteria. e.g. ⋆Penicillin, ⋆streptomycin, ⋆actinomycin D, etc.

**antibody** A special protein produced by vertebrate white blood cells (⋆leucocytes) that react with, and help make harmless, foreign substances within the body

(★antigens), thus helping to protect the organism. Antibodies are globular proteins (★immunoglobulins) produced by ★B-cells. Compare antigen. See ★immune system, clonal selection and Fig. A7.

**anticoagulant** A substance that stops blood clots from forming. e.g. ★Heparin is a naturally occurring anticoagulant. Leeches and other blood sucking organisms produce anticoagulants to help keep the blood flowing.

**anticodon** A group of three bases next to each other on a transfer RNA molecule that pairs with a ★complementary ★codon of three bases on a messenger RNA molecule. See translation.

**antidiuretic hormone** Called **ADH** for short, or ★**vasopressin**.

**antigen** A foreign substance that, when it invades an organism, stimulates the formation of specific ★antibodies against it. Such a substance, usually a protein or polysaccharide, may be separate (e.g. toxins etc. in the bloodstream), or on the surface of larger particles. e.g. Bacteria, virus. See also immune system.

**antigenic determinant** The part of an ★antigen that combines with the active portion of an ★antibody. Applies especially to that part of a bacterium or virus (e.g. surface molecule) with which the antibody reacts. See immune system.

**antigenic drift** A gradual change in the ★antigens on the surface of some viruses as a result of small genetic changes. e.g. The ★influenza virus continually changes so that we are not ★immune to new strains that arrive each season.

**antihistamine** Any medicine that is used to reduce the effects of ★histamines in the body, especially those associated with ★allergies such as hay fever and asthma.

**antisense** Used to describe a strand of RNA or DNA that is a ★complement of the strand that contains the genes to be ★transcribed. e.g. The antisense DNA is the strand not read during ★transcription of DNA into RNA. With RNA, antisense strands can be used to combine with the normal mRNA and thus stop it being ★translated into protein.

**antiseptic** A substance that destroys microorganisms when used on the surface of the body, or on objects with which the body could come in contact. Compare disinfectant.

**antitoxin** A mixture of antibodies produced in response to a ★toxin. See also serum and antivenene.

**antivenene** An ★antitoxin ★serum used to treat ★venom from snakes, spiders, etc.

**antivivisection** Against ★vivisection or the cutting into, or dissecting of, a living organism, especially for research purposes. See animal rights and bioethics.

**antlion** The ★larvae of lacewings, insects in the order Neuroptera. Antlions are well known for building conical shaped traps of sand in dry areas. Any insect, often ants, that stumbles into the trap cannot escape because of the moving sand on the sides, and falls into the waiting jaws of the larva. The antlion actually throws sand at its hapless victim to help confuse it, and to make the sides of the cone even more unstable. See Appendix 1.

**Anura** An order of amphibians that includes frogs and toads. See Appendix 1.

**anus** The opening at the end of the

★digestive system through which solid wastes (faeces) are removed. See Fig. D2.

**aorta** The large blood vessel (★artery) that carries blood away from the heart. It then splits into many other arteries which carry the blood to other parts of the body. See Fig. C8.

**ape** One of the large, tail-less ★anthropoid apes, family Pongidae. e.g. Chimpanzee, gorilla, orang-utan, gibbon. See Appendix 1.

**apex** The top or point of an object.

**aphid** An insect of the family Aphididae that sucks juices from plants.

**Apiaceae** The parsley family (carrots, parsley, celery) of dicotyledons, which also includes the ★flannel-flower. See Appendix 1.

**apical** Of the apex or tip of something.

**apical dominance** The influence of a terminal (endmost) bud of a plant in suppressing the growth of lateral (side) buds further down the stem or branch. Often due to the release of ★growth substances such as ★auxins.

**apical meristem** A growing point at the tip of the stem or root of vascular plants. It is the point at which cell division occurs, allowing stems and roots to grow in length.

**apiculate** A term used to describe a leaf having a small broad point at the apex. See Fig. L1.

**apocarpous** A flower having separate ★carpels, not united. Compare syncarpous.

**Apoda** An order of limbless worm-like amphibians that inhabit tropical areas. Not found in Australia. See Appendix 1.

**apoptosis** Cell death under the control of so-called 'suicide genes'. The genes produce enzymes that cut the genetic material into segments that ★phagocytes of the immune system remove. Compare necrosis.

**appendage** Any part of a body attached to the main axis or ★trunk. e.g. Arms, legs, antennae and wings.

**appendicitis** An inflammation of the ★appendix.

**appendix** A small hollow worm-like structure attached to the start of the large intestine. It contains some lymphatic tissue. See vestigial organ and Fig. D2.

**appetitive behaviour** Any pattern of animal behaviour where a specific need is satisfied, usually by an ★instinctive (innate) act. e.g. Searching for a mate leading to innate courtship behaviour and mating.

**apple** 1 The fleshy fruit (★pome), or the tree in genus *Malus* upon which it grew. See Fig. F4. 2 See *Angophora* above.

**apposition** The act of adding to, or fitting together. e.g. In plant cells, the adding of extra layers of cellulose to thicken the wall.

**appressed** Pressed together, but not fused with. e.g. Hairs pressed against a leaf surface.

**apterous** Without wings or wing-like structures.

**aquaculture** The cultivation of the natural produce of water, such as fish or shellfish.

**aquatic** Relating to water, fresh or salt.

**aqueous** Containing water. e.g. aqueous solutions are made from substances dissolved in water.

**aqueous humour** In the ★eye, the transparent fluid that lies between the cornea and the iris and lens. Compare vitreous humour. See Fig. E9.

**Arachnida** A class of arthropods with eight legs and two body parts (*cephalothorax, *abdomen). e.g. Spiders, scorpions, ticks, mites. Compare insects. See Appendix 1.

**arbor** A tree.

**arboreal** 1 Treelike or relating to trees. 2 Living in or attached to trees.

**arbovirus** A large group of RNA-containing *viruses that are transmitted to humans by the bites of mosquitoes and ticks. e.g. Encephalitis, various fevers.

**Archaeobacteria** Any of a class of primitive bacteria including methane-producing forms, some forms that only live in very salty habitats, and others that live in harsh hot acidic environments. See Appendix 1.

***Archaeopteryx*** The genus of a fossil classified as a bird due to its feathers, but with many reptile-like features e.g. teeth. It is said to be 'intermediate' (or 'transitional') between dinosaurs and modern birds.

**archegonium** The organs that produce the female gamete (the egg) in lower plants such as bryophytes (mosses, liverworts, etc.), pteridophytes (ferns, etc.), and some gymnosperms. It consists of a neck and a swollen base part called the venter. Compare antheridium.

**Arecaceae** The palm family of monocot flowering plants, including the cabbage and date palms. See Appendix 1.

**arginine (arg)** One of the 20 common *amino acids found in proteins. It is also used in the converting of nitrogen-containing wastes into *urea. (See Fig. A6).

**aril** A hairy or fleshy covering of a seed outside the *testa. e.g. Mace is the aril or covering of nutmeg seeds; the covering of certain waterlily seeds to help them float.

**arm** The forelimb of a vertebrate. e.g. The upper limb of humans, excluding the hand.

**arrowworms** A small phylum of marine worm-like invertebrates. They occur free-swimming or floating in the plankton. Their thin bodies have small fins, no segments, but have separate mouth and anus on opposite ends of the body. See also Chaetognatha and Appendix 1.

**artefact** A feature or structure of a tissue that has been produced by humans and is not naturally present. A major problem with the preparation of specimens for microscopy. e.g. Air bubbles in student-prepared microscope slides.

**arteriole** A small *artery.

**arteriosclerosis** Disease of the arteries where the walls become thickened and inelastic, thus reducing the flow of blood. See heart attack. Compare atherosclerosis.

**artery** One of many tube-shaped vessels that carries blood away from the heart to the tissues of the body. e.g. *Aorta. Arteries usually have thicker, more muscular and more elastic walls than do *veins. See circulation and Fig. C8.

**arthritis** *Inflammation of the *joints. Compare rheumatism.

**Arthropoda** A phylum of invertebrates that contains animals with jointed appendages (such as legs) and a stiff external skeleton (exoskeleton). e.g. insects, spiders, crabs, millipedes. See Appendix 1.

**articulation** The attachment of two bones at a moveable *joint. e.g. The thigh bone (femur) articulates with the pelvic girdle at the hip.

**artificial insemination** The placing of *semen, using a syringe, at the opening to the uterus (womb) to make *conception possible. e.g. Used widely in animal breeding, and in certain cases of human ⋆infertility.

**artificial selection** The control by humans as to which organisms will be allowed to reproduce. In this way desirable features of domesticated plants and animals can be retained during breeding programs. e.g. The breeding of cattle, racehorses or show dogs. Compare natural selection.

**Artiodactyla** An order of even-toed herbivorous placental mammals that includes deer, antelope, cattle, goats, camels, hippopotamuses and pigs. Their 2 or 4 toes have hooves on their ends. Most have complex stomachs, and often horns or antlers. See also ungulates and Appendix 1.

**ascending** Rising, going upwards. e.g. Plants, or their parts, that grow at an angle at first, but then grow upwards.

**Ascomycetes** A class of fungi (phylum Eumycota) which includes the yeasts, 'true' moulds, mildew, cup fungi, truffles, etc. See Appendix 1.

**ascorbic acid** Also called **vitamin** C, a water soluble vitamin present in citrus fruits (e.g. lemon, orange) and some other vegetables. As humans cannot make their own vitamin C, it must be included in our diets. The disease ⋆scurvy occurs if there is insufficient vitamin C. It acts as a ⋆cofactor in a number of reactions, including the synthesis of cartilage and bone. It is needed for the making of the protein ⋆collagen, especially during wound healing. See Fig. A8.

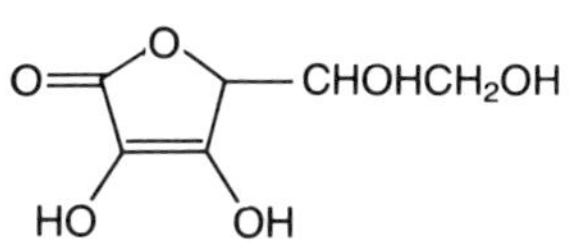


**Fig. A8 Ascorbic acid (vitamin C)**

**ascus** Plural is **asci**. A flask-shaped structure that produces spores in the ⋆Ascomycetes fungi.

**aseptic** Free from the micro-organisms that cause disease. Compare antiseptic.

**asexual** Not sexual. e.g. Asexual reproduction does not involve the combining of gametes (e.g. sperm and ova) as in ⋆fertilisation. See also grafting, spore. Compare sexual.

**ash 1** In Australia, used for tall, straight trees with smooth bark on most of the trunk and branches, rough bark being confined to the lower trunk. The spectacular stands of Mountain Ash (*Eucalyptus regnans*) and Alpine Ash (*E. delegatensis*) in the high country of Victoria and southern New South Wales are well known. **2** Northern hemisphere trees of the genus *Fraxinus*. **3** The remains of biological material after burning, especially carbon and minerals.

**asparagine (asn)** One of the 20 common ⋆amino acids found in proteins. Also used in making ⋆purines. See Fig. A6.

**aspartic acid (asp)** One of the 20 common ⋆amino acids found in proteins. It is also used in the making of ⋆pyrimidines, and the reactions that produce ⋆urea from nitrogen wastes. See Fig. A6.

**asperate** Rough with short stiff hairs or points.

**aspiration** The removal from the body of fluids such as pus or blood serum.

**aspirin** A drug (acetylsalicyclic acid) mostly used to relieve pain and reduce inflammation. Acts by suppressing ⋆prostaglandin production. See Fig. A9.

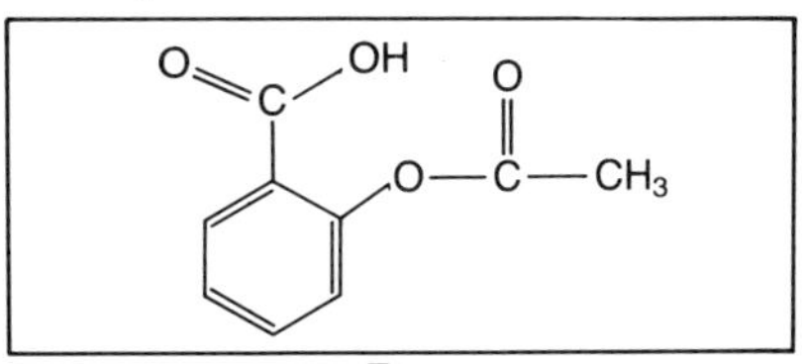


**Fig. A9 Aspirin (acetylsalicylic acid)**

**assay** To analyse a drug or other chemical for its strength or purity.

**assimilation** The process of converting food into the living material of the body. Thus the simple molecules absorbed into the blood after digestion are converted into the complex molecules of the body. Compare biosynthesis, anabolism, catabolism.

**association** A group of ecosystems that have the same dominant trees (after which the association is named), but different species in the lower strata. Compare alliance.

**association area** Those areas of the ⋆brain's ⋆cerebrum that lie between ⋆motor and ⋆sensory areas, and are concerned with memory, emotions, reasoning, will, judgement, personality traits and intelligence. See Fig. B7.

**Asteraceae** The daisy family of dicots, often called composites because what appears to be one flower is actually a head or composite of many small flowers. It also includes the dandelion and fireweed. See Appendix 1.

**Asteroidea** The class of echinoderms that includes the ⋆sea-stars (*wrongly* called star-fish). See also crown-of-thorns and Appendix 1.

**asthma** A usually ⋆allergic reaction that results in wheezing and difficulty in breathing. Attacks occur when smooth muscle in the smaller ⋆bronchi and ⋆bronchioles contract uncontrollably, partly closing off the passageways. The walls of such passageways also secrete excessive amounts of ⋆mucus that further clog the air passage ways. Usually asthma victims are ⋆allergic to substances in food or in the air, but there is sometimes a ⋆psychosomatic origin as well. Compare hay fever.

**astigmatism** A visual defect due to the unequal curvature of the ⋆eye's lens. Thus rays from a point do not focus at the same points in each eye.

**atavism** The reappearance in an organism of a trait or character typical of an ancestral form, usually due to genetic recombination. e.g. Sperm whales born with hind limbs; horses born with three toes.

**atherosclerosis** The thickening of artery walls due to the build-up of fatty deposits and growth of large numbers of smooth muscle cells (plaque), which limits and eventually blocks the flow of blood. Compare arteriosclerosis.

**athlete's foot** A ⋆contagious fungus disease of the feet, that grows well on moist skin.

**atlas** The ring-shaped first ⋆vertebral bone of the backbone of four-limbed vertebrates. It allows movement to occur between the skull and the backbone. See also spinal cord and Fig. S3.

**atom** The smallest particle of an element that cannot be split into smaller particles by chemical means. In the centre is a nucleus containing ⋆neutrons and ⋆protons, while orbiting around this nucleus are the ⋆electrons. Each chemical type of atom has its own number of electrons (called atomic number), there being equal numbers of electrons and protons in uncharged atoms. Chemical properties are due to the number of electrons on the outside of the atom. See also ion, molecule, isotope.

**atomic mass** The mass of an atom compared to carbon which is set at 12. e.g. Hydrogen is 1; iron is 56; uranium is 238. Compare molecular mass.

**ATP** Short for **adenosine triphosphate**. A common energy donating molecule in the functioning of cells.

Classed as a ★nucleotide, where ★adenine and ★ribose sugar are joined to three phosphates. During a ★hydrolysis reaction, ATP loses one phosphate and one hydrogen to become adenosine diphosphate (★ATP), thus releasing energy that can be used by various chemical reactions of the cell. When a second phosphate is removed, ADP becomes adenosine monophosphate (★AMP). During ★respiration, ADP is converted back to ATP, thus effectively storing energy. See Fig. A10.

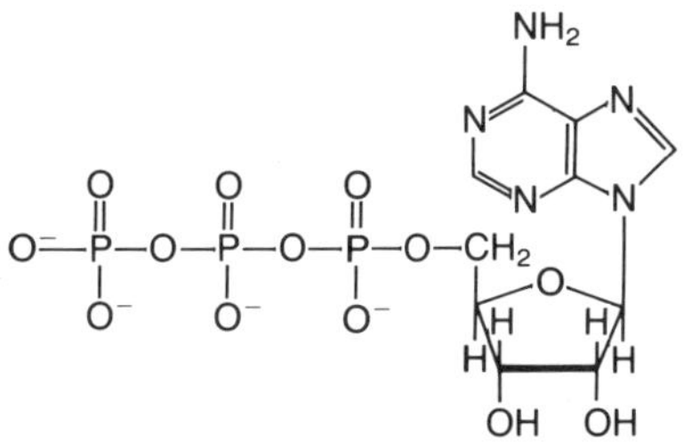

**Fig. A10 ATP (adenosine triphosphate)**

**atrio-ventricular node** See A-V node below.

**atrium** Plural is **atria**. It is also called the **auricle**. A chamber. e.g. The thin-walled chamber in the heart that receives blood from the body and passes it onto the thick-walled ★ventricles which pump it out to the body again. See also heart, veins and artery. See Fig. H2.

**atrophy** The degeneration or wasting away of organs or various parts of the body.

**attenuation** 1 The process of reducing the ability of an organism to cause a disease. e.g. During the production of ★vaccines. 2 The conversion, by yeasts, of carbohydrates to alcohol during the making of wine and beer. See fermentation.

**auditory** Concerned with hearing or the organs of hearing. See ear.

**aural** Perceived by the organs of hearing.

**auricle** See atrium.

**Australasian** The ★biogeographic realm that includes the Australian mainland, Tasmania, New Zealand, New Guinea, and the islands as far as ★Wallace's line. Characterised by the marsupials. See Fig. B5.

**Australoid** The race of peoples including the Australian Aborigines and certain peoples from Asia and various Pacific islands.

***Australopithecus*** A genus of chimpanzee-sized primates with skulls like those of apes, but with jaws having features similar to those of humans. There were two main species living 1.3 to 4 million years ago. *A. africanus* is often called 'gracile' and was around 1.35 metres tall, while *A. robustus* was up to 1.5 metres tall with more massive bones. ★*A. afarensis*, commonly called 'Lucy' was also gracile in form. See *Homo* and Oldowan culture.

**autogamy** In plants, ★self-fertilisation.

**autoimmune disease** A disease resulting from the body's ★immune system attacking the body's own tissues. e.g. Rheumatoid ★arthritis, allergies and asthma.

**autolysis** The breakdown of plant or animal tissues by enzymes that are contained within those tissues. See lysosomes.

**autonomic nervous system** That part of a vertebrate's nervous system over which the animal usually has no control, and which is important for controlling the heart, smooth muscle and internal organs in general. Divided into the ★sympathetic and ★parasympathetic systems. See also nervous system.

**autoradiograph** A photographic image showing the distribution of radioisotopes within a specimen.

**autosome** Any ★chromosome that is not a sex chromosome. Thus humans have 22 pairs of autosomes and 1 pair of sex chromosomes.

**autotomy** The casting off of damaged or trapped body parts. e.g. Tails of lizards, legs of spiders.

**autotrophs** Means 'self-feeding'. Organisms that are able to synthesise organic compounds from inorganic raw materials. e.g. Photosynthetic plants and bacteria which make complex chemicals from carbon dioxide, water and sunlight. Some bacteria can obtain energy from other chemicals and not use photosynthesis. See chemosynthesis. Compare heterotrophs.

**auxiliary** Giving support or helping.

**auxin** A group of plant ★growth substances that are produced in the growing tips of roots and shoots. It temporarily softens the plant cell walls, allowing the cells to enlarge. In association with other plant growth substances, it helps maintain ★apical dominance, and starts the formation of roots in cuttings. The only known natural auxin is ★IAA. Synthetic auxins include ★2,4–D and ★2,4,5–T that are used as weed killers. See also growth substance and Fig. A11.

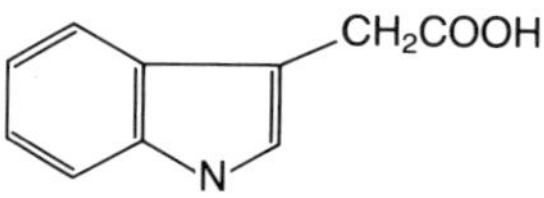


**Fig. A11 Auxin (indole acetic acid)**

**A–V node** A compact mass of nerve cells in the ★heart that detect the impulses from the ★S–A node, and then cause the ventricles to contract after the atria. See also pacemaker.

**Aves** The class of vertebrates that contains the ★birds. See Appendix 1.

**axil** The angle between the stem axis and a leaf, stem, or branch attached to it. See Fig. L1.

**axillary** Growing from the ★axil, between the leaf and the stem.

**axis** An imaginary line passing through a body around which the parts are symmetrically distributed. See Fig. S11.

**axolotl** The ★larva of a type of ★salamander, commonly called the Mexican Walking Fish. They are capable of breeding while still in the larval stage. See neoteny.

**axon** That part of a ★neurone that carries impulses away from the cell body. See also giant axon and Fig. N1. Compare dendrite.

AZT Short for ★**zidovudine**.

**azygous** Unpaired; existing singly.

# Bb

**B-cell** Also called **B-lymphocytes**. A type of ★lymphocyte (white blood cell) of the ★immune system, produced in the bone marrow and found in the blood. Unlike T-cells, B-cells do not have to go to the ★thymus gland to become active. When B-cells meet an ★antigen they change into cells that produce ★antibodies (★plasma cells). Some B-cells do not become plasma cells but circulate in the blood for years as B-memory cells. Compare T-cells. See also lymphocytes, immunity.

**B-lymphocyte** See B-cell.

**bacillus** Plural is **bacilli**. **1** A general term for any rod-shaped bacterium. **2** A genus of rod-shaped ★bacteria, occurring singly or in filaments or clusters, and able to produce ★spores that can survive hard times. e.g. *Bacillus cereus* lives in the soil. *B. anthracis* causes the disease anthrax in mammals including humans. Compare coccus, spirillum, vibrio, spirochaete. See also bacteria, antibiotic, and Fig. B1.

**backbone** The ★spine or spinal column made up of many ★vertebrae. It grows from the ★notochord in the ★embryo. See Fig. S3.

**backcross** **1** A cross between a hybrid and one of its parents. **2** In particular, a cross between an F1 hybrid (heterozygote) and an individual with the parental genotype (usually homozygous). See also test cross.

**backswimmer** A water-living bug of insect order Hemiptera, which swims on its back using long legs as paddles. Often seen in freshwater pools. See Appendix 1.

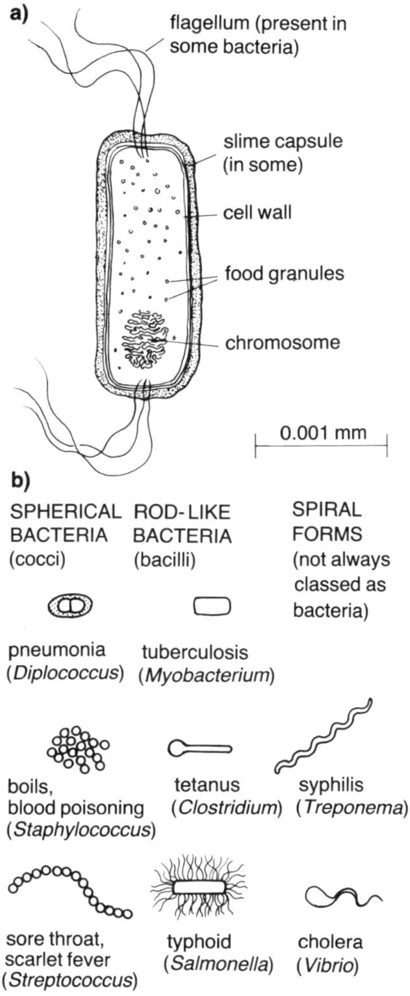


**Fig. B1 Bacteria (a) A bacterium (b) Types of disease-causing bacteria**

**bacteria** Singular is **bacterium**. Microscopic single-celled organisms that do not have a nuclear membrane around the 'nuclear region' which contains the DNA. See Fig. B1.

Each bacterium is surrounded by a ★glycoprotein cell wall, and some by a further ★capsule outside of that. They are classified by their reaction with ★Gram's stain (positive or negative), their type of respiration (★aerobic, ★anaerobic, ★chemo-synthetic), and their shape. A bacteria may be spherical (*coccus), rodlike (*bacillus), spiral (*spirillum), comma-shaped(*vibrio), or corkscrew-shaped (*spirochaete). Many have *flagella, and some can produce *spores to survive hard times. Bacteria normally reproduce by cell division (*mitosis), but a complex form of sexual reproduction is possible (*conjugation). Most bacteria are *saprophytes or *parasites. Those bacteria that are saprophytes are largely responsible for the decomposition and decay of dead plants and animals, thus returning valuable nutrients to the ecosystem (see nitrogen cycle, carbon cycle). Those bacteria that are parasitic usually harm their host by the production of *toxins. There are a few photosynthetic bacteria, while some obtain their energy from compounds such as hydrogen sulfide ($H_2S$), and are called chemosynthetic. There are very few places on Earth where there are no bacteria. For example, they live in the gut of most animals, within the vagina, and on the skin. A gram of soil may contain from a few thousand to a few hundred million. They are the source of some *antibiotics, are one of the most important organisms used in *genetic engineering, help clean up sewage, and help produce yoghurt. Compare virus. See also phage.

**bacteriophage** A *virus that infects bacteria, also called a phage. See Fig. V4.

**bag of waters** The fluid-filled sac which helps protect the baby (as a *foetus and *embryo) when it is in the womb. It bursts before or during childbirth. See amnion, amniotic fluid, and Fig. E4.

**balance** 1 Environmental balance is the state of natural equilibrium where organisms coexist without endangering the survival of any species. 2 Sense of balance. See vestibular apparatus.

**baleen** Also called **whalebone**. Bony plates which occur in the mouth of certain whales and are used to filter food (e.g. krill) from the oceans.

**bandicoot** Any of the small nocturnal, *omnivorous, rat-like marsupials of the family Peramelidae that live in Australia, New Guinea and Melanesia. Their forelimbs are shorter than their hindlimbs. They dig in the soil for insects and other arthropods, but will feed on fruit and soft tubers when available. See Appendix 1.

**banding** 1 The banded pattern on certain chromosomes. e.g. (a) Giant chromosomes from *Drosophila* salivary glands (see polytene). (b) Chromosomes during metaphase of cell division when treated with certain staining techniques. 2 The placing of coloured and/or numbered bands on the legs of birds to be used for identification, especially when tracing *migration routes.

**bangalay** The red-timbered tree *Eucalyptus botryoides* growing along the coast of New South Wales and eastern Victoria and widely used commercially.

***Banksia*** A genus of Australian dicots, family Proteaceae, with tough leathery leaves, and dense flower-heads. Sometimes called bottlebrush (see *Callistemon*). Banksia are a major part of most sand-dune *successions. See Fig. S10 and Appendix 1.

**baobab** Thick trunked, bottle-like tree of genus *Adansonia* found in north-western Australia.

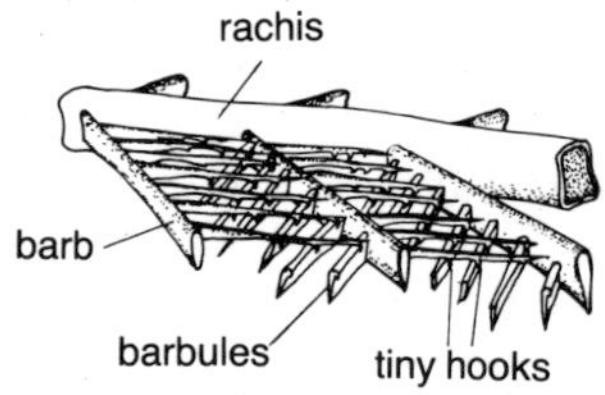


**Fig. B2 Barbs**

**barb** 1 In birds, one of the hairlike structures attached on either side of the shafts of feathers. (See Fig. B2) Compare barbules. 2 On plants, a stiff, hooked hair.

**barbiturate** Any of the many chemicals made from barbituric acid and used as ⋆sedatives to calm and soothe.

**barbule** Any of the minute filaments attached along either side of the ⋆barb of a ⋆feather. Tiny hooks on the barbules of flight feathers help hold the filaments in place. (See Fig. B2).

**bark** 1 A non-technical term describing the external covering of woody stems and branches. Often easily separated from the wood, it helps protect the phloem just below the surface of the plant. 2 All tissues outside the cambium in a woody stem. See also wood and Fig. W2.

**barnacle** Members of the marine crustacean order Cirrepedia, which settle onto solid objects and secrete shells around themselves after a period drifting with the currents (a free-living ⋆larval stage). They feed by poking feathery appendages out of a valve in the top of their shell. See Appendix 1.

**Barr body** See sex chromatin.

**barren** Unable to reproduce. e.g. Barren woman cannot have children. Barren ground cannot grow new plants.

**basal ganglia** Paired masses of grey matter containing the cell bodies of the ⋆neurones found within the brain. They connect with other brain centres and help with ⋆motor control, especially inhibiting unwanted behaviours. See also ganglion.

**basal metabolism** The ⋆metabolism of an organism that is using just enough energy to maintain life while awake. Expressed as kilojoules per square meter of body surface (or per unit weight) per hour.

**base** 1 That part of an organ or organism at or near its bottom, or point of attachment. e.g. base of a tree. 2 See purine and pyrimidine and Fig. B3. 3 See alkali.

**base pairing** When the strands of DNA pair with other DNA, or with RNA, they do so such that adenine

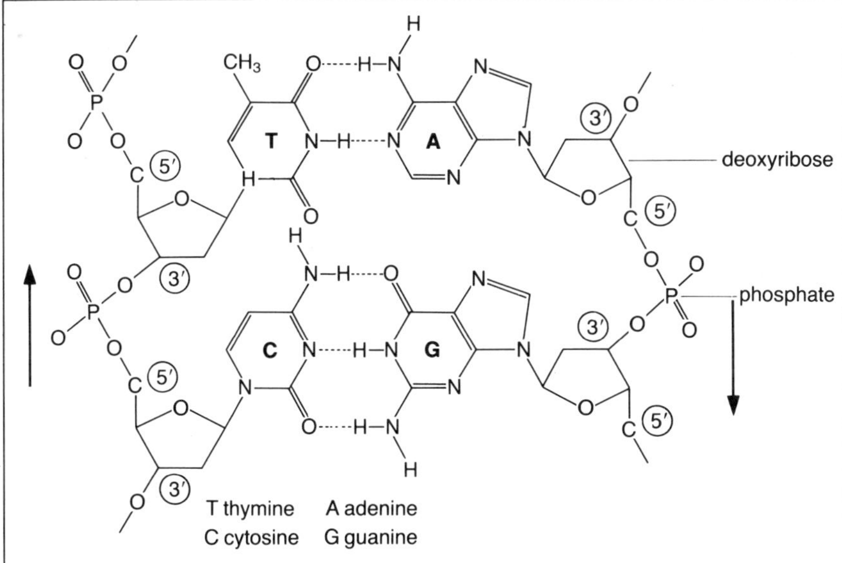


Segment of the DNA molecule showing the sugar-phosphate backbones running in opposite directions (arrows) and hydrogen bonding between complementary base pairs.

**Fig. B3 Base pairing**

must always pair with thymine (uracil in RNA), and guanine with cytosine. See DNA, RNA, translation, transcription and Fig. B3.

**base ratio** The ratio of the number of adenine and thymine bases (A + T) to the number of guanine and cytosine bases (G + C) in the DNA of a particular species. This ratio varies widely between different species. Compare base pairing.

**Basidiomycetes** A class of fungi (phylum Eumycota) in which the spores are produced on specialised reproductive structures called basidia, which are often shaped like a club. Hence they are also called the club fungi. Includes the rusts, smuts, mushrooms and toadstools, bracket fungi, puffballs, dry rot. See Appendix 1.

**bat** Any of the flying placental mammals of the order Chiroptera. Their front limbs (*pentadactyl) are highly modified and covered by a membrane-like skin for flight. See Appendix 1.

**beak** 1 The bill of a bird, made of horny material. 2 Similar structures in turtles, platypus etc.

**bee** A fly-like insect of the order Hymenoptera, superfamily Apoidea, including both the many Australian solitary bees, as well as the introduced social honeybee *Apis mellifera*. See also castes and Appendix 1.

**beetle** Insects of the order Coleoptera having its rear flying wings protected by hard horny coverings when not in use. See Appendix 1.

**behaviour** The actions or pattern of activity of an organism.

**behavioural adaptations** Any evolutionary change in the way an organism behaves that makes it better able to survive in its environment. e.g. burrowing in mud helps desert dwelling frogs avoid *desiccation.

**'bends'** Also called **caisson disease**. The sometimes fatal formation of nitrogen bubbles in the bloodstream when the pressure to which the body has been subjected is reduced too rapidly. e.g. Underwater divers surfacing too quickly.

**Benedict's test** A biochemical test for glucose sugar, the blue solution turning greenish with low concentrations of glucose, yellowish at higher concentrations, and brick-red at high concentrations. Benedict's solution is a mixture of copper sulfate solution and sodium citrate, which *reducing sugars (such as glucose) convert to copper (I) oxide.

**benign** Kind or gentle. e.g. Benign tumour is a *cancer that is not *malignant.

**Benson-Calvin cycle** See photosynthesis.

**benthos** All the organisms living on the bottom of the oceans and lakes, including the *littoral and sublittoral organisms, and those living on the floor of the *abyss. See Fig. O1.

**Bergmann's rule** Within a species, the average size of individuals tends to be larger in colder climates. Compare Allen's rule.

**beri-beri** A disease of the *peripheral nerves resulting from a lack of vitamin $B_1$ (thiamine). Symptoms include pain and paralysis of the extremities, and the body being abnormally thin, or severely swollen.

**berry** 1 In botany, a fleshy fruit with few to many seeds, and that does not open. e.g. currants, tomatoes, grape, gooseberry. Compare drupe. See fruits (Fig. F4). 2 Many so-called berries (e.g. strawberry, raspberry, blackberry) are *compound fruits. Strawberries have *achenes attached to an enlarged fleshy part. Raspberries and blackberries are composed of a number of *drupes.

**bettong** A shortnosed ★rat-kangaroo of the genus *Bettongia*. See Appendix 1.

**bicarbonate ion** The hydrogen carbonate ion ($HCO_3^-$). It has an important part to play in the way in which the blood carries carbon dioxide, and helps control ★blood acidity (pH). It is also added to the small intestine to neutralise acid from the stomach, thus allowing the digestive enzymes produced by the pancreas and intestine walls to work. See also buffer.

**bicep** Having two heads. e.g. Bicep brachii is the muscle on the front of the upper arm used to bend the forearm. Compare tricep. See Fig. B4.

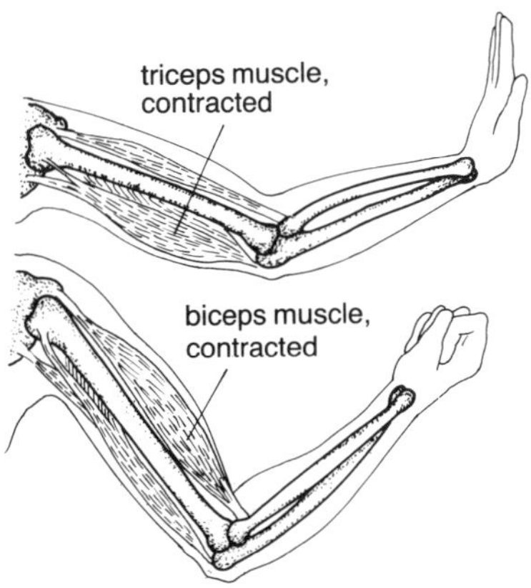


**Fig. B4 Bicep and tricep**

**bicuspid** 1 ★Teeth having two points. There are four in each human jaw between the incisors and molars. Also called pre-molars. 2 The mitral valve on the left side of the mammalian heart between the atrium and the ventricle. Made of two membrane-like flaps, it prevents fresh (oxygenated) blood from flowing back into the left atrium when the muscular ventricle contracts. See heart and Fig. H2.

**biennial** Occurring once in two years. e.g. A plant that requires two years to complete its life cycle. Commonly leaves, stems and roots grow the first year while it reproduces in the second year. See also perennial, annual, ephemeral.

**bilateral** Involving two sides or parts. e.g. **Bilateral symmetry**: An organism where its right and left halves are approximately mirror images of each other. See Fig. S11.

**bilby** One of two marsupial ★bandicoots, *Macrotis lagotis* or *M. leucura*, with rabbit-like ears. See Appendix 1.

**bile** Also called **gall**. A greenish-yellow alkaline solution produced by the liver and secreted into the small intestine through the ★bile duct. If not immediately needed, bile is stored in the ★gall bladder. Bile has two main parts: ★Bile salts and ★lecithin help in digestion by emulsifying fats so that they form smaller globules (with a larger surface area for digestive enzymes to act upon); ★bile pigments and excess ★cholesterol are ★excreted. See Fig. D2.

**bile duct** The duct that carries bile from the liver to the small intestine. See also bile and Fig. D2.

**bile pigment** Coloured compounds, found in the ★bile, that are the products from the breakdown of ★haemoglobin of the blood.

**bile salts** Organic salts produced by the liver, that help emulsify fat in the small intestine (turns large globules into small ones) thus increasing the surface area of the fat for digestive juices to act upon.

**bilharziasis** A disease due to a parasitic blood ★fluke of the flatworm genus *Schistosoma*. See schistosomiasis.

**billabong** A waterhole in a river or creek that dries up outside the wet season, especially if it is an ★anabranch or part of the stream now cut off from the main flow except at flood time.

**bilobed** Having or being divided into two lobes. e.g. Leaves with two rounded parts.

**binary fission** Division of a single celled organism into two cells. e.g. *Paramecium, Amoeba*. See mitosis.

**binaural** Concerning both ears.

**binocular vision** Involving the use of two eyes. Since the field of view is overlapping, it results in the ability to see in three dimensions.

**binomial nomenclature** The scientific naming of organisms with two terms, the first being the ⋆Genus (spelt with a capital first letter), and the second the ⋆species (spelt with a small first letter). Printed in either italics or underlined. e.g. *Homo sapiens* or Homo sapiens.

**bioassay** A quantitative (numerical) estimation of the strength of a biologically active substance, when measured under certain standard conditions, as it acts on living organisms.

**biochemical oxygen demand** See BOD.

**Biochemistry** The study of the chemical reactions that occur in living things.

**biocide** A substance that is able to kill living organisms.

**biodegradable** A material able to be broken down by the action of living things, especially bacteria.

**biodiversity** The numbers of different species of plants and animals living in an environment. Most often used in the context of the biodiversity crisis — the accelerating rate at which species are becoming ⋆extinct due to human activities.

**bioengineering** The manufacture and use of artificial body parts needed to replace those lost or which no longer function. e.g. Artificial limbs, heart pacemakers. Compare biotechnology.

**bioethics** The standards of conduct used during the study of biology, medicine and related areas. See also animal rights and antivivisection.

**biofeedback** Using information from the senses of sight and hearing to help control body functions normally unable to be consciously controlled. e.g. Pulse rate, blood pressure.

**biogenesis** The principle that all living organisms come from pre-existing living organisms. It is also expressed as 'organisms only produce other organisms much like themselves' (i.e. like produces like). Compare spontaneous generation, evolution.

**biogeographical realms** Regions of the world containing characteristic plant and animal life. The regions are ⋆Palaearctic, ⋆Nearctic, ⋆Ethiopian, ⋆Madagascan, ⋆Oriental,⋆Neotropical, ⋆Australasian. See Fig. B5.

**biological amplification** See biological magnification.

**biological clock** Mechanism(s) built into living things which control their normal growth, and patterns of activity (⋆biological rhythms), regardless of external time or short-term changes in the environment. For example, yearly ⋆hibernation, daily sleep, and many other less obvious changes in physiology. However, if organisms are kept away from their normal environment for extended periods of time, they gradually readjust to the new conditions. See also acclimation.

**biological control** The control of pests by biological rather than chemical means. e.g. (a) The use of natural predators to control pests, such

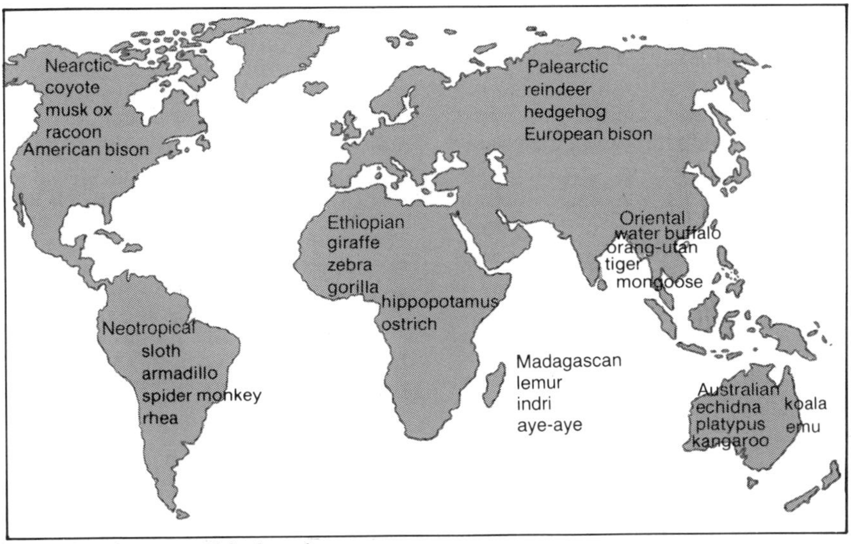


**Fig. B5 Biogeographical regions**

as the introduction of the cactus moth ⋆*Cactoblastis cactorum* to control the ⋆Prickly Pear cactus. (b) The release of males of the pest species that have been sterilised, so that many matings result in no offspring, thus reducing the size of the population.

**biological evolution** See evolution.

**biological magnification** The making of a biological quantity larger or greater. e.g. The concentration of insecticides etc. in the tissues of living things increases along a ⋆food chain. See also pyramid.

**biological oxygen demand** See BOD.

**biological rhythm** Innate patterns of growth and activity that appear to be under the control of a ⋆biological clock. e.g. ⋆Circadian rhythms, ⋆hibernation, ⋆migration.

**biological warfare** 1 Warfare involving the use of living organisms (e.g. disease-causing bacteria) or their toxic products as weapons against humans. 2 Warfare involving the use of ⋆herbicides against crops or vegetation that conceals enemy troops.

**biology** The scientific study of living things. Some of the divisions of biology include ⋆botany, ⋆zoology, ⋆microbiology, ⋆ecology, ⋆cytology, ⋆anatomy, ⋆physiology, ⋆genetics, ⋆taxonomy, ⋆biochemistry, etc.

**bioluminescence** The ability of some living things to give off light. e.g. *Firefly, glow-worms (an insect larvae), some fungi. Bioluminescence can be used for finding mates or attracting food. See also luminescence.

**biomass** The total mass of living matter in a given population of organisms. Usually expressed as the dry weight per unit area. See also pyramid and standing crop.

**biome** A major and distinctive community with its characteristic climate, plants and animals. e.g. Coral reef, tropical rain forest, desert. Compare ecosystem, habitat.

**biometry** The use of mathematics, especially statistics, to study living things.

**biopesticide** A ⋆pesticide derived from natural biological sources.

**biopsy** The removal and ★diagnostic study of a piece of tissue from a living organism.

**biosensor** A device that uses enzymes or other biological material to detect molecules in medicine or industrial samples. They can be used in medical diagnosis (e.g. detect glucose in the blood of people with diabetes) or in the monitoring the production of chemicals (e.g. making of medicines).

**biosphere** The parts of the Earth (air, land and water) where living organisms can be found. See also ecosphere.

**biosynthesis** The turning of elements and simple compounds into the usually more complex organic compounds of living organisms. e.g. Photosynthesis. Compare catabolism and synthesis. See assimilation and anabolism.

**biota** The living organisms (plants and animals) in an area.

**biotechnology** The use of biological processes by industry. Some uses have existed for centuries. e.g. Using yeasts in making bread, and in wine and beer making during ★fermentation; bacteria in making yoghurt. Other uses are much more recent. e.g. Fungi and bacteria used to make ★antibiotics, ★genetic engineering. Compare bioengineering.

**biotic** Having to do with living organisms; the living components of an environment. Compare abiotic.

**biotin** One of the B group ★vitamins. Biotin acts as a ★coenzyme in chemical reactions that result in carbon fixation—it helps carry the carbon dioxide needed for various steps of the ★Krebs' cycle. Usually the bacteria in the intestines can produce enough for most vertebrates, but foods such as cereals, vegetables, milk, egg yolk, and yeast are particularly rich in biotin.

**bipedal** Tendency to walk on two legs. e.g. Humans, birds.

**bipinnate** A leaf that is twice ★pinnately divided, so that the first side divisions themselves have side divisions. e.g. Many fern fronds. (See Fig. L1.)

**biradial symmetry** A creature with both ★radial and ★bilateral symmetry. See Fig. S11.

**birds** Any of the ★Aves, a class of homoiothermic (warmblooded) vertebrates at least partly covered with feathers. See Appendix 1.

**birth** The act of being born. e.g. A baby leaving the birth canal. See also parturition.

**birth canal** The ★vagina. See Fig. R4.

**birth rate** Usually expressed as the number of births per thousand population at a given time. Compare death rate.

**bisexual** See hermaphrodite.

**Biuret's test** A biochemical test for proteins in solution. Copper sulfate in alkaline solution turns from blue to purple if protein (and a few other substances) are present. Compare Millon's test.

**bivalent** Any paired ★homologous chromosomes during ★meiosis, especially prophase I.

**Bivalvia** Also called **Pelecypoda**. Any members of the marine and freshwater class of molluscs with two shells hinged together. The animal is laterally compressed inside, and filters food from the water as it passes through its gills. e.g. Clam, oyster, mussel. See Appendix 1.

**blackbutt** Any of a number of tall straight eucalyptus species with rough bark at the base of the trunk, but smooth barked branches above. Especially applied to *Eucalyptus* *pilularis*, common in coastal regions of New South Wales.

**blackheads** A small blacktipped, fatty mass in the pore of a ⋆sebaceous (oil) gland on the face. See also acne and pimples.

**bladder** Sac-like structures for storing fluids. **1** A sac found in the pelvic region where urine is stored after it is filtered from the blood by the kidneys—the urinary bladder (Fig. E8). **2** See swim bladder, gall bladder, pneumatocyst.

**blade** The flat part of a plant leaf or large frond of an algae, where photosynthesis occurs.

**blastocyst** The stage in the life of a mammal ⋆embryo when it is implanted into the walls of the womb (⋆uterus). Roughly equivalent to a ⋆blastula. See also gastrula.

**blastula** A stage in the life of an animal ⋆embryo, just before the changes occur that produce a ⋆gastrula. Usually a hollow ball of cells.

**Blattodea** The insect order that includes the ⋆cockroaches. See Appendix 1.

**bleeder** Someone suffering from ⋆haemophilia.

**blight** A destructive widespread plant disease causing sudden leaf damage. Many organisms can produce such effects, but especially famous is potato blight caused by ⋆fungi, and fireblight of pears caused by a bacterium.

**blind spot** The small area of the ⋆retina where the optic nerve leaves, and as there are no rods or cones, it is unable to detect light. See also eye and Fig. E9.

**blindsnake** A genus, *Typhlina*, of small, worm-like, burrowing snakes with eyes only small spots under the scales of the head. See also Squamata and Appendix 1.

**blood** The mixture of ⋆blood cells and liquid ⋆plasma that is pumped through the ⋆blood vessels by the ⋆heart. Blood carries nutrients to the body tissues, and wastes away again.

**blood cells** Also called **blood corpuscles**. The red cells (⋆erythrocytes) carry oxygen to the body tissues. The white cells (⋆leucocytes) help remove dead cells and unwanted microbes and other

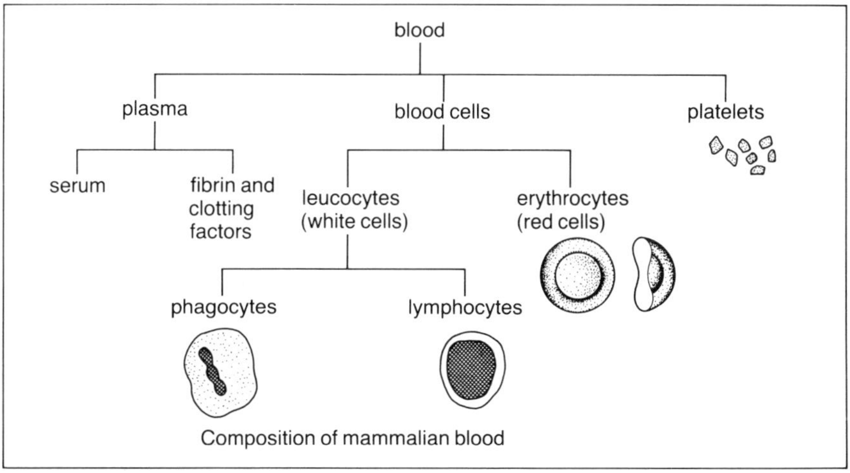


**Fig. B6 Blood**

invaders of the body. Sometimes the ⋆platelets are counted as cells, but they are really just cell fragments. See Fig. B6.

**blood clotting** The forming of a jellylike mass (a blood clot) to stop the escape of blood from an injury. If at the surface, this may dry to form a scab. Blood clotting is a complex process with at least 15 different enzymes or ⋆clotting 'factors' involved. 'Factor VIII' is often missing in those suffering from ⋆haemophilia. Some of the major stages are: Damage to tissues releases an enzyme from those tissues and from ⋆platelets that converts ⋆prothrombin into the enzyme ⋆thrombin, which in turn converts the water soluble protein ⋆fibrinogen into a tangle of ⋆fibrin threads. These threads trap a mass of platelets, blood cells and serum to form the clot.

**blood corpuscle** See blood cells above.

**blood count** The number of red or white blood cells in a particular volume of blood.

**blood groups** The classification of blood into groups according to its clotting reactions. The main groups are A, B, AB, and O. Each one of the groups listed can also be $Rh^+$ or $Rh^-$. There is also another 15 or more blood groups such as Hh, Lewis, Sese, etc. See also transfusion.

Blood groups arise because of the chemicals, called ⋆antigens, on the surface of the blood cells. Thus blood cells of group A have the A antigen on their surface, cells of group B have B antigen, while cells of blood group AB have both A and B antigens on their surface. Blood group O has neither of these antigens. If you mix blood cells of group B with a lot of blood from group A, then the group B cells form lumps (see agglutination). This is because ⋆antibodies in the serum of blood type A attack the 'invading' antigen B, making the cells clump together. The reverse would also occur. If some AB cells are mixed with a lot of blood of group A, then as B antigen is still present, antibodies will cause these blood cells to also coagulate. You can, however mix type O cells with type A blood, as there is no A or B antigen on the surface of O for antibodies to attack, and thus it does not coagulate. This is why group O is considered the universal donor blood. Group AB is called universal receiver, as it can be mixed with a little A, B, or O type blood. However, this mixing can only be done in emergencies as there are many other blood groups that could cause an immune response.

**blood plasma** See plasma.

**blood pressure** The pressure of the flowing blood against the walls of the arteries, measured in millimetres of mercury (mmHg). The pressure is greatest when the heart contracts (⋆systolic pressure) to pump blood into the arteries, and least when it relaxes and fills with blood (⋆diastolic pressure). Normal blood pressure is expressed as systolic/diastolic, and for adults is around 120/80 mm Hg. This value can vary considerably. See hypertension and carotid body.

**blood serum** See serum.

**blood sugar** Glucose dissolved in the blood. Its concentration is kept fairly constant by the hormones ⋆insulin and ⋆glucagon. See also diabetes.

**blood vessel** Any of the tubes (⋆arteries, ⋆veins or ⋆capillaries) through which the blood flows.

**bloodwood** A group of eucalypts having characteristic bark with short fibres that form fairly loose rectangular pieces. Many such trees produce a reddish gum. e.g. Red bloodwood *Eucalyptus gummifera,* common in coastal areas of New South Wales.

**bloom** 1 The flower of a plant. 2 The opening of a flower bud. 3 A rapid increase in population, especially of algae so that they discolour the water in which they are growing. e.g. Along the Murray–Darling river system. **4** The delicate powdery or waxy coating on the fruit or leaves of certain plants.

**blossom** **1** The flower of a plant. **2** To ⋆bloom or produce blossoms.

**blowfly** Any of the true flies, order Diptera, which deposit their eggs which hatch into larvae which feed on carcasses (meat), or in sores and wounds. Especially the blue blowfly *Lucila cuprina*. See Appendix 1.

**blubber** The insulating layer of fat under the skin of whales and similar aquatic mammals.

**blue-green algae (bacteria)** See Cyanobacteria.

**bluebottle** **1** A cnidarian (or coelenterate) of genus *Physalia* found in warm seas, and having a floating blue gas-filled bag from which long tentacles hang. These tentacles can inflict a painful sting. **2** Any of the large metallic blue and green flies from the order Diptera, family Calliphoridae.

**bluebush** A herbaceous blue-grey appearing shrub from genus *Kochia* that grows on the dry and often ⋆saline plains of central Australia. Compare saltbush.

**BOD** Short for **biochemical (or biological) oxygen demand**. The oxygen needed by microorganisms such as bacteria as they remove the organic matter present in a water sample. Measured in milligrams per litre (mg/L) or parts per million (ppm). Often used to measure levels of water pollution such as sewage, as a high BOD indicates high levels of organic matter.

**body** **1** All the physical parts of an organism. **2** The parts of an animal less the head and limbs. The ⋆trunk or ⋆torso.

**body cavity** The internal cavity of the body of an animal which contains the major organs. In vertebrates, the body cavity is divided by a transverse ⋆septum, called the ⋆diaphragm in mammals. This divides the body cavity into the ⋆thorax and ⋆abdomen. See also coelom and haemocoel.

**bog** Wet spongy earth mainly made up of decayed vegetable matter.

**boil** A painful, pus filled, inflamed sore caused by microbial infection.

**bole** The main trunk of a tree, from ground level to the first branches.

**bolus** A lump of chewed food mixed with saliva that enters the food pipe (⋆oesophagus) as one swallows. The saliva helps lubricate the movement of the food along the food pipe, as it is pushed by a muscular contraction called ⋆peristalsis.

**bond** Something that binds, fastens or holds together. e.g. ⋆Chemical bonds hold atoms together in groups. See covalent and ionic.

**bond energy** The energy needed to break a particular chemical bond. Usually expressed as joules per mole of bonds.

**bonding** The process which occurs immediately after birth which establishes the relationship between parent and offspring. e.g. mother and child; bird and nestling.

**bone** The hard tissue from which the ⋆skeleton of most vertebrates is made. Its fibres are the protein ⋆collagen (30% by weight), and the hard material is mainly salts of calcium and phosphate (70%) in which are embedded bone cells (⋆osteocytes).

During the development of an embryo, the final shape of the bone is first made in cartilage, which is later relaced by bone. There are two types of bone. **a** Compact bone forms the main cylinder section, and is made from layers (lamellae) of bone around tiny ★Haversian canals, in which are embedded the osteocytes, etc. **b** Spongy bone is chemically the same, but forms bony bars within the cylinder of compact bone, especially at the ends. Bone ★marrow may fill these spaces in spongy bone, or they may be hollow to conserve weight, as in birds. Compare cartilage. See also parathyroid and calcitonin hormones.

**bone marrow** See marrow.

**boomer** Any large male kangaroo, but especially the great grey or forester kangaroo, *Macropus giganteus*.

**boongary** Lumholtz's ★tree kangaroo, *Dendrolagus lumholtzi*, found only in a small rainforest area near Cairns.

**boreal** Northern. e.g. Aurora borealis or northern lights.

***Boronia*** A genus of Australian dicot shrubs in family Rutaceae, including species such as *Boronia serrulata*, the Native Rose, and *B. megastigma*, the scented brown boronia.

**botanical garden** A usually large garden where plants from many countries and/or parts of Australia are grown and studied. Compare herbarium.

**botany** The scientific study of plants and plant life.

**bottlebrush** Any species of the Australian genus *Callistemon* (Family Myrtaceae), whose flower spikes look like a cylindrical brush. Sometimes certain other plants, such as types of *Banksia* are also called bottlebrush.

**botulism** A serious form of food poisoning. See toxin.

**bough** One of the main branches of a tree.

**bovine somatotrophin** Called BST for short. A ★genetically engineered ★hormone that increases milk yields in cows.

**bovine spongiform encephalopathy** Called BSE for short and commonly called **mad cow disease**. Thought to be caused by a ★prion, it leaves parts of the brain with a spongy look. The cows lose coordination of the limbs and may become very aggressive. See also kuru and scrapie.

**bowel** The gut or ★intestine.

**bowerbird** A group of birds in Australia and New Guinea, order Passeriformes and family Paradisaeidae, which build structures called bowers where the male courts the female.

**Bowman's capsule** In the kidneys of vertebrates, it is the bulb-like end of the ★nephron where the liquid parts of the blood are filtered from the blood cells and large molecules. Compare glomerulus. See also kidney and Fig. E8 and K1.

**box** Any of a group of eucalypts with a characteristic short fibred bark, often in the form of small rectangular plates, and usually extending to the small branches. e.g. *Eucalyptus microcarpa* which occurs over much of central New South Wales and Victoria.

**box jellyfish** Tropical group of cnidarians (coelenterates) where the 'head' is box shaped. The Australian species *Chironex fleckeri*, or sea wasp, is extremely venomous, and is dangerous in tropical waters during the summer months.

**brachial** Belonging to the forelimb of a vertebrate. See also ★brachiation.

**brachiation** Movement by swinging the arms from one hold to another as when primates move through trees.

**Brachiopoda** A small phylum of marine invertebrates called lampshells, with two shells like a bivalve, but with a different arrangement of internal organs. They possess a lophophore, a special filter-feeding organ. Compare Bivalvia. See Appendix 1.

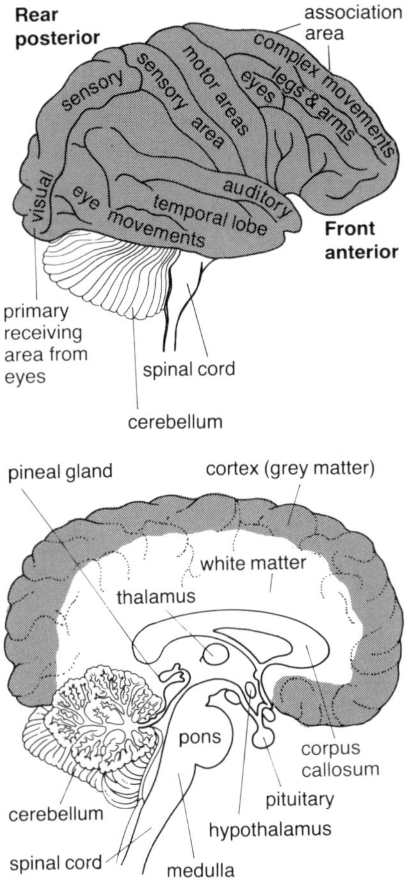


**Fig. B7 Brain**

**brachyurous** Short tailed. e.g. crabs. Compare macrurous.

**bracken** *Pteridium esculentum*, a ground fern not restricted to moist habitats that grows to a height of a metre or more, and is native to Australia. Can cause poisoning of cattle if eaten at certain times of the year.

**bract** A leaflike structure found at the base of a flower-stalk or an ⋆inflorescence of flowers. Compare bracteole and sepal.

**bracteole** Small leaflike structures or bracts (usually a pair) on a flower stalk or at the base of the ⋆calyx (full set of ⋆sepals). Compare bract.

**brain** The mass of whitish and greyish nervous tissue within the skull, that is the centre of ⋆nervous and ⋆hormonal control of the body of a vertebrate. It is connected to the ⋆spinal cord, and together with it, is called the ⋆central nervous system. In humans, the location of mental processes such as thought and emotion. In invertebrates, the brain is a concentration of nerve ⋆ganglia at the front (anterior) end. See Fig. B7.

The main part of the human brain is called the ⋆cerebrum, which is divided down the center into two parts called the ⋆cerebral hemispheres. There are a number of areas of ⋆grey matter such as the surface layer of the cerebrum, called the ⋆cerebral cortex, which is where most mental activity occurs. The grey matter inside includes the various ⋆nuclei, and the ⋆thalamus and ⋆hypothalamus. Connecting the areas of grey matter are tracts of ⋆white matter. The ⋆pituitary hangs from under the brain. At the rear of the brain is the ⋆cerebellum, while the ⋆brainstem (⋆medulla oblongata, ⋆pons, ⋆midbrain) connects to the spinal cord. See also grey matter and white matter.

**brain wave** The electrical activity of the brain as recorded on an ⋆electroencephalogram and/or a cathode ray oscilloscope (CRO).

**brainstem** The part of the brain connecting cerebrum and spinal cord which helps coordinate and integrate many automatic functions of the body e.g. breathing, some aspects of heart beat. Includes the medulla oblongata, pons and midbrain. See brain and Fig. B7.

**Brassicaceae** Also called **Cruciferae**, or the cabbage family of dicots. Other examples include cauliflowers, radish and stocks. See also Appendix 1.

**breast** 1 The front part of the body from neck to waist. The chest. 2 ★Mammary or milk glands of females.

**breastbone** The ★sternum, found at the front between the upper ribs.

**breath** The air inhaled or exhaled from the lungs during breathing. While in the lungs, oxygen is removed from the air and carbon dioxide is added.

**breech birth** A birth during which a baby's posterior (rather than the head) leaves the birth canal (vagina) first.

**breed** 1 To produce offspring. 2 To carefully control the reproduction of organisms. e.g. Breed stud cattle. 3 A subdivision of a species. e.g. breeds of cattle. See also variety, race.

**brigalow** 1 The wattle species *Acacia harpophylla* growing in large areas of New South Wales and Queensland. 2 Plant communities in which brigalow is the dominant plant.

**brittle star** A group of echinoderms similar to seastars, but with long thin arms that break off easily; hence the name brittle. See Ophiuroidea and Appendix 1.

**brolga** An Australian bird of the crane family Gruidae, *Grus rubicundus*, renowned for its 'dancing' as part of courtship and the maintaining of pair ★bonding.

**bronchioles** The fine tubes in the lungs that are at the end of each ★bronchus, and connect to the ★alveoli. See also lungs and Fig. L4.

**bronchitis** Disease due to the inflammation of the linings of the bronchial tubes.

**bronchus** Plural is **bronchi**. One of the pair of tubes at the end of the windpipe (trachea) that carry air to the two halves of the lungs. Subdivides further into ★bronchioles and then ★alveoli. See Fig. L4.

**Brownian motion** The continuous movement at random of microscopic particles when viewed through a microscope. It is caused by the particles being hit by even smaller, and thus not visible, moving particles.

**brucellosis** A disease often found in cattle where it causes abortions. Due to bacteria called *Brucella abortus*.

**bruise** To injure by striking or pressing such that a blue colour arises because blood vessels are broken and blood seeps into the surrounding tissue without reaching the surface.

**brushborder** An area on surface cells, such as those in the small intestine, which are covered by ★microvilli.

**Bryophyta** A Phylum (Division) of non-vascular plants that includes the ★mosses, ★liverworts and hornworts. See Appendix 1.

**Bryopsida** Also called **Musci**. A class of bryophytes also called ★moss. See Appendix 1.

**Bryozoa** A phylum of mostly marine invertebrates, often called moss animals due to the way they live in colonies. The polyps often produce calcium carbonate cases, and have tentacles around their mouth, like corals. However, they have a U-shaped gut rather than a bag-shaped cavity for digesting food. See Appendix 1.

**BST** See bovine somatotrophin.

**bubonic plague** The 'plague' or 'black-death' of history. Contagious epidemic disease spread by rat-fleas, and caused by the bacterium *Yersinia pestis*.

**buccal** Having to do with the mouth and/or cheeks.

**bud** 1 A growth on a plant protecting developing leaves or flowers. 2 Asexual reproduction that occurs when a new organism develops from an outgrowth of the parent. e.g. Sea anemone young grow out of the side of the adult before dropping off to begin life on their own. See Fig. P8.

**budding** See bud (2).

**budgerigar** A small green and yellow parrot-like bird, *Melopsittacus undulatus*, found wild in inland regions of Australia. Domesticated birds have been bred with a wide variety of colours.

**buffer** A substance that helps prevent changes in the ⋆pH of a solution when small amounts of acid or base are added. e.g. ⋆Bicarbonate ions help buffer the blood to keep its pH constant even if, say, foods are digested and acidic molecules enter the bloodstream.

**bug** Members of the insect order ⋆Hemiptera. e.g. Green vegetable bug, assassin bug. See also Appendix 1.

**bulb** A usually underground bud that has fleshy thickened leaves for the storage of food. e.g. Daffodil, onion, tulip. Compare corm.

**Bunya pine** See pine.

**buoyancy** The upward force of water or other medium on a body, helping it to float. If the body is less ⋆dense than water it will float, but if its density is greater than that of water, it will sink. Larger surface areas tend to increase buoyancy, as do layers of fat or blubber under the skin.

**burr** A rough, prickly case found around many seeds allowing the seed to be attached to and thus ⋆dispersed by various animals. e.g. Bathurst burr. See also dispersal.

**Burramyidae** The family of mouse-sized marsupials that includes the pygmy-possums and pygmy-gliders (feathertailed glider). These mostly tree-living marsupials feed on insects, nectar and pollen. See Appendix 1.

**Burramys** A rare marsupial pygmy possum, *Burramys parvus*, found in high mountain areas of Victoria. See Burramyidae.

**bush** 1 A relatively small woody plant with many branches usually arising near the ground. Compare tree, shrub, herb. 2 A non-technical term for an area covered with bushy plants, scrub or low trees. Compare forest, mallee.

**bustard** Also called the Plains turkey, the bustard, *Ardeotis australis*, is a large turkey-like bird that inhabits grassy plains and open country throughout much of Australia.

**butterfly** Insects of the order Lepidoptera that can be told apart from ⋆moths as they fly by day using colourful and conspicuous wings, and by having club-shaped antennae. They feed using a ⋆proboscis. See Appendix 1.

**buttresses** Flat extensions of trunks and roots in shallow rooted rainforest trees that help with support. e.g. figs.

# Cc

**$C_3$ plants** Any plant that in the first step of *photosynthesis produces the three-carbon molecule phosphoglyceric acid. This type of photosynthesis is common amongst plants of the temperate regions, and they use *photorespiration. $C_3$ plants produce less glucose for a given area of leaf, but can continue to carry out photosynthesis at much lower light intensities than can *$C_4$ plants. See Fig. P5.

**$C_4$ plants** Any plant that in the first step of *photosynthesis produces the four-carbon molecule oxaloacetic acid. See Fig. P5. This type of photosynthesis, using the Slack-Hatch pathway, is found in many tropical species of grass including maize, sugar cane, sorghum, etc. $C_4$ plants produce more glucose for a given area of leaf, can continue to carry out photosynthesis at high light intensities but low carbon dioxide concentrations. Compare $C_3$ plants.

***Cactoblastis cactorum*** A small moth introduced to Australia as a method for controlling the numbers of prickly pear cactus. See biological control.

**caddis-fly** Insects of the order Trichoptera, caddis flies look like small drab-coloured wasps, the two pairs of wings covered by hairs or scales. Their larvae and pupae are often found in freshwater streams and lakes, hiding in cases made from pieces of leaf, stick, etc. See Appendix 1.

**caecum** A one-ended pouch. Especially the sac at the beginning of the large *intestine to which the *appendix is attached. In herbivorous animals like rabbits and cows, the caecum is quite large and filled with cellulose-digesting *bacteria. Small in humans. See Fig. D2.

**Caesalpiniaceae** The Cassia family of dicots. They are so similar to the pea family that they are often included in the Fabaceae. See Appendix 1.

**caesarean** The removal of a baby from its mother by cutting an opening through the walls of her abdomen and womb (*uterus).

**caffeine** A stimulant drug found in coffee, tea and kola nuts. Also a *diuretic.

**Cainozoic** The most recent geological era. See Appendix 2.

**calcareous** Rich in calcium carbonate. e.g. Shells, coral, and certain algae.

**calcitonin** A hormone produced by the *thyroid gland that inhibits the release of calcium from bone, and reduces its concentration in the blood. Compare parathyroid hormone.

**calcium (Ca)** An element (metal) important in the bones, and is a vital regulator of many cell reactions. e.g. Clotting of blood.

**calibration** The checking and/or adjusting of an instrument or machine to ensure the scales are accurate.

**calicivirus** A virus that causes death of rabbits. Released by accident in South Australia in 1995. A rabbit haemorrhagic virus, rabbits die from heart and lung failure within 40 hours of infection.

***Callistemon*** A genus in the dicotyledon family Myrtaceae, often called ⋆bottlebrush.

**Callitrichidae** A family of monkeys in the suborder Anthropoidea, the marmosets. See Appendix 1.

**callosity** A ⋆callus.

**callus** 1 The hard bone-like material that serves to unite fractured bone. 2 A thickened or hardened area of skin, due to friction or pressure. 3 Protective tissue that forms over the wounds of plants to allow healing. 4 Undifferentiated tissue, especially in ⋆tissue culture.

**calmodulin** A calcium-binding protein, present in most cells, that stimulates a wide variety of enzymes to act. It carries the signal from the cell membrane receptors to the sites of enzyme action within the cell.

**calorie** A (now rarely used) measure of the amount of heat needed to raise the temperature of 1 gram of water by one degree Celsius. 1 calorie = 4.2 Joules.

**Calvin cycle** A set of chemical reactions in the dark phase of ⋆photosynthesis during which a series of enzymes converts (⋆reduces) carbon dioxide to complex molecules from which carbohydrates can be made. Occurs in the ⋆stoma of the chloroplasts. See photosynthesis and Figs C1 and P5.

**calyx** The ⋆sepals collectively.

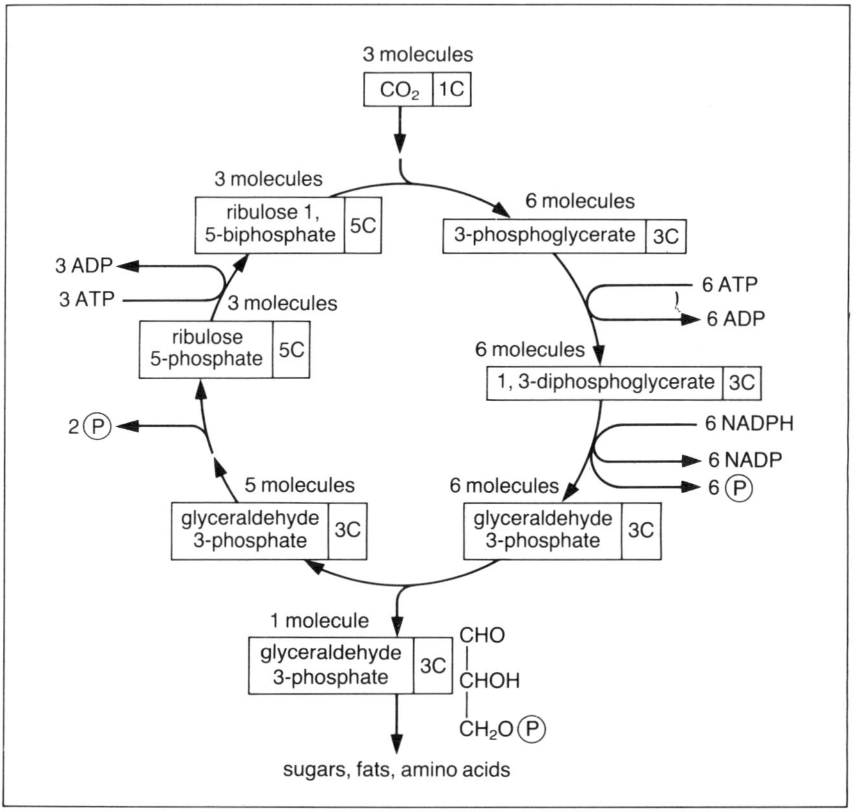


**Fig. C1 Calvin cycle (photosynthesis)**

Usually the outermost set of ⋆flower parts. Compare corolla. See Fig. F2.

**calyx-tube** The tube formed when the sepals of the ⋆calyx are fused together.

**cambium** 1 Vascular cambium is a layer of cells (⋆meristem) found around the edges of mature roots and stems of many dicots and conifers. Mitosis in this cambium produces cells to form new ⋆xylem and ⋆phloem tissue, thus increasing the diameter of the plant. See also growth rings, and Figs S7, V1 and W2. 2 The cork cambium (phellogen) is just under the surface of many woody plants producing a protective layer of cork (⋆phellem). See also vascular tissue, wood, and Fig. W2.

**Cambrian** The oldest geological period whose rocks contain numerous marine fossils. First period in the Palaeozoic era. See Appendix 2.

**camouflage** The way in which an organism makes itself indistinguishable from its background using colour patterns or shape. e.g. Stick insects resemble sticks and twigs. Moths, such as the Peppered Moth, resemble bark. Compare protective colouration, cryptic colouration, and mimicry.

**Canada balsam** A yellowish resin obtained from the balsam fir. It is used to mount specimens on microscope slides, as it has similar optical properties as glass.

**cancer** An uncontrolled growth of cells. A ⋆malignant and invasive growth or ⋆tumour, that is often able to spread via the bloodstream or lymphatic system to other parts of the body (⋆metastasis). Cancers are sometimes named after the tissue they invade. e.g. ⋆Carcinoma, ⋆sarcoma's, ⋆lymphomas, etc.

**cane toad** The introduced ⋆toad *Bufo marinus*, now a widespread pest in Queensland, Northern Territory and New South Wales. See also Amphibia.

**canine** In mammals, the four pointed teeth next to the front incisors, one on each side of each jaw.

**cannabis** Formed from the dried leaves of the Indian hemp plant, *Cannabis sativa*. See marijuana.

**canopy** The outermost or topmost layers of branches and leaves of a forest. Compare emergents.

**capillary** The smallest thin-walled blood vessels connecting ⋆arteries to ⋆veins. The capillaries pass through the tissues of the body, so that no cell is more than about 130 µm from a capillary. Food and oxygen pass through the capillary walls from the blood to the cells of the body, while wastes move in the reverse direction. See circulatory system and Fig. C8.

**capitulum** A type of flower head (racemose inflorescence) as seen on daisy and dandelion, where there are many small stalkless flowers (florets) attached to a flat fleshy disc, and surrounded by a ring of bracts. It looks like a single flower, but is in fact a head of many flowers. See Fig. I4.

**capping** The modification of the 5′ end of ⋆eukaryotic ⋆messenger RNA after ⋆transcription by adding a methylated guanosine group. This cap remains after the ⋆introns have been removed during ⋆editing. Compare tailing.

**capsid** The protein coat of a ⋆virus, protecting the RNA or DNA inside when the virus is outside the host cell. See Fig. V4.

**capsule** 1 A dry split open (⋆dehiscent) ⋆fruit that develops from two or more carpels. See Fig. F4. 2 The spore case of bryophytes (moss, liverwort, hornwort). 3 A slimy gelatinous layer around the cells of some ⋆bacteria. See Fig. B1.

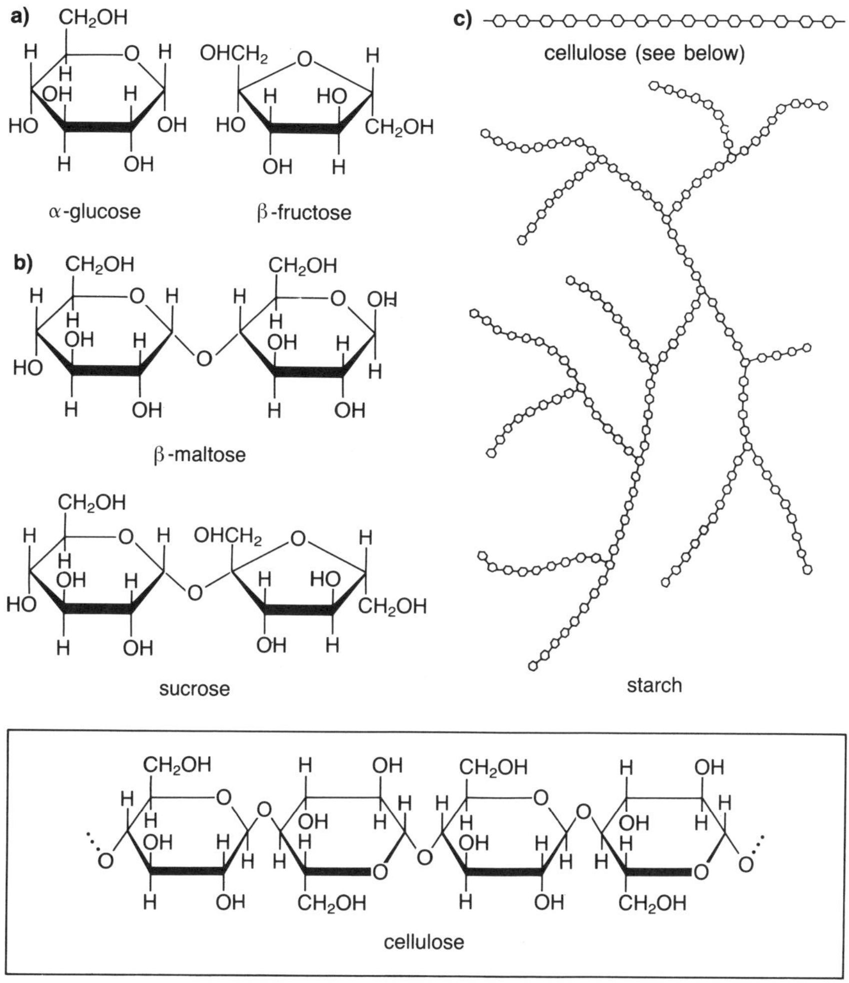


**Fig. C2 Carbohydrates (a) Single sugars (b) Doubles sugars (c) Multiple sugars**

**carapace** 1 A shield (part of the ⋆exoskeleton) that covers several segments of the body of some ⋆arthropods. e.g. crabs. 2 The top (dorsal) part of the 'shell' of tortoises and turtles (Chelonia).

**carbohydrate** Any of a number of organic compounds containing carbon, hydrogen and oxygen, and consisting of one or more simple sugars. Carbohydrates are classed as ⋆monosaccharides (simple sugars including glucose and fructose), ⋆disaccharides (double sugars including sucrose), and ⋆polysaccharides (multiple sugars, including starch, glycogen, and cellulose). See Fig. C2.

**carbon** (C) The primary element of ⋆organic compounds. Its ability to bond chemically to itself in long chains allows it to form a greater variety of compounds than any other element.

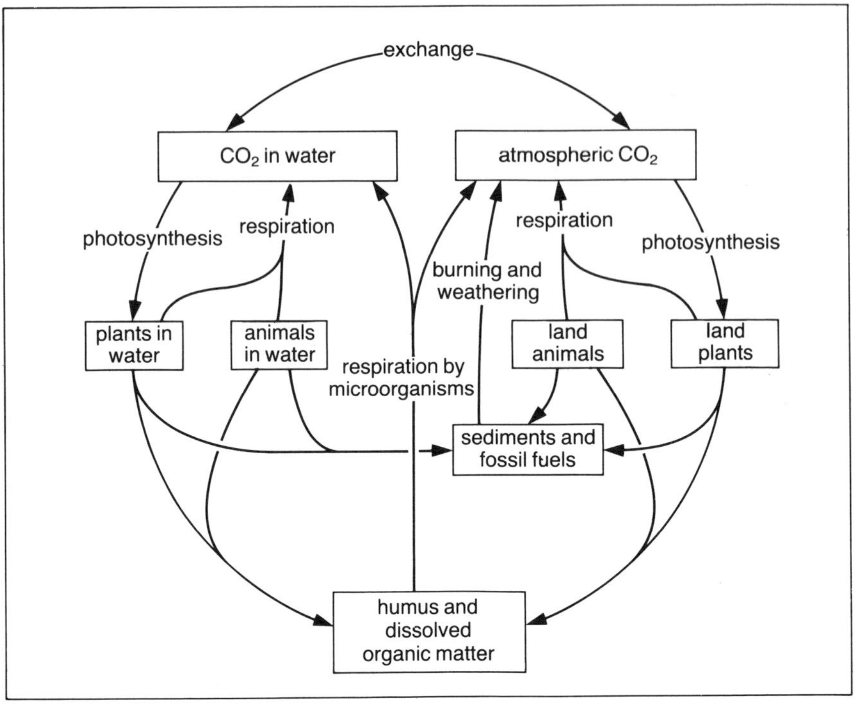


**Fig. C3 Carbon cycle**

**carbon cycle** The movement of carbon atoms from carbon dioxide in the atmosphere, to living things, and then back to the atmosphere. See Fig. C3.

Carbon enters the carbon cycle during ⋆photosynthesis when carbon dioxide is taken from the air and incorporated into the living material of plants. This carbon is transferred to herbivores and carnivores as it passes along the ⋆food chain. At all stages, some carbon re-enters the atmosphere when carbon dioxide is produced during ⋆respiration. Even when the organism dies, microbes (e.g. bacteria) and fungi begin decay, returning more carbon to the atmosphere as carbon dioxide. Compare nitrogen cycle.

**carbon dating** See radiocarbon dating.

**carbon dioxide** A colourless, odour-less gas (formula $CO_2$), normally present in the atmosphere. ⋆Respiration (or combustion) adds carbon dioxide to the atmosphere, which is later removed by photo-synthesis. In recent years, the activities of humans have been increasing the concentration of carbon dioxide in the atmosphere, with possible adverse environmental effects. See greenhouse effect.

**carbon fixation** The removal of carbon dioxide from the atmosphere during ⋆photosynthesis to produce carbohydrates and other organic chemicals. See Calvin cycle.

**carbon monoxide** A colourless, odourless gas (formula CO) that is highly poisonous as it binds more strongly to the haemoglobin of the blood than does oxygen, thus preventing oxygen from reaching the cells of the body.

**Carboniferous** Geological period of the Paleozoic era between the Permian and Devonian. See Appendix 2.

**carboxyl group** A functional group (-COOH) that is present in organic acids. e.g. Acetic acid is $CH_3COOH$.

**carboxylic acid** Organic acids that include the ★carboxyl group (-COOH). Various long chain carboxylic acids occur in fats and oils, and are then called ★fatty acids.

**carcass** The dead body of an animal.

**carcinogen** A substance that causes cancer in a body. e.g. The chemical TCA in cigarette smoke, ionising ★radiation, ★dioxin.

**carcinoma** A cancer of tissues covering internal or external surfaces (★epithelial tissue). See cancer.

**cardiac arrest** When the heart stops beating.

**cardiac muscle** The special type of muscle found in the vertebrate heart. It is made from branched and interlocking fibres that contract without the need of nerve signals. However, the vagus nerve can change the basic rate of contraction. See also muscle. Compare smooth and striated muscle.

**cardiac sphincter** The ring of muscle at the entrance to the stomach that prevents food in the stomach returning up the food pipe (★oesophagus). See Fig. D2.

**cardiovascular** Concerning the heart and blood vessels. e.g. Cardiovascular disease. See also circulation.

**Carnivora** An order of placental mammals that feed on the flesh of other animals. It includes dogs, bears, weasels, badgers, skunks, otters, mongooses, hyenas, and cats. They have relatively large canines, and their molars and pre-molars are shaped for shearing. Claws are usually sharp. See Appendix 1.

**carnivore** Animal that feeds on the flesh of other animals. Compare herbivore and omnivore.

**carnivorous plants** See insectivorous plants.

**carotenoids** Pigments that colour many plants (e.g. red tomatoes, autumn leaves), and that capture light and pass it to ★chlorophyll-*a*. Includes the carotenes (yellows, oranges, reds) and xanthophylls (yellow). See also photosynthesis.

**carotid** 1 Either of the two large arteries, one on either side of the neck, which carry blood to the head. 2 A body consisting of ★chemoreceptors lining a section of carotid artery close to where it starts near the neck. The carotid body helps measure changes in oxygen and carbon dioxide concentrations in the blood, and the ★pH of the blood. Nearby are receptors that measure ★blood pressure.

**carpal bones** The bones that form the wrist (or corresponding part of the forelimb) of terrestrial vertebrates. See pentadactyl limb and Fig. P2.

**carpel** The female part of a ★flower (★gynoecium), in which are contained one or more ★ovules. It is usually able to be divided into the ovary, style and stigma. A flower may have one or more carpels, either singly or joined together (fused). Compare pistil. See also apocarpous and syncarpous, and Fig. F2.

**carrier** 1 A living thing which carries disease organisms, and although it may sometimes be healthy itself, is able to transmit the disease organisms to others. 2 An organism which has a detrimental recessive gene, causing little or no harm to itself, but which can be genetically transmitted to the next generation.

**carrion** A dead, rotting animal.

**carrying capacity (K)** The number of organisms of a species that a particular environment can support indefinitely.

**cartilage** Also called **gristle**. Skeletal tissue composed of scattered cells surrounded by tough, flexible protein fibres. e.g. In humans, found at the end of nose, top of ear, and between joints. Makes up the whole skeleton of ⋆cartilaginous fish, ⋆jawless fish, etc. Compare bone.

**cartilaginous fish** The group of fish that have skeletons of cartilage. e.g. Sharks and rays. Cartilage is more flexible and lighter than bone giving cartilaginous fish increased buoyancy and easier movement in marine environments. See Chondrichthyes and Appendix 1.

**casein** One of a group of phosphorus-containing proteins found in milk. It can be precipitated from milk by the action of the enzyme ⋆rennin, a process that is used in cheese making.

**Casparian strip** In plant roots, a band-like thickening (containing ⋆lignin and ⋆suberin) within the primary cell walls. A strip of such cells forms a cylinder around the central bundle of vascular tissue. This prevents the movement between the cell walls of solutes absorbed from the soil, forcing them through the cells, which can then control the nutrients entering the vascular tissue. See roots and Fig. R8.

**cassowary** A large flightless bird, *Casuarius casuarius*, found in the tropical rainforests of north Queensland. A similar species exists in Papua-New Guinea.

**caste** A type of organisation found amongst social insects where individuals have specific roles. e.g. Among honey bees, there are three castes: Queens (fertile females), drones (males), and workers (sterile females). Castes also exist in wasps, ants and termites.

**castration** The removal of the ⋆testicles or ⋆ovaries.

**Casuarina** Any member of the family Casuarinaceae of dicot trees, generally referred to as oaks. e.g. ⋆river-oaks, ⋆she-oaks. Most Australian species are in the genus *Allocasuarina*. Instead of normal leaves, they have long needle-like ⋆cladodes or photosynthetic stems with scale leaves at the nodes, and resembling segmented pine needles.

**cat** 1 Domesticated carnivore *Felis domesticus*. 2 One of the marsupial native 'cats'. See Quoll.

**catabolism** Chemical reactions where complex molecules are broken down into simpler molecules. e.g. respiration. Compare biosynthesis and anabolism. See also metabolism.

**catalase** An enzyme that catalyses the conversion of hydrogen peroxide to water and oxygen. Hydrogen peroxide is produced as a by-product of metabolism, and must be removed before it can cause damage.

**catalyst** A substance that speeds up chemical reactions, but is not itself used up during the reaction. ⋆Enzymes are biological catalysts.

**cataract** Eye problem caused by the lens becoming cloudy and hard to see through.

**catecholamines** A group of chemicals that includes ⋆adrenalin (epinephrine), ⋆noradrenaline (norepinephrine) and ⋆dopamine. They are important in the functioning of the hormonal and nervous systems.

**caterpillar** The ⋆larval stage of a butterfly or moth.

**cation** A positively charged atom or group of atoms. e.g. Sodium ion

(Na⁺), ammonium ion ($NH_4^+$). Compare anion.

**catkin** A flower head where the ★inflorescence is a spike of unisexual flowers without petals, and which hangs in such a way as to aid in wind pollination. Compare spike. See Fig. I4.

**caucasian** The so-called 'white-race' characterised by paler skin and without the mongolian eye-fold. Typically people native to Europe. See race.

**caudal** Towards the tail end of an animal.

**caulescent** Having a plant stem above ground.

**cauterise** To burn, in the process of treating a medical condition. Usually achieved with a special electric iron, lasers, etc. e.g. Warts can be removed by burning them off.

**CD-4 cells** A type of ★T-cell with a protein called CD-4 on its surface. Helper T-cells are CD-4 cells. Since AIDS-causing ★HIV attacks cells at the CD-4 protein, it is the helper T-cells that are primarily infected by this virus. This causes a change in the ratio of CD-4 to ★CD-8 (T-4 to T8) cells which is examined to detect AIDS.

**CD-8 cells** A type of ★T-cell with a protein called CD-8 on its surface. Killer T-cells and suppressor T-cells are CD-8 cells. Compare CD-4 cells.

**cecum** See caecum.

**cell** The structural unit of all living things (except viruses), being composed of ★protoplasm enclosed in a ★membrane. Plant cells also have a cellulose ★cell wall outside the membrane.

Cells usually range in size from 0.5 to 20 μm. ★Prokaryote cells lack a nuclear membrane, and have no ★organelles. ★Eukaryote cells have their nucleus and organelles enclosed in membranes, and are generally larger and more complex. **a Plant cells** (Fig. C4): Plant cells differ from animal cells

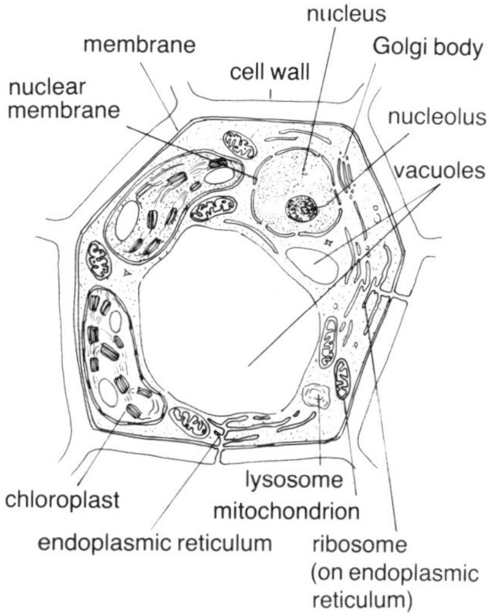


**Fig. C4 Plant cell**

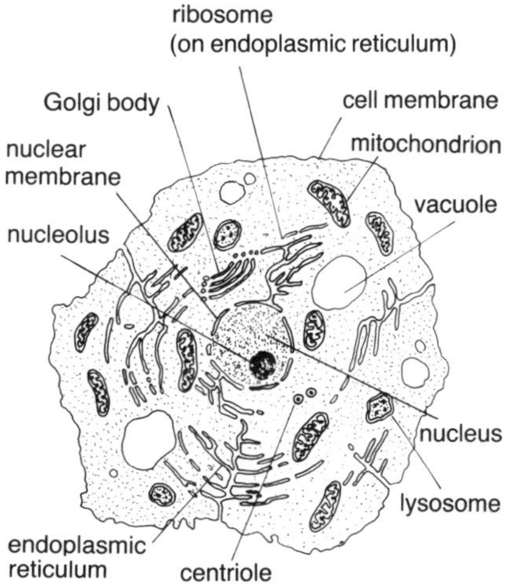


**Fig. C5 Animal cell**

because they have cells walls usually composed of cellulose, ★chloroplasts that contain ★chlorophyll, and large ★vacuoles. Also present are ★mitochondria, ★golgi bodies and ★lysosomes. Spread throughout the ★cytoplasm is the ★endoplasmic reticulum, and many ★filaments and ★microtubules to help organise the contents of the cell. Along the surface of the endoplasmic reticulum are the ⋆ribosomes where protein synthesis takes place. Besides these many structures, there are a number of small bodies containing pigment granules, fat or oil globules, etc. Inside the membrane enclosed ⋆nucleus are the ⋆chromosomes and ⋆nucleolus. **b Animal cells** (Fig. C5): Animal cells contain most of the same structures as plant cells, but lack the cell wall, chloroplasts, and large vacuoles. Animal cells have a ⋆centriole.

**cell body** That part of a ⋆neurone that contains the nucleus and much of the cell machinery. Extending from the cell body are the ⋆dendrites and ⋆axon. See Fig. N1.

**cell differentiation** The process of cells becoming ⋆specialised during development of an ⋆embryo. See differentiation.

**cell division** The splitting of a cell, both ⋆cytoplasm and ⋆nucleus, into two functioning cells. See meiosis and mitosis.

**cell membrane** See plasma membrane.

**cell plate** The plate-like structure that is produced by dividing plant cells at the position where the new cell wall is to be formed. It develops across the spindle at the equator of the cell during telophase of cell division.

**cell wall** 1 The cellulose boxlike structure that encloses the cell of green plants, and some fungi and algae. The primary cell wall is composed of cellulose ⋆microfibrils, thin fibres that run through a mass of complex polysaccharides (including ⋆pectic substances), some protein, and some space filled with water. Pores in the cell wall, called ⋆plasmodesmata, allow cytoplasm to connect between cells. When a secondary cell wall is deposited, extra layers of cellulose are laid down, and sometimes strengthened by ⋆lignin e.g. found in xylem vessels, tracheids and sclerenchyma cells. ⋆Pits are left in areas where there is little or no secondary thickening. See Fig. C4. 2 The cell walls of bacteria and blue-green algae are strengthened by ⋆mucopeptide rather than cellulose. There is often a ⋆capsule surrounding the cell wall. The cell walls of many fungi are strengthened by ⋆chitin. (See Fig. B1.)

**cellulose** An insoluble ⋆carbohydrate made from long chains of glucose molecules. Forms the main part of plant ⋆cell walls. Cotton is 98% cellulose. See also polysaccharide and Fig. C2.

**cement** Also called **cementum**. The hard substance that covers the surface of the roots of vertebrate teeth. See also tooth.

**Cenozoic** See Cainozoic.

**centipede** Any member of the arthropod class Chilopoda, where each body segment has one pair of appendages. The tail of these fast moving carnivores looks dangerous, but the unpleasant bite comes from the jaws on its head. See Appendix 1.

**central nervous system** Called **CNS** for short. In vertebrates, the ⋆brain and ⋆spinal cord. Compare reflex arc.

**centrifuge** A fast spinning machine used to separate liquid substances of different density, and to separate cell organelles or their fragments from the cytoplasm.

**centriole** A pair of tiny structures found just outside the nucleus of many cells (absent in higher plants). In normal cells, ⋆microtubules run from the centriole to the cell membrane. The pair of centrioles doubles before cell division, the two centrioles then separate to either ends of the cell, and the spindle (of microtubules) forms between them. See also cytoskeleton and centrosome. See Fig. C5.

**centromere** The point where the two ⋆chromatids formed from a chromosome are held together, and where the ⋆spindle fibre is attached. See also mitosis.

**centrosome** A dense zone surrounding the ⋆centriole of a cell.

**cephalisation** The concentration of nervous tissue and sense organs in the head.

**Cephalochordata** A subphylum of the chordates, also called the ⋆lancelets. See Appendix 1.

**Cephalopoda** A class of molluscs which includes ⋆squid, ⋆octopus, ⋆cuttlefish and ⋆nautilus. All have a ring of tentacles around the mouth, a complex nervous system, and a siphon with which they direct jets of water as they move off in the opposite direction. See Appendix 1.

**cephalothorax** The combined head and ⋆thorax found in ⋆arachnids and many ⋆crustaceans. Compare abdomen.

**cerci** Singular is **cercus**. A pair of appendages found at the rear end of many insects. e.g. Cockroach, silverfish, stonefly, mayfly, earwig.

**Cercopithecidae** A family in the order Primates which includes the old world monkeys and baboons. Their tail (if present) is not prehensile, and the nostrils are flattened. See Appendix 1.

**cerebellum** A part of the ⋆brain of vertebrates, relatively large in humans and located under the rear of the ⋆cerebrum (Fig. B7). It helps coordinate muscular activities and maintain equilibrium (balance). Recent research indicates it may also have a role in processing information from the senses.

**cerebral cortex** The layer of ⋆grey matter (neurones) covering the surface of the brain of vertebrates (Fig. B7). Substantial only in mammals, especially humans, where it is much folded to increase the surface area. The location for conscious sensations (⋆sensory area) and voluntary muscular activity (⋆motor area). It is also involved with memory, language, thought, and intelligence (⋆association area). See also brain.

**cerebral hemisphere** One of the two halves of the vertebrate ⋆cerebrum, connected by a band of nerve fibres called the ⋆corpus callosum.

**cerebrospinal fluid** A clear fluid, derived from the blood, that fills the cavities in, and surrounds, the brain and spinal cord. Helps cushion the brain.

**cerebrum** The major part of the vertebrate brain (excluding the cerebellum) that is divided into two ⋆cerebral hemispheres that are connected by a band of nerve fibres (corpus callosum). See also brain and Fig. B7.

**cervix** **1** The neck or opening into the womb (⋆uterus) from the birth canal (⋆vagina). See also pap smear and Fig. R4. **2** The region of the body connecting the head to the trunk. **3** The cervical or neck vertebrae in the neck region of the backbone.

**Cestoda** A class of flatworms (phylum ⋆Platyhelminthes), the parasitic ⋆tapeworms. See also Appendix 1.

**Cetacea** The order of marine mammals, including ⋆whales, dolphins and porpoises. Front limbs are shaped as flippers, and at the rear are horizontal tail flukes. They have large brains, and the adults have little hair. Their bodies are insulated by layers of blubber (fat). See also Appendix 1.

**CFC** Short for ★chlorofluorocarbon.

**chaeta** A bristle made of ★chitin found on the body segments of earthworms and marine worms (polychaetes). Compare seta.

**Chaetognatha** A small phylum of small marine wormlike organisms called ★arrowworms. See Appendix 1.

**chancre** The initial ulcer or sore of ★syphilis, often on the genitals.

**cheek** 1 Either side of the face below the level of the eyes. 2 Either side of the buttocks.

**cheek pouch** Sacs or 'bags' in the cheeks of some rodents in which they carry food.

**chelicera** Plural is **chelicerae**. Piercing and sucking mouthparts of arachnids such as spiders, ticks, etc. Associated with poison glands in spiders.

**Chelonia** The order of reptiles that includes ★turtles and ★tortoises. See Appendix 1.

**chemical bonds** The forces that hold two atoms together in molecules or crystals. See also covalent and ionic.

**chemical coordination** See hormones and endocrine system.

**chemical reactions** The rearranging of the atoms in one or more chemicals to produce new chemicals. See also activation energy, catalyst.

**chemical receptor** See chemoreceptor below.

**chemoreceptor** A sensory cell or organ that responds to specific chemicals. e.g. ★Smell and ★taste receptors, ★carotid receptors.

**chemosynthesis** The production of organic molecules from inorganic molecules, but not using photosynthesis. e.g. Hot springs in rift valleys on the floor of the ★abyss where there is no sunlight, and where hydrogen sulfide ($H_2S$) supplies energy for metabolism of certain bacteria.

**chemotaxis** The response of a cell or organism as it follows the increasing concentration of a chemical to its source. e.g. (a) Movement of certain white blood cells (★polymorphs) towards a bacterium in response to chemicals produced when special blood proteins (called ★complements) react with the outer coat of the bacterium. (b) The swimming of a fern's male gamete towards the female gametes, following the increasing concentration of a chemical produced by the female sex organs. See also pheromones and taxis.

**chemotropism** A plant's behaviour in turning towards or away from a chemical stimulus. e.g. The growth of a ★pollen tube down the style of a flower in response to the presence of sugars. See tropisms.

**Chenopodiaceae** The ★saltbush family of dicots which includes many saltmarsh plants as well as the ★saltbush and ★bluebush of the drier regions of Australia. See Appendix 1.

**chest** The part of the body between the neck and the abdomen.

**chiasma** Plural is **chiasmata**. The X-shaped figure formed when two homologous ★chromatids cross one another. It is the site of ★crossing over where chromatid segments are swapped. Compare centromere. See Fig. H3.

**chickenpox** A common skin disease caused by a herpes-type virus. People with chickenpox have a pink rash with tiny pus-filled blisters that later form a scab.

**Chilopoda** The class of arthropods that includes the ★centipedes. See also Appendix 1.

**chimaera** 1 An organism composed of genetically different tissues. e.g. A mutation in the development of a leaf may produce a discoloured area. 2 An artificially produced creature having tissues from two or more different sources. e.g. Chimaeric rats and mice are produced by mixing together cells of very early embryos.

**chimpanzee** Anthropoid apes in the mammal family ⋆Pongidae, the common chimpanzee *Pan troglodytes* and the pygmy chimpanzee *P. paniscus* found living in equatorial Africa. See also Appendix 1.

**Chiroptera** The order of flying mammals called ⋆bats, which includes insect-eating and fruit-eating (⋆flying fox) varieties in Australia. There are no vampire bats native to Australia. See also Appendix 1.

**chitin** A tough, resistant polysaccharide that forms the ⋆exoskeleton of arthropods and the ⋆cell walls of many fungi.

**chiton** A group of sluggish molluscs of class ⋆Amphineura, its 'shell' having 8 overlapping plates. See also Appendix 1.

**Chlamydia** A ⋆sexually transmissible disease (STD) caused by the bacterium *Chlamydia trachomatis*. It causes inflammation of the urethra, and in females may lead to sterility.

**chlorenchyma** ⋆Parenchyma or packing tissue which contains ⋆chloroplasts. e.g. palisade and spongy mesophyll inside a leaf.

**chlorofluorocarbon** Called CFC for short. Any of several simple gaseous compounds that contain carbon, chlorine, fluorine, and sometimes hydrogen, that are used as refrigerants, cleaning solvents, and aerosol propellants and in the manufacture of plastic foams. They are suspected to be a major cause of ⋆ozone depletion in the upper atmosphere.

**chlorophyll** A group of pigments which give the green colour to plants. Chlorophyll is found in the ⋆chloroplasts of nearly all plant cells and the autotrophic blue-green bacteria, and is responsible for trapping light energy during ⋆photosynthesis. See Fig. C6.

Chemically, chlorophyll is a ⋆porphyrin containing the metal magnesium, and there are a number chemically similar forms. Chlorophylls absorb mainly red and blue light (see action spectrum), and the green light that is reflected gives plants their green colour. See also photosynthesis, photosynthetic pigments, photosystems 1 and 2.

**Chlorophyta** A Phylum (Division) of plants also called the green algae. Many are microscopic single cells or groups (colonies) of cells. The majority are freshwater forms, but they can occur on soil, snow, bark, etc. Marine forms include the green seaweeds such as sea-lettuce (*Ulva*). See also Appendix 1.

**chloroplast** A small body (⋆plastid) found in the cells of green plants which contains the green pigments called ⋆chlorophyll. A typical higher plant will have a hundred or more chloroplasts in a leaf cell. They are the site of the light reaction and part of the dark reaction of ⋆photosynthesis. In higher plants, the chlorophyll is organised in plate-like ⋆grana within the chloroplasts. Chloroplasts also contain a little DNA and some ribosomes, although they don't seem to be able to produce all their own proteins. See Fig. C6.

**cholera** An acute infection of the digestive tract due to the ⋆bacterium, *Vibrio cholerae*, resulting in extreme diarrhoea, vomiting, cramp, etc. Often fatal. The antibiotic tetracycline can be used to kill the bacterium and thus control the diarrhoea and vomiting. However, far more important is the replacement of the large amounts of fluid and salts that are lost (Oral Rehydration Therapy).

a)

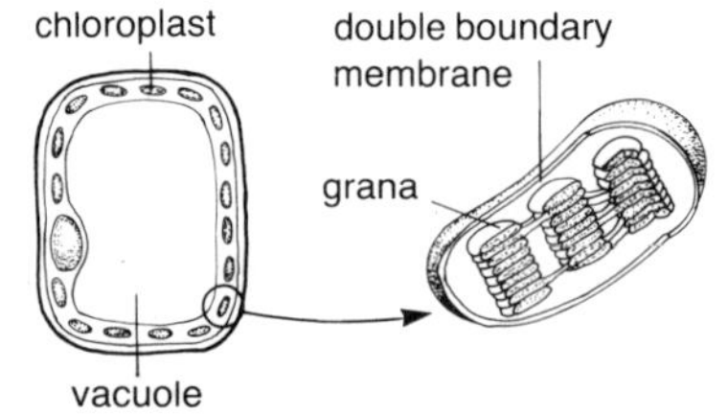


**The green plant cell contains chloroplasts which contain grana**

**cholesterol** The most abundant complex lipid (a ⋆steroid) in the body. An important part of cell membranes (see plasma membrane). Found in bile, blood plasma, blood cells. Common in foods such as animal fats and egg yolk. Cholesterol is converted to a number of other important chemicals including various sex hormones such as oestrogen. See Fig. C7.

b)

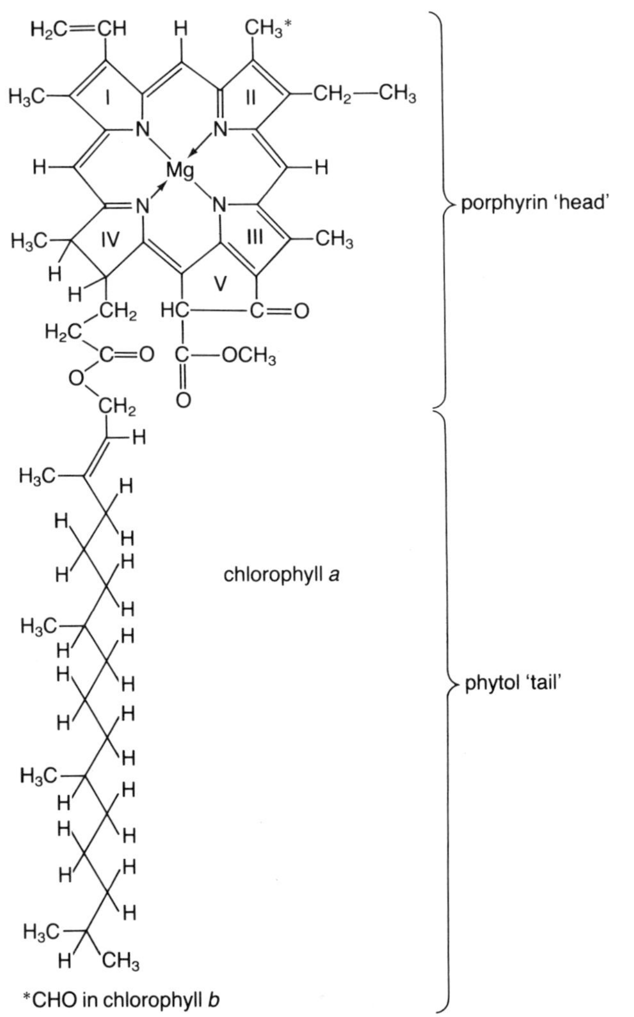


**Fig. C6 Chloroplasts (a) Chloroplast structure (b) Chlorophyll molecule**

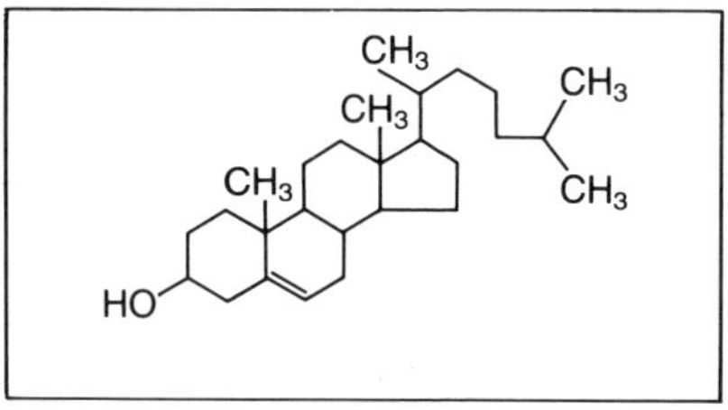


**Fig. C7 Cholesterol**

**Chondrichthyes** A class of vertebrates, also called ★cartilaginous fish as their skeletons are made from cartilage. Includes the sharks, skates and rays. See Appendix 1.

**Chordata** A major animal phylum all members of which possess, for at least some stage of their life, a flexible rod (★notochord) just below a dorsal (top or back) nerve cord, and structures that look like gill slits near the pharynx. In vertebrates the notochord is replaced by the vertebrae as the embryo develops. e.g. ★Tunicates, ★lancelets, ★acorn-worms, and ★Vertebrata. See also Appendix 1.

**chorion** 1 A membrane lining the inside of the eggs of birds and reptiles, and important in respiration and excretion. 2 A membrane that surrounds the embryo of vertebrates, and is involved in the formation of the ★placenta in mammals. See Fig. E4.

**Christmas bell** A spectacular bell-shaped flower produced by any of the species of *Blandfordia*, a genus of monocots (family Liliaceae) found in eastern Australia. See Appendix 1.

**Christmas bush** Any of various trees and shrubs that flower near Christmas. e.g. In New South Wales, *Ceratopetulum gummiferum*, a bush with red 'flowers' which are really coloured sepals that remain after the whitish flower falls off. In Victoria, *Prostanthera lasianthos*, with white more-or-less bell-shaped flowers.

**chromatid** Either of the two strands of a duplicated ★chromosome while still joined at the ★centromere. Chromatids are produced during interphase, and are identical in mitosis, but ★crossing over may make them different in ★meiosis. See Fig. H3.

**chromatin** The thin strands of DNA and protein that condense to form ★chromosomes during cell division (except bacteria and blue-green bacteria). Not readily visible during the cell resting stage (interphase), but are readily stained during cell division. See also histones and nucleosomes.

**chromatography** Any of several techniques for separating or analysing mixtures of liquids or gases by selective adsorption as they pass through a column of adsorbent material (column chromatography), or along a strip of suitable paper (paper chromatography). Compare electrophoresis.

**chromatophore** In animals, a pigment-containing cell, especially those of certain crustaceans and lower vertebrates that allow them to change colour. See also melanin.

**chromoplast** See plastid.

**chromosome** Threadlike structure made of condensed ★DNA and protein, and found in the nuclei of cells only during cell division (not at all in bacteria and blue-green bacteria). The number of chromosomes in the nucleus is usually constant for each species. They occur in pairs (e.g. Humans have 46, or 23 pairs), except for sex cells where the number is halved (e.g. Humans 23). During early stages of cell division, the duplicated chromosomes are still held together by a ★centromere, each of the four threads being called a ★chromatid. Compare chromatin. See also histones, nucleosomes, diploid, haploid, homologous, sex chromosomes, autosomes.

**chromosome mapping** Determining the position of genes along a chromosome by techniques such as measuring the frequency of crossing-over, examining chromosome banding patterns, etc. See also linkage.

**chromosome mutation** A visible change in the chromosomes of an organism best seen under a microscope. e.g. The numbers of chromosomes can change (*polyploidy, *Down's syndrome), or the relative position of sections of chromosome as indicated by banding can change (*inversion, *translocation, *deletion, *duplication). Compare gene mutation. See also mutation.

**chronic** Constant or continuing for a long time.

**chrysalis** The hard *pupa of butterflies and moths.

**Chrysophyta** A Phylum (Division) of microscopic plants, also called *golden algae and *diatoms, that exist as single cells, filaments, or colonies. See Appendix 1.

**chylomicrons** Small membrane-covered globules of *lipid produced by the cells in the walls of the small intestine and released into the *lymphatic system for transport to the liver. See also lacteal.

**chyme** The partly digested and semi-fluid contents of the stomach as it passes to the intestine. See digestion.

**cicada** Any member of the insect family Cicadidae, order Hemiptera. Their loud 'song' in summer months, is used by the males to attract females. See Appendix 1.

**cilia** Singular is **cilium**. Short rhythmically beating hair-like structures, usually arranged in groups on various surfaces. (a) Helps propel many single celled organisms (*Ciliata). e.g. *Paramecium*. (b) Line the *trachea of humans and help remove mucus from the lungs. Compare flagella.

**ciliary muscle** Small muscles of the *iris that help change the size of the *pupil. See also eye.

**Ciliata** A class of single celled protozoans that move using *cilia. e.g. *Paramecium. See also Appendix 1.

**circadian rhythms** Also known as **diurnal rhythms**. Cycles of approximately 24 hours that commonly occur in organisms. e.g. Leaf movements in certain plants. Sleep in humans and other mammals. See also biological rhythms.

**circulation** The movement of fluid through a system of tubes and spaces. e.g. Flow of blood (*circulatory system) and lymph (*lymphatic system).

**circulatory system** Also called the **cardiovascular system**. The movement of blood and lymph through the body of an animal by which the tissues are supplied with oxygen and nutrients, and wastes are removed. See Fig. C8

In insects and other invertebrates, the blood leaves the heart and arteries to flow through body spaces before re-entering the heart (see Fig. I5 and haemocoel). In vertebrates the flow of blood is confined to tubes. The *heart pumps used (deoxygenated) blood through the *pulmonary circuit to the *lungs where carbon dioxide is lost, and oxygen is added. After returning to the heart, the now fresh (oxygenated) blood is pumped through the *systemic circuit to the tissues of the body. Blood leaves the heart through *arteries, and enters the tissues through a fine network of *capillaries. In the tissues, oxygen and nutrients are lost to the cells, and carbon dioxide and wastes are returned. The blood is then collected into *veins and returned to the heart. During its time in the systemic circuit, some of the blood is sent to the ★kidneys to have wastes removed, and to the ★liver for the concentrations of the many chemicals present to be adjusted. See also blood, lymphatic system.

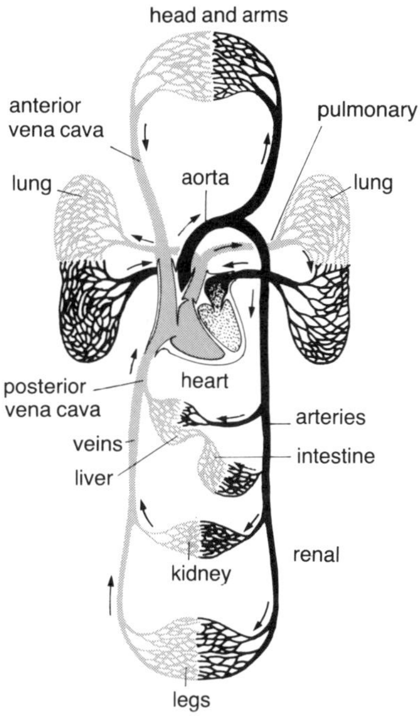


**Fig. C8 Circulatory system**

**circumcision** The removal of the foreskin from the male penis.

**cirrhosis** Damaged liver due to chronic ★inflammation. The liver attempts to repair itself by replacing damaged liver cells by fibrous or ★adipose connective tissue. Cirrhosis can be caused by hepatitis, liver parasites, and alcoholism.

**cirri** Singular is **cirrus**. **1** In animals, hair-like or filament-like structures. e.g. Hairs on the antennae or legs of certain insects. Copulatory organs of certain flatworms and molluscs. **2** In plants, ★tendrils.

**cisterna** Plural is **cisternae**. Flattened sac-like cavities (★vesicles) in the endoplasmic reticulum of cells, especially the ★Golgi apparatus. Often involved in cell transport and storage. See Fig. E7.

**cistron** A length of DNA that codes for a particular protein or enzyme using messenger RNA. Genes that code for transfer RNA or ribosomal RNA, or act as regulators, are not called cistrons. Compare operon.

**citric acid** An organic acid common in citrus fruits such as lemons and oranges. Present in all cells as part of the ★citric acid cycle.

**citric acid cycle** Also called the ★Krebs' cycle (respiration).

**CJD** See Creutzfeldt–Jakob disease.

**cladistics** A system of classification where living things are grouped according to their assumed common ancestry. It assumes that new species evolve suddenly (e.g. ★punctuated equilibrium), and not gradually (e.g. ★neo-Darwinism). See also classification.

**cladode** The green stem of a plant that looks like a leaf, but has the internal structure of a stem, and fewer stomata. e.g. *Allocasuarina*.

**clam** A ★bivalve mollusc, especially the larger varieties found on tropical reefs.

**clasper** A structure which helps a male organism to hold on to the female during ★copulation. e.g. Certain insects and fish.

**Class** In classification, the major subdivisions of a Phylum (or Division). A Class is itself divided into a number of orders.

**classical conditioning** A type of simple learning where an organism learns to respond to a ★stimulus with a ★conditioned response. It differs from ★operant conditioning in that the learner has no control over the leaning situation. e.g. A dog that has learned to salivate at the sounding of a bell has no control over when the bell is rung.

**classification** The systematic arrangement of living things into groups based on common (usually structural) characteristics and evolutionary descent. The common groupings from the most general to the most specific are: phylum (animals) or division (plants), class, order, family, genus, species. e.g. Humans are in kingdom Animalia, phylum Chordata, class Mammalia, order Primates, family Hominidae, genus *Homo*, and species *sapiens*. See also binomial nomenclature, cladistics, and taxonomy. Appendix 1 provides a classification of living things with an emphasis on Australian examples.

**clavicle** In people, the collar bone. In general, a bone of the ⋆pectoral girdle extending from the breastbone (sternum) to the shoulder blade (scapula). See Fig. S3.

**claw** The usually curved, and sharp, nail on the foot of an animal.

**clay** A fine-grained deposit of clay ⋆minerals, left after the ⋆weathering of ⋆igneous rocks such as basalt. It is like plasticine (plastic) when wet, and water cannot pass through it (impermeable).

**cleavage** A series of cell divisions of the fertilised egg (⋆zygote), ending in the formation of the hollow ball of cells called the ⋆blastula. See also development.

**cleft** A narrow, crack-like slit or opening. e.g. Cleft palate, a birth defect where the two sides of the roof of the mouth (palate) have not closed, leaving a gap (cleft).

**climate** The generally prevailing weather conditions, as measured over a number of years. Compare weather.

**climax** In ecology, the final stage in a ⋆succession of a ⋆community, which in some cases is thought to be stable as long as the environment remains unchanged. e.g. In a sand dune succession, the eucalypt forest behind the dunes is usually considered to be the climax community. See Fig. S10.

**climber** A plant that grows upwards using other plants or objects for support. e.g. Grape and passionfruit vines. See also tendril.

**cline** A relatively smooth gradation of characteristics from one end of an organism's range to the other, usually reflecting a gradient in environmental factors such as climate and soil. e.g. Gradual increase in size and darkening of coat colour of koalas as one travels from Queensland through New South Wales to Victoria.

**clitellum** The smooth band around a section of the body of certain annelid worms, especially ⋆earthworms, where a ring of frothy material is produced that contains the eggs. The worm then moves forward to deposit the sperm obtained and stored during mating.

**clitoris** The touch-sensitive organ of the female ⋆vulva, located above the ⋆urethra, which becomes erect on stimulation. See Fig. R4.

**cloaca** The common exit chamber for faeces, urine, and gametes found in all vertebrates except placental mammals, jawless fish, and the ray finned fish.

**clonal selection** In an ⋆immune response, the way ⋆antibodies are produced in response to any possible ⋆antigen that invades the body. The body contains a huge variety of ⋆B-cells, each type producing only one type of antibody. When a B-cell meets an antigen it recognises, it multiplies to produce a large ⋆clone of ⋆plasma cells that produce antibodies against that antigen. See also immunity and immuno-globulin.

**clone** 1 A population of cells, all of which have arisen from the same single cell by cell division (*mitosis). 2 A population of organisms that are descended by asexual reproduction from a single ancestor. e.g. Reproducing plants from cuttings, bulbs, or tissue culture. In both cases, the offspring are genetically identical to the parent.

**cloning DNA** Lengths of DNA are said to be cloned when they are inserted into *plasmids, which are then introduced into a suitable bacterium by the techniques of *genetic engineering. As the bacterium reproduces, the length of DNA is also copied.

**clot** A semi-solid mass of blood. See also blood clotting.

**clotting factors** See blood clotting.

**clover** Any of the various *legume dicots of the genus *Trifolium*, their leaves usually having three lobes. Like other legumes, they form *symbiotic associations with *nitrifying bacteria, and so are often grown to be ploughed into the soil, thus enriching it with nitrogen.

**clubmosses** Vascular plants, similar to but usually larger than mosses, in the Class Lycopsida. They have roots, and many small leaves spiral around their stems. e.g. *Lycopodium*, *Selaginella*. See Appendix 1.

**clutch** The set of eggs laid by one bird in its nest.

**Cnidaria** Also called **Coelenterata**. Animals built on a 'bag-like' structure, with sting-containing tentacles around an opening that acts as both the mouth and anus. e.g. Jellyfish, hydra, corals and sea anemones. See Appendix 1.

**cnidoblast** A specialised stinging cell amongst the surface cells of Cnidaria (Coelenterates). Each cnidoblast contains a fluid-filled capsule called a *nematocyst, inside of which is a coiled hollow thread with a dart on the end.

**co-dominance** Similar to *incomplete dominance. In genetics, when two forms of a gene (*alleles) are present and both are equally dominant. The resulting offspring have detectable characteristics of both alleles. e.g. In human *blood groups, type A is co-dominant with type B, resulting in blood type AB when both alleles are present. Molecules for both type A blood and type B blood are on the surface of the red blood cells. In cattle, genes for red and white hair are co-dominant. The genes produce separate red and white hairs resulting in a roan colour.

**cocaine** A white crystalline powder obtained from coca leaves. Used as a local *anaesthetic. Misused as a drug of addiction.

**coccus** Plural is **cocci**. A spherical *bacterium. Cocci can occur singly (monococcus), in pairs (diplococcus), in groups of four or more as cube-shaped packets (*Sarcinae*), in grapelike clusters (**Staphylococcus*), or in chains (**Streptococcus*). e.g. *Gonorrhoea and *pneumonia. See Fig. B1.

**coccyx** A small bone at the bottom end of the backbone (vertebral column or spine) in humans and apes. Formed by the fusing together of four smaller vertebrae. See also tail and Fig. S3.

**cochlea** The spiral (snail) shaped inner *ear of humans and many other animals where the vibrations caused by sound are detected by fine hairs and nerve endings. See Fig. E1.

The vibrations are carried by small bones called *ossicles, connected to the oval window of the cochlea. As this window vibrates, it causes vibrations within the fluid of the coiled cochlea. Sensitive cells with tiny hairs move with the vibrations of the fluid and generate nerve impulses that are sent by the auditory nerve to the brain. High notes are detected at the start of the tube, while low notes are detected at the far end.

**cockatoo** Any of the crested parrots in the genera *Probosciger* (Palm), *Cacatua* (Corella, Galah, Pink, Sulphur-crested), *Callocephalon* (Gang-gang), or *Calyptorhynchus* (Black). Fairly large seed eating birds that are basically white, pink, or black in colour.

**cockroach** Any of the flattened, nocturnal insects of Order Blattodea, most of which feed on rotten wood and other vegetable matter. Household pests such as the oriental cockroach (*Blatta orientalis*) are in the family Blattidae. See also Appendix 1.

**cocoon** Any of the various coverings provided by invertebrates to protect their eggs or larvae. e.g. The silken covering of the larva or *pupa of an insect (such as silkworms). The membrane-like sac that encloses the developing larvae of earthworms and leeches. The silken sac that encloses the eggs of spiders.

**codeine** A white powder obtained by processing crude opium and used in medicine as an *analgesic (removes pain) and *sedative (calms, relaxes and puts to sleep).

**codon** Also called a **triplet**. The basic instruction ('letter') of the *genetic code. Made up of three bases that are next to each other in a molecule of DNA, or messenger RNA. In messenger RNA they code for a particular amino acid during the production of protein e.g. CGA codes for Arginine. See also amino acids, anticodon, transcription, translation. See Figs G1 and T4.

**coelacanth** See Latimeria.

**coelom** The main body cavity of many animals in which the internal organs are suspended. See also body cavity.

**coenzyme A** Called **Co A** for short. A complex organic compound that helps enzymes in the transfer of *acetyl and other groups during the *Krebs' cycle of respiration, and during the synthesis and oxidation of fatty acids. Chemically it is a *nucleotide made from the vitamin B called *pantothenic acid, the purine base adenine, and ribose sugar and phosphate. See also coenzyme, and acetyl Co A.

**coenzymes** An organic compound that must be present before an enzyme can become active. Often acts as a donor or acceptor of a substance involved in the reaction. e.g. *NAD, *NADP, and *coenzyme A. Many *vitamins are essential for the production of coenzymes. See also cofactors.

**coevolution** The simultaneous evolution in two or more populations that interact so closely that each is a strong selective force on the other. e.g. Flowers and the insects that pollinate them.

**cofactors** A non-protein substance that must be present before an enzyme can become active. e.g. Metal ions ($Na^+$, $K^+$, $Mg^{2+}$); *Coenzymes.

**cohesion** The sticking together of molecules of the same or different substances. e.g. Cohesion theory helps explain *transpiration pull, the movement of water up the xylem tubes of plants. It is partly due to the strong forces between the water molecules (due to *hydrogen bonds), and between the water and the xylem cell walls (capillary effect). Compare adhesion.

**cohort** A group used in classification; between a class and an order.

**coitus** Sexual intercourse, where fertilisation is internal. Compare amplexus.

**cold blooded** More correctly called ⋆poikilothermic.

**Coleoptera** An order of insects with chewing mouthparts, that includes the ⋆beetles, ⋆fire-flies, weevils, wood-borers, etc. The front wings form a hard covering for the rear wings which fold beneath them when not in use. See Appendix 1.

**coleoptile** The protective leaf-like sheath that covers the first (embryonic) shoot of a germinating seed of the grasses. Because of its rapid growth rate, coleoptiles are suitable tissues for ⋆bioassays. e.g. Wheat, corn and lawn-grass. See also auxin. Such shoot tips respond by moving towards light. See tropisms.

**colic** A sudden pain in the abdomen or bowels.

**coliform bacteria** A group of bacteria that mostly inhabit the human intestine (e.g. *Escherichia coli*), and thus their presence in water is a good test to see if the water is contaminated by human wastes (faeces). Such wastes may also contain more dangerous, but less common, bacteria such as *Salmonella*.

**colitis** An inflammation of the mucus membrane of the colon.

**collagen** A fibrous ⋆protein contained in ⋆connective tissue, ⋆tendons and ⋆bones. Becomes ⋆gelatin when boiled. About 40% of the body's protein is collagen. Compare elastin.

**collarbone** Also called the ⋆clavicle.

**Collembola** An order of insects that includes the small, soft bodies springtails common in leaf litter. See Appendix 1.

**collenchyma** Supporting plant tissue composed of living cells whose walls are thickened at the corners. It is common in young plant stems, leaves and flowers, but is replaced by ⋆sclerenchyma in mature trees and shrubs.

**colloid** A substance whose particles range from 1 nm to 100 nm in size. Colloidal solutions are intermediate between true solutions and suspensions. See also sols and gels.

**colon** The ⋆large intestine or bowel. See Fig. D2.

**colony** **1** A group of animals or plants, of the same kind, living together in close association. e.g. Colony of bats. **2** Colonial organisation of organisms such as sponges (⋆Porifera) and many algae where groups of cells live together without the organisation of tissues. **3** The growth of bacteria on an agar plate is often in circular patches called colonies.

**colostrum** The first 'milk' produced by the breast after the birth of a baby. This slightly cloudy liquid is not true milk, but a mixture of fats and sugars that have a slight laxative effect on the newborn baby. Colostrum also contains antibodies from the mother to help the baby resist infection until its own immune system has developed. See also lactation.

**colour blindness** The inability to distinguish colours. There are many types and degrees of colour blindness, with inability to distinguish red-green being the most common. Colour blindness is nearly always due to genetic causes, and is ⋆sex linked (it affects mostly men as it is carried on the X-chromosome). Colour blindness does not influence the ability to see light, shade or with clarity.

**colour vision** Ability to see colour due to the presence of ⋆cones in the ⋆retina. The three types of cones are sensitive to red, green and blue light respectively.

**Colubrid** Any snake of the family Colubridae, which in Australia includes the solid-toothed non-venomous snakes, and the venomous rear-fanged forms. e.g. Various tree snakes and water snakes. See Appendix 1.

**commensalism** A close living association between two different species where one species is benefited while the effect on the other is neutral or unknown. e.g. Epiphyte plants living on trees, and sucker fish (remoras) attached to sharks. See also symbiosis.

**community** The ⋆populations of plants, animals and microbes found living together in a given area (⋆habitat) and often interacting with one another. See also ecosystem, biome, association.

**companion cell** A specialised ⋆parenchyma cell containing a nucleus, and found in association with the sieve tubes of the ⋆phloem in flowering plants. Companion cells probably have a role in controlling the phloem cells which lack their own nucleus. See Fig. P4.

**compensation point** The intensity of light at which the rate of production due to ⋆photosynthesis just balances the rate of ⋆respiration.

**competition** The 'struggle' between organisms within a community, of the same or different species, for limited or hard-to-reach resources. e.g. Plants may compete for water, minerals or light. Animals may compete for food and shelter. If the community is made unstable and a new competitor is introduced, one of the competing organisms is usually displaced.

**complement** 1 A family of proteins (20 or more) that as part of the ⋆immune system, attacks and kills microbes. This complex series of events creates holes in their surface and allows water and salts to enter which kills the microbe. Complement proteins are also involved in the ⋆inflammation response, including enlarging the capillaries and making them more permeable so that chemicals can reach surrounding tissues. 2 See complementary bases and complementary DNA.

**complementary bases** The bases that normally pair on the opposite strands of DNA or RNA: Guanine and cytosine; adenine and thymine; adenine and uracil. See also base pairing and Fig. B3.

**complementary DNA** Called **cDNA** for short. Complementary DNA is produced from messenger RNA using the enzyme ⋆reverse transcriptase. This technique can be used to produce ⋆gene probes.

**Compositae** See Asteraceae and Appendix 1.

**composite fruit** Also called **multiple fruit**. A fruit that has formed from the individual ovaries of many flowers clustered together. e.g. mulberry, pineapple. See fruit and Fig. F4.

**compost** A mixture of various types of decaying organic matter, such as leaves, manure etc. Often used to fertilise land.

**compound** Chemical substance that can be broken down into elements. The atoms in a compound are joined together by ⋆chemical bonds. The numbers of each type of atom in a compound are always in the same ratio.

**compound eye** The type of eye found in insects and crustaceans, made up from many column-shaped units called ⋆ommatidia. Such eyes form a mosaic-like image that is not useful for seeing detail, but detects movement easily. See Fig. O3.

**computerised axial tomography** See CT scanner.

**concave** Hollow like the inside

of a saucer. Curving inwards. Compare convex.

**conception** Fertilisation. The joining of the ovum (egg) and sperm cell. The act of becoming pregnant.

**conditioned reflex** See conditioned response below.

**conditioned response** Also called a **conditioned reflex**. The learning of a response to a stimulus that is not a natural response to that stimulus, or, if it is natural, is not as intense. e.g. A dog trained to respond by producing saliva for a normally unrelated stimulus such as the ringing of a bell.

**conditioned stimulus** A stimulus that does not naturally produce a particular response. e.g. A dog which does normally respond to a stimulus such as a bell by producing saliva, will do so with training.

**conditioning** The process by which an organism learns to respond to a particular stimulus. In ⋆classical conditioning, the learning involves the ⋆conditioned response. e.g. When a dog responds by salivating after it hears a stimulus such as a bell. In ⋆operant conditioning, the response is learnt by ⋆reinforcement. e.g. When desired behaviour is rewarded with food.

**conduction** 1 The movement of heat or electricity through a solid substance. Note that nerve ⋆impulses do not travel in the same way as electricity in a wire. 2 Also used to describe vascular tissue (conduction tissue).

**cone** The reproductive structure of gymnosperms such as conifers. The male cone is often comparatively small but produces enormous amounts of wind-carried pollen. The larger female cone produces winged seeds between the hard shelf-like layers of the cone.

**cone cell** A light-sensitive cell in the ⋆retina of most vertebrates. These cells are only roughly cone shaped. Because of the way they are connected to neurones, cone cells are more sensitive to details of an image than are ⋆rod cells. They contain pigments that makes each cell sensitive to one of the three basic colours—red, green and blue. See rhodopsin.

**cone sticks** Common name of plants in the genus *Petrophile*, dicot Family Proteaceae, with their flowers on cone-like structures. See Appendix 1.

**conformation** The three-dimensional folding pattern of a protein or other ⋆macromolecule. See proteins.

**congenital** Existing from one's birth. Thus a congenital defect is a birth defect, an abnormality present at birth.

**conifer** Short for **Coniferopsida**. See below.

**Coniferopsida** A Class of mostly evergreen gymnosperms in which reproduction involves ⋆cones. Includes pines, spruces and firs. e.g. Hoop pine, Bunya pine, Cedar. See Appendix 1.

**conjugated protein** A complex organic molecule where a non-protein molecule is attached to a relatively simple protein. The non-protein may be a nucleic acid (⋆nucleoprotein), lipid (⋆lipoprotein), polysaccharide (⋆glycoprotein), etc.

**conjugation** The union of two organisms for the exchanging of genetic material. Usually applied to sexual reproduction in certain protozoa (e.g. ciliates such as *Paramecium*), algae (e.g. *Spirogyra*) and bacteria (e.g. *E. coli*, *Salmonella*) where a tube connects the two cells, and genetic material is transferred. See also F factor, pili.

**connate** United, especially from birth. Often describes similar organs that are joined together, such as petals or leaves.

**connective tissue** Supporting and packing tissues which lie between groups of blood vessels, glands and muscle cells. The cells are spread amongst larger amounts of jelly-like and fibrous material (e.g. ⋆collagen, ⋆elastin). Bone, cartilage, blood and lymph are also classed as connective tissue.

**conservation** The act of preserving for future generations our natural resources, especially plants and animals, and their environments. See also national parks, wilderness areas, pollution.

**constipation** A condition of the bowels when it is difficult to eliminate faeces.

**constriction** Made narrow, made smaller, compressed. Compare dilation. See also vasoconstriction.

**consumer** Organism that uses other organisms as a source of food (⋆heterotrophic). Primary consumers (⋆herbivores) feed on plants; secondary consumers (⋆carnivores) feed on herbivores and other carnivores. See also food chains.

**contagious** Easily spread from one to another by physical contact. e.g. Contagious diseases can be carried and spread to others. Compare infectious.

**continental drift** Better called **plate tectonics** (Fig. P7). The gradual shifting of the world's continents that has occurred over hundreds of millions of years. This movement is believed to have greatly influenced the distribution of plants and animals amongst the continents. See also biogeographical realms (Fig. B5) and seafloor spreading (Fig. S1).

**continental shelf** The ocean floor under the shallow seas near the edges of the continents, where the rocks are more similar to those on the nearby continent, than to those under the floor of the ⋆abyss. See also littoral zone and Fig. O1.

**continuous variation** The variation that exists within a population of organisms for one feature when it is controlled by many genes and/or environmental factors. When graphed, such variation often forms a bell-shaped or ⋆normal curve (Fig. N3). e.g. Human height. See also variation.

**contractile fibres** Also called **contractile fibril** and **⋆myofibril**. The structural unit of ⋆striated muscle cells that allows the cells to contract. See also muscle fibre.

**contractile vacuoles** A small cavity in some protozoa, sponge cells and algae that is able to contract and discharge its contents through the cell's outer membrane. Mainly used as a pump to remove excess water that has entered the cell by ⋆osmosis. See also vacuoles.

**control** A standard against which observations and results can be checked in order to help test their validity. Used during experimental work.

**convergent** Coming together, meeting at a point; tending towards a common result or conclusion. e.g. Convergent ⋆evolution is the independent evolution of similar features by unrelated organisms (such as sharks and porpoises) due to adaptation to similar environments.

**convex** Shaped more or less like the outside surface of a saucer. Curved outwards. Compare concave. See also eye and lens.

**coolabah** A species of eucalypt (*E. microtheca*) that is common in arid and semi-arid parts of Australia, wherever the land is occasionally flooded, and around the edges of swamps and lagoons.

**coordination** 1 The action of the

⋆nervous and ⋆endocrine (hormone) systems in ensuring the different parts of the body act as a united whole. 2 The action of different or opposing muscles to produce a desired movement. e.g. A bird catching an insect in flight.

**copepod** Any crustacean of the subclass Copepoda, mostly microscopic organisms such as *Daphnia* and *Cyclops* that are often used as fish food. Copepods are an important part of the ⋆zooplankton. See Appendix 1.

**copulation** The physical act that allows sperm to pass from one organism to another. The act of sexual intercourse.

**coral** 1 A cnidarian (coelenterate) in the class Anthozoa. This small ⋆polyp secretes a small calcium carbonate (calcareous) cup in which it lives either alone or in a colony. Mainly inhabits warm, shallow and clean waters. See Fig. P8. 2 The mass of coral skeletons that form a ⋆coral reef.

**coral reef** A marine reef formed by calcium carbonate extracted from the sea by living organisms. Calcifying algae, as well as coral, play important roles in this process. But it is the coral building on top of the previous generation which gives the reef its characteristic appearance. Often the reef partly or completely encircles an island, or runs along the coastline for a considerable distance. e.g. Great Barrier Reef.

**coral tree** Any of several tropical ⋆deciduous trees in genus *Erythrina*, family Papilionaceae, with large red or orange pea-shaped flowers that appear before the leaves.

**cord** A sting or rope. e.g. ⋆Spinal cord is a long rope-like structure made up of billions of nerve fibres found inside the backbone. See also umbilical cord.

**cordate** A heart-shaped ⋆leaf that is broadest and indented at the base. See Fig. L1.

**corepressor** A small molecule that joins with a ⋆repressor molecule to block the expression of a gene (i.e. stops the gene being read to produce messenger RNA).

**coriaceous** Leathery, especially of leaves.

**corium** See dermis.

**cork** 1 The outer protective covering of stems and roots of woody plants. It is a ⋆secondary tissue produced by the cork cambium. These air-filled dead cells have had their walls impregnated by a water-repellent fatty substance called ⋆suberin. 2 A tree called cork oak produces a very thick cork that can be peeled off and used to make bottle corks, floats etc. See also phellem, wood, cambium.

**corm** A short, underground, upright and swollen base of a plant stem. e.g. gladiolus, crocus. The swollen stem stores food during winter for the next growing season. Around the outside of the corm are protective scale leaves. Shoots grow from the base of these leaves in the next growing season, a method of vegetative (asexual) reproduction. Compare bulb.

**cornea** The outer, exposed, transparent part of the vertebrate ⋆eye that is in front of the ⋆iris (Fig. E9). In many vertebrates the curvature of the cornea helps refract light and thus affects the focusing of the eye.

**cornification** Also called **keratinization**. See keratin.

**corolla** The collective name for the ⋆petals of a ⋆flower. Compare calyx. See Fig. F2.

**coronary artery** An artery that supplies blood to the ⋆heart muscles. A blockage of this artery is a frequent cause of heart attacks. Oxygen carrying blood cannot reach the muscles of the heart, resulting in their death. See also arteriosclerosis.

**corpse** A dead body, usually that of a human being.

**corpus callosum** The bundle of nerve fibres that connect the two halves (*cerebral hemispheres) of the *brain in mammals.

**corpus luteum** The 'yellow body' that forms from an egg follicle (see Graafian follicle) in an *ovary after the ovary has ruptured, and released the egg into the *oviduct. It is the main source of *progesterone until the *placenta develops, and may also be a source of *oestrogens.

**corpuscle** A very small body. **1** The cells (e.g. red or white *blood cells) that float in the blood. **2** The small swellings on the ends of some sensory nerves. e.g. *Meissner's corpuscle detects light pressure or touch.

**correlation** The relationship between two sets of measurements.

**cortex** **1** An outer layer. **2** The outermost layers of organs such as the adrenal gland, kidney and ovary. **3** The grey matter that forms the outermost layer of the surface of the brain. See cerebral cortex. **4** The tissue in a plant stem or root that lies between the vascular tissue (e.g. xylem, phloem) and the epidermis (e.g. bark).

**corticosteroid** Any of several hormones made by the cortex of the *adrenal glands. Chemically, they are *steroids made from cholesterol. Some help control the use of carbohydrates, proteins and fats by the body (e.g. *cortisone), while others help ensure a proper balance of salts and water in the body fluids (e.g. aldosterone).

**corticotrophin** See ACTH.

**cortisone** An anti-inflammatory hormone produced by the cortex of the adrenal glands. Used to treat rheumatoid arthritis. See also corticosteroids for its more general role.

**corymb** A type of *racemose *inflorescence in flowers in which all the flowers are at the same final level, even though their stems grow from different positions on the main axis. See Fig. I4.

**costate** Ribbed.

**cotyledon** Also called a **seed leaf**. A leaflike structure of the embryo of a seed plant. In some plants it forms the first leaf after the seed germinates. In other plants it is swollen, and acts as a food source for the germinating seed. Flowering plants (angiosperms) are divided into two main groups according to the number of cotyledons in the seeds: *monocotyledons have one cotyledon per seed and *dicotyledons have two cotyledons per seed. See Fig. S2.

**courtship** A type of animal behaviour that occurs before mating, often involving *display behaviour by the male. Such behaviour ensures that mating is between the same species, that they are of different sexes, and that the female is receptive to the male. It also helps establish the pair bond in many species.

**covalent bond** A *chemical bond between two atoms that is formed as the result of the sharing of a pair of electrons. Two atoms may be held together by more than one covalent bond (see saturated). This is the most common bond in carbon-based organic molecules. Compare ionic bond.

**Cowper's gland** One of two small glands near the *prostate gland of males that produce part of the *seminal fluid during sexual excitement. This fluid is the first to be excreted, cleansing the urethra and acting as a lubricant for the penis. See also seminal vesicle, and Fig. R3.

**crab** 1 One of the many ★crustaceans of the order ★Decapoda. e.g. blue swimmers, mud crabs, most rock crabs. 2 One of many crustaceans similar to true crabs. e.g. Hermit crabs, king crab. See also Appendix 1.

**cranial nerves** A series of 12 paired nerves (10 in some vertebrates) that start in the brain and supply the head, neck and some internal body organs. The ★sensory and/or ★motor nerves include those connected to the eyes, ears, nose, tongue, and pharynx. The ★vagus nerve connects to the heart, lungs and other internal organs.

**cranium** The part of a vertebrate ★skull that encloses the brain. See Fig. S3.

**crayfish** In Australia, various large ★crustaceans (order ★Decapoda) such as yabby and marron that have large claws. Compare lobster. In other parts of the world crayfish refers to different crustaceans.

**Creation** The original bringing into existence of the universe, including life, by God.

**creature** An animal, as distinct from humans.

**creeper** A plant that grows along, or just under the surface of the ground, or over the surface of another object, and which sends down roots at various intervals. e.g. Couch grass, ivy.

**crenate** Describes a structure such as a leaf or petal with margins that have small rounded teeth or lobes. See Fig. L1.

**crenulate** Slightly ★crenate. See Fig. L1.

**crepuscular** Active at dawn or dusk when light is dim. e.g. Bats, rabbits. Compare nocturnal, diurnal.

**Cretaceous** The third and last ★period of the Mesozoic era. See Appendix 2.

**Creutzfeldt–Jakob disease** Called CJD for short. A rare disease believed caused by a ★prion that results in the human brain becoming spongy, and a person suffering ★dementia and loss of muscular coordination. Compare kuru, scrapie and mad cow disease.

**cricket** True crickets are members of the insect order Orthoptera, and family Gryllidae, being more often heard than seen due to their ★stridulation or 'chirp'. Closely related are the tree crickets, mole crickets, cave crickets, etc. See Appendix 1.

**Crinoidea** A class of echinoderms which includes the sea-lilies and feather-stars. The sea-lilies live in deep water and, as adults, are usually attached to the bottom by a stalk. They use their feather-shaped arms to catch prey. The feather-stars also have feathery arms, but as adults are not attached by a stalk, and are free-swimming in tropical waters. See Appendix 1.

**crista** 1 A crest or ridge. 2 A fold on the inner surface of a ★mitochondrion, along which are arranged many of the enzymes needed for respiration. See Fig. M2.

**Cro-magnon** A race of humans whose fossils were first found in caves in southern France in 1868. Their skeletons are almost the same as those of modern humans, except they have, on average, a larger brain volume and are taller. They are believed to have lived during the upper Pleistocene period 20 000 to 50 000 years ago. See also Magdalenian culture. Compare Neanderthal Man.

**Crocodilia** An order of large aquatic reptiles, including *Crocodylus porosus*, which lives in tropical areas of Australia. Distinguished from ★alligators by having a narrower snout, and the large tooth visible towards its tip when the mouth is closed. See also Appendix 1.

**crop** 1 Cultivated produce such as grain (e.g. wheat), fruit (e.g. cherries) etc. 2 A part of the oesophagus in many birds where food is stored.

**cross** 1 A usually controlled mating between two organisms. e.g. Controlled crosses are used in genetics experiments, and in breeding livestock and crops to improve their quality. 2 The plant or animal that results from such a mating.

**cross fertilisation** The fertilisation of an ovum by a sperm produced by a different individual. This is usual in animals.

**cross pollination** See pollination.

**cross-breed** To produce a hybrid by mating two different varieties, races, breeds, strains etc.

**crossing over** The swapping of pieces of chromosomes during the special type of cell division that produces sperm and ova. Thus during ⋆meiosis, genetic material is exchanged between paired ⋆chromatids of ⋆homologous chromosomes (Fig. H3). During the last stages of prophase I, the homologous chromosomes remain in contact at corresponding points called ⋆chiasma which allows pieces of chromatid to be exchanged, thus establishing new ⋆linkage groups (new combinations of alleles). See variation, recombination.

**crow** A large black scavenging bird, the crows are hard to distinguish from similar looking ravens. Both are members of the genus *Corvus*, and are distributed throughout Australia.

**crown** 1 The top part of a structure. 2 Tooth crown is the top exposed part of a tooth covered in enamel. 3 The top leaves and branches of a tree. Compare bole. 4 The point where the root of a seed plant joins the stem.

**crown-of-thorns** A large ⋆sea-star (starfish), *Acanthaster planci*, its body being covered with sharp spines. It periodically reaches plague numbers on the Great Barrier Reef. See also Asteroidea.

**Crustacea** A class of arthropods with at least 10 appendages. They have two pairs of antennae, and usually obtain their oxygen with the use of ⋆gills. e.g. Lobsters, crayfish, crabs, shrimp, prawns, barnacles, water fleas, krill, and fossil trilobites. See also Appendix 1.

**crustose** Describing ⋆lichens that lie flat and close to the rocks, wood, etc. to which they are attached. Compare foliose, fruticose.

**cryptic** Hidden, secret. e.g. cryptic colouration is a form of ⋆camouflage that helps hide an organism from predators. Compare mimicry.

**CT** Short for **computerised axial tomography**, where X-ray scanners are used to obtain cross-sectional images of an individual's internal organs. Compare MRI, PET and ultrasound.

**Ctenophora** A small phylum often called comb-jellies, that look like jelly-fish but lack stinging cells. See Appendix 1.

**cud** The mass of food which a ⋆ruminant animal (e.g. cow, deer) returns to the mouth from the first stomach. It is then chewed a second time before being re-swallowed.

**cull** To remove certain animals from a herd or flock. e.g. To remove those of lesser quality; to reduce numbers to prevent overcrowding, etc.

**cultivar** Short for a cultivated variety, or a variety of a plant species that has been bred by humans to enhance some desirable characteristic, and does not occur in nature.

**cultural evolution** The gradual change of knowledge, beliefs, ideas, customs etc. through non-genetic means as they are transferred from one generation to the next. e.g. Art, dance or communication. Compare biological evolution.

**culture** 1 The growing of ⋆microbes (e.g. bacteria, yeast, moulds) or ⋆tissue cells on specially prepared media such as ⋆agar. 2 The way in which a group of people live, including work styles, social structure, religious practices, etc. The main cultures associated with the fossil record are Oldowan (*Australopithecus*), Acheulian (*Homo erectus*), Mousterian (*H. sapiens*, Neanderthal Man), and Magdalenian and other upper ⋆Palaeolithic cultures (*H. sapiens*, Cro-magnon Man).

**cuneate** Wedge shaped, being narrowest at the base. See Fig. L1.

**cunjevoi** A sea-squirt, or tunicate, such as *Pyura*, a member of the small subphylum Urochordata, and common along the edge of the rock platform. See Appendix 1.

**cuscus** Either of two species of usually slow moving, nocturnal, and tree-living marsupials found in the rainforests of Cape York, *Phalanger maculatus* or the Spotted cuscus, and *P. orientalis*, or Grey cuscus. See Appendix 1.

**cusp** A point on the top surface (crown) of a tooth.

**cuticle** 1 An outer skin. 2 A layer of waxy substance (⋆cutin) on the outer surface of plant (epidermis) that helps prevent loss of water. 3 A noncellular, outermost layer found on many invertebrates. In some it prevents damage by chemicals or abrasion. In others it helps provide strength to the exoskeleton, and helps reduce water loss. 4 The outermost layer of each hair. 5 The dead outer layer of human skin, especially that at the base of the nails.

**cutin** A waxy, waterproofing material present in plant ⋆cuticles, especially the surface cells of leaves, flowers and fruits. It also helps cut water loss through the surface of the leaves. See xeromorphic.

**cutting** A part of a plant, such as stem, root, or leaf, that when removed from the parent, and planted, grows into a new plant. It is thus a type of ⋆vegetative reproduction.

**cuttlefish** Along with ⋆squid, part of the Order of ten-armed molluscs called ⋆Decapoda. The internal skeleton of a cuttlefish is often washed ashore (called cuttlebone), and is broader than that of a squid (called a pen).

**Cyanobacteria** Also called **blue-green bacteria** or **blue-green algae**. A Phylum (Division) of often aquatic bacteria containing chlorophyll, but not in chloroplasts. The cells also lack a nucleus, and are grouped with ⋆bacteria as ⋆prokaryotes. See also Appendix 1.

**cycad** See Cycadopsida below.

**Cycadopsida** A Class of gymnosperms, many with the general appearance of palms, and with about 100 species at present. They were very common during the ⋆Mesozoic era. Sometimes called a 'living fossil'. See Macrozamia (Burrawang) and Appendix 1.

**cycle** A repeated sequence of events. e.g. ⋆Life cycle, ⋆nitrogen cycle, ⋆carbon cycle, ⋆water cycle.

**cyclic AMP** A cyclic form of the ⋆nucleotide adenosine monophosphate (⋆AMP) that helps regulate many metabolic processes in animal and bacterial cells. AMP helps in the action of hormones such as ⋆vasopressin, ⋆adrenaline, ⋆ACTH and the ⋆prostaglandins.

**cymose** An arrangement of flowers (⋆inflorescence) on a plant stem in which flowers at the end of the stem stop the growth of the stem. One or two lateral buds then grow, and the process is repeated. Compare racemose. See inflorescence and Fig. I4.

**Cyperaceae** The sedge family of monocotyledons, being grass-like or rush-like plants. See Appendix 1.

**cypsela** A small achene-like ⋆fruit characteristic of the daisy family (Asteraceae). See Fig. F4.

**cyst** 1 A thick-walled resistant stage in the life-cycle of many organisms (e.g. bacteria, algae, protozoa, flukes). 2 An abnormal sac-like structure in animals that is filled with fluid or pus. e.g. Ovarian cyst.

**cysteine** (cys) One of the 20 common ⋆amino acids found in proteins, and is important in maintaining the three-dimensional shape of protein molecules. It is also used to make ⋆coenzyme A. See Fig. A6.

**cystic fibrosis** An ⋆hereditary disorder where glands secrete a very thick ⋆mucus, leading to blockage of the ⋆pancreas and chronic infections of the lungs, which generally cause death in childhood or early adulthood. The ⋆recessive gene responsible for cystic fibrosis has now been identified, and tests have been developed to detect the most common types. These tests can identify unaffected ⋆carriers of the disorder.

**cystitis** An inflammation of the urinary bladder.

**cytochrome** One of several iron-containing proteins. Those found in mitochondria transfer electrons during the chemical reactions of cellular ⋆respiration. Those found in chloroplasts transfer electrons during ⋆photosynthesis. See also haem, electron transport chain.

**cytokinesis** The division of a cell's cytoplasm into two during ⋆mitosis and ⋆meiosis, as distinct from division of the nucleus.

**cytokinins** Plant ⋆growth substances (hormones) that stimulate cell division (mitosis), and have other effects. They are naturally occurring ⋆kinins.

**cytology** The study of cells, especially when using light and electron microscopes.

**cytoplasm** A general term for all the contents of a ⋆cell outside the nucleus and within the cell membrane. Compare protoplasm.

**cytoplasmic streaming** The stream-like movement of cytoplasm within a cell.

**cytosine** One of the nitrogen-containing organic bases that make up the 'letters' of the genetic code of DNA and RNA. Chemically cytosine is a pyrimidine, and helps form the ⋆nucleotides that make up the nucleic acids. See Fig. B3.

**cytoskeleton** The system of fibres within a cell that helps establish its shape, and plays a role in movement and cell division. There are three systems of fibres: **a** The finest are the ⋆microfilaments, averaging 6 nm in diameter and made of the protein ⋆actin; **b** the ⋆microtubules are 22 nm in diameter and consist of ⋆tubulin; **c** the ⋆intermediate filaments are between 7 and 11 nm in diameter, and the protein that makes them up varies with the type of cell. See also spindle. Compare endoplasmic reticulum.

**cytosol** The liquid part of the ⋆cytoplasm, not counting the particles (⋆organelles).

**cytotoxic T-cells** See T-cells.

# Dd

**dactyl** A part of the body such as a finger or toe, e.g. A ⋆pentadactyl limb is a limb with five fingers or toes.

**damselfly** Long thin insects in the order Odonata which differ from dragonflies by holding their membranous wings above their body when they have landed. See Appendix 1.

**dargawarra** A native Australian rodent (⋆placental), also called the spinifex hopping mouse, *Notomys alexis*.

**dark reaction** See photosynthesis.

**Darwinism** Evolutionary explanations of the origin of species according to the theory of ⋆natural selection as originally proposed by Charles Darwin (1809–1882). Compare Neo Darwinism and punctuated equilibrium.

**Dasyuridae** A family of marsupial carnivores including the ⋆native 'cats', tiger 'cats' (⋆quoll), ⋆dunnart, ⋆antechinus and the ⋆Tasmanian devil. See Appendix 1.

**daughter cell** A cell produced by ⋆mitosis or ⋆meiosis from the 'parent' cell.

**day-neutral** Plants whose flowering appears to be independent of the length of daylight or darkness. Compare long-day and short-day plants. See also photoperiodism.

**DDT** Short for **dichloro-diphenyl-trichloroethane**. A powerful insecticide that is very difficult for bacteria to break down (i.e. not very ⋆biodegradable). Thus DDT tends to remain in the environment long after it is used and is often ⋆biomagnified in the food chain. Now banned in many countries. See Fig. D1.

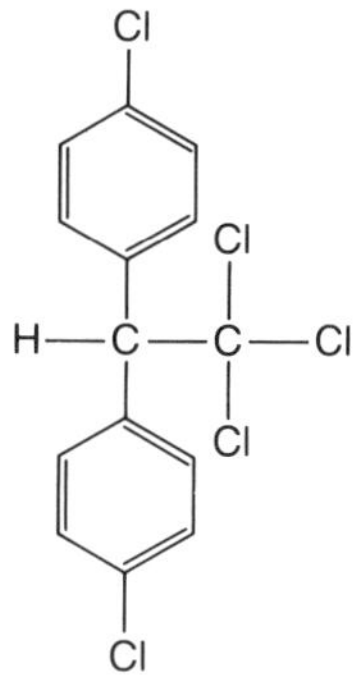


**Fig. D1 DDT**

**dead** No longer living.

**deadwood** Dead tree or dead branches on a tree.

**deamination** The removal of an amino group ($NH_2$–) from an organic compound. In humans this reaction occurs in the liver where ammonia ($NH_3$) forms before being excreted as ⋆urea.

**death** The end of life when vital functions (e.g. heartbeat, brain activity) can no longer be detected.

**death rate** The number per thousand of a population that die each year. Compare birth rate.

**Decapoda** 1 An order of ⋆crustaceans, including crabs, lobsters, crayfish, prawns and shrimp. All have five pairs of legs, two pairs of antennae, and a ⋆carapace over the front part of the body. 2 An order of ⋆cephalopod ⋆molluscs with ten arms. e.g. ⋆squid, ⋆cuttlefish. See Appendix 1.

**decarboxylation** The loss or removal of carbon dioxide from an organic acid, especially the amino acids.

**decay** 1 To decompose or rot. 2 The decomposition of organic matter through the actions of decomposers such as bacteria and fungi. 3 Tooth decay (caries) arise when the tooth enamel is eaten away and bacteria invade the *dentine inside the tooth. If they reach the tooth pulp, then persistent pain (tooth ache) is felt.

**deciduous** The shedding or falling of leaves each year (annually) from trees and shrubs, usually in autumn. See also abscission. Compare evergreen.

**decompose** To decay, rot or break down. Usually due to the action of decomposers such as bacteria and fungi.

**decomposer** An organism that helps break down and rot the dead bodies (plant or animal) or wastes (excreta) of other organisms, thus returning nutrients to the soil. e.g. bacteria and fungi (yeasts, moulds, etc.).

**decorticate** To shed outer bark, usually in long strips.

**decurrent** Running down along a stem, after joining with it, as with leaf-bases or petioles.

**decussate** The arrangement of leaves where successive pairs of *opposite leaves are at right angles to each other along the stem. See Fig. L1.

**defence** 1 Resisting attack; protection. 2 **Territorial defence**: The aggressive behaviour of animals to drive out competitors for food, mates, etc. See agonistic. 3 **Defence mechanisms**: The production of *antibodies and *antitoxins to protect an organism from disease. See immune system.

**deficiency disease** Sickness caused by an insufficient supply of one or more of the essential dietary chemicals. e.g. Lack of vitamin C helps cause scurvy.

**deforestation** The process of clearing of forests; the state of having been cleared of forests. e.g. The deforestation in tropical areas is causing great concern amongst environmental experts.

**dehiscence** The natural bursting or splitting open of a part of a plant when ripe or mature. e.g. Capsules on moss plants, fruits, seed pods, or anthers.

**dehydrate** To remove water.

**dehydrogenase** An enzyme that catalyses the removal of hydrogen atoms from a molecule, thus oxidising the molecule.

**deleterious** Hurtful, harmful.

**deletion** A type of *chromosome mutation where a section of chromosome is lost during mitosis or meiosis.

**delirium** 1 A more or less temporary loss of normal mental functions. e.g. During fevers. 2 **Delirium tremens**: Called **DTs** for short. The effect of long term abuse of alcohol, including trembling and terrifying hallucinations.

**deltoid** Triangular in shape. e.g. A shoulder muscle that helps move the arm, connecting the humerus to the clavicle and scapula.

**dementia** 1 A condition of deteriorated mental ability sufficient to interfere with work and social life. Emotional expression is also often changed. 2 Senile dementia is associated with old age and begins with failing attention and memory, loss of mathematical ability, irritability and loss of sense of humour, and poor orientation in space and time. See also Alzheimer's disease.

**denaturation** The altering of the

three-dimensional shape of a molecule, and thus its physical and biological properties, due to heat treatment, too much acid or alkali (base) etc. See also conformation.

**dendrochronology** The use of the pattern of annual ★growth rings in trees to help date archaeological, fossil and climatic changes.

**dendrite** The usually branched part of a ★neurone that receives impulses from other neurones and conducts the impulse towards the cell body. See Fig. N1. Compare axon.

**dendroid** Tree-like. Branching like a tree in shape.

**denitrification** The action of certain soil bacteria in changing (reducing) nitrates to simpler compounds (e.g. nitrites, ammonia) and eventually to nitrogen gas.

**denizen** An inhabitant or resident.

**density** 1 The mass per unit of volume. 2 Population density: The numbers of organisms per unit volume or per unit area.

**dental pulp** The living tissue inside the ★tooth that contains the nerves and blood vessels.

**dentate** Describing leaves with a toothed edge. See Fig. L1.

**denticulate** Describing a leaf edge with small teeth. See Fig. L1.

**dentine** The hard and dense material of mainly calcium carbonate, forming the major part of the ★tooth. Dentine surrounds the dental pulp, and is covered by the even harder enamel of the crown.

**dentition** 1 Teething; the growing teeth breaking through the gums. 2 The kind, number and arrangement of the teeth in the tooth row.

**deoxygenate** To remove oxygen from a material.

**deoxyribonucleic acid** Called ★**DNA** for short.

**deoxyribose sugar** A five-carbon sugar important in ★DNA. See Fig. R6.

**dependence** 1 The state of relying on one particular thing for aid or support, especially with ★drugs. 2 **Physical dependence**: Extensive use of drugs causes changes to the nervous system resulting in severe ★withdrawal symptoms if the drug is not continually used. 3 **Psychological dependence**: A strong emotional desire or craving to continue using a drug, often without any physical need to do so. See also tolerance.

**depolarisation** The decrease of electrical potential difference (volts) across a cell membrane. e.g. When a nerve ★impulse passes along a neurone, the movement of sodium and potassium ions across the cell membrane causes the voltage to drop. See also action potential.

**depressant** A drug that lowers the activity of the central nervous system. e.g. alcohol, barbiturates.

**Dermaptera** An order of insects that are often called ★earwigs. These relatively long flattened insects have a pair of large, but harmless, ★cerci at the end of their abdomen. See Appendix 1.

**dermatitis** An inflammation of the skin.

**dermis** The layer of the ★skin found under the ★epidermis.

**desert** Land where rainfall is so low (often less than 250 mm per year), and irregular, and covered with only sparse, widely spaced vegetation (or none at all). Most desert-growing plants have special adaptations to reduce loss of water.

**desert pea** Also called **Sturt's desert pea**, *Gossypium sturtianum*, a spectacular scarlet and black flower found in inland Australia.

**desertification** The process of becoming ★desert due to land mismanagement and/or climate change.

**desiccation** The process of thoroughly drying out as water is lost.

**detritus** Pieces of dead organic matter.

**Deuteromycetes** A class of fungi (fungi imperfecti) many of whose members cause serious plant diseases. Other examples include *Candida* which causes the disease thrush in humans, *Botrytis* which gives special flavours to wine, ★*Penicillium*, and *Aspergillus flavus* (see flavotoxins). Fungi are placed in this class because they have not yet been found to produce sexual spores and most will be reclassified in time. See also Appendix 1.

**development** The series of orderly changes that occur to a fertilised ovum (★zygote) as it divides and matures into tissues, organs and finally the complete organism (Fig. R4).

With humans, the stage after the zygote is called an ★embryo. At nine weeks it is called the ★foetus. In the early stages of embryo development, the cells begin to ★differentiate. The zygote undergoes cell division (mitosis) to form a solid ball of cells, and then develops into hollow ball called a ★blastocyst. At about the seventh day of development, the blastocyst attaches itself (implanted) to the soft walls of the ★uterus. The blastocyst continues to develop, being called a ★gastrula when three cell layers (★germ) have formed around the hollow. Each of these three layers develops into the various organs as the embryo matures. In particular, the ★placenta develops from the outer layers of the embryo. By the end of the first month, the embryo has developed limb buds, and the start of the nervous system, eyes, heart, etc.

**Devonian** A period in the Palaeozoic era, after the Silurian but before the Carboniferous. See Appendix 2.

**dextrose** See glucose.

**diabetes** A disease caused by an imbalance in the amount of the hormone ★insulin in the body. **a Diabetes mellitus**, or **sugar diabetes**, is caused by too little insulin resulting in a build-up of sugar in the blood (and hence urine). **b Diabetes insipidus** is caused by too little ★vasopressin in the blood, resulting in an enormous thirst and large urine flow.

**diagnosis** In medicine, the process, or its result, by which a doctor decides the nature and causes of a disease.

**diakinesis** The final stages of ★prophase during the first ★meiotic division, just before the nuclear membrane disappears. See Fig. M1.

**dialysis** 1 The separation of small particles from larger ones by allowing the smaller ones to ★diffuse through tiny pores (holes) in a ★semi-permeable membrane. e.g. Large molecular proteins and polysaccharides (such as starch) are trapped in a dialysis tube while the smaller sugar and salt particles can pass through. 2 When kidneys are not working correctly, a machine can remove waste products using the process of dialysis.

**diapause** A resting condition (type of ★dormancy) in which the metabolic rate is very low. Commonly occurs in insects. Compare hibernation, aestivation.

**diaphragm** 1 A thin membrane or partition used to divide two spaces. 2 The muscular division between the ★abdomen (containing the intestines) and the thoracic cavity (containing the heart and lungs). See thorax. 3 On a ★light microscope, the device for controlling the amount of light reaching the specimen on the stage. 4 A device used in contraception, being placed over the neck of the cervix to prevent the passage of sperm.

**diarrhoea** Highly frequent loss of very liquid ⋆faeces.

**diastase** See amylase.

**diastole** The relaxation of the heart muscles following the muscular contraction which actually pumps the blood. During diastole the heart fills with blood. Compare systole. See blood pressure.

**diatom** Microscopic single celled and colonial algae with silica cell walls. Sometimes classified with the Chrysophyta, but often regarded as a separate Phylum (Division). Diatoms are an important part of the phytoplankton in both lakes and oceans. See ⋆Chrysophyta and Appendix 1.

**dibbler** A variety of marsupial 'mouse', ⋆*Antechinus apicalis*.

**dichotomous** 1 Divided into two parts. 2 **Dichotomous key**: A method for identifying/classifying organisms where each feature selected to help identification requires a choice between two alternatives.

**dicot** Short for ⋆**Dicotyledonae**.

**Dicotyledonae** Called **dicot** for short. One of the two major divisions of flowering plants where the seeds contain two cotyledons, and a number of other distinguishing features. e.g. Rose, peach, eucalypt. Compare monocotyledon. See Appendix 1.

**Didelphidae** A family of marsupials also called opossums, living in north and south America. See Appendix 1.

**dieback** A disease of shrubs and trees often caused by a fungus. The plant progressively dies, starting at the tips of the branches and extending down to the trunk.

**differentiation** The process during ⋆development whereby newly formed cells become more specialised as they mature. e.g. **a** In root tips of plants, the newly formed cells differentiate into phloem, xylem, root hairs and packing cells, etc. **b** In animals, the ⋆germ layers differentiate into various organs of the body. See also embryo.

**diffusion** Also called **passive transport**. The movement of atoms, ions or molecules from regions of high concentration to regions of lower concentration. Compare active transport. See osmosis, dialysis, semipermeable.

**digestion** The process whereby the large, complex food molecules are broken down into molecules simple and small enough to be absorbed into the blood stream. See digestive system.

**digestive juices (enzymes)** The mixture of enzymes and other chemicals (e.g. acid, bicarbonate ions) secreted into the ⋆alimentary canal to break down food into its components. See saliva, gastric juices, pancreatic juices, bile, intestinal juices, and digestive system.

**digestive system** The alimentary canal and its associated organs where ⋆digestion occurs. See Fig. D2.

**a** Food taken into the mouth (⋆ingestion) is mechanically broken down by the combined action of the ⋆tongue and ⋆teeth. This helps increase the food's surface area, allowing the ⋆digestive juices to work more quickly. **b** The ⋆saliva mixed with the food helps lubricate it so that it moves as a lump called a ⋆bolus down the food pipe (⋆oesophagus) more readily. The saliva also contains a digestive enzyme (salivary ⋆amylase) that starts to break down complex ⋆carbohydrates such as starch. **c** The food travels to the stomach via the oesophagus. The muscular cardiac ⋆sphincter at the bottom end of the oesophagus helps prevent food returning to the mouth. **d** In the ⋆stomach the food is churned by the muscular walls, and mixed with digestive juices and hydrochloric acid to form ⋆chyme. These juices, such as pepsin, start to breakdown the ⋆protein molecules in the food. **e** The muscular ⋆pyloric sphincter at the bottom of the stomach releases food into the start of the ⋆small intestine, called the ⋆duodenum. Here it is mixed with digestive juices from the ⋆pancreas, ⋆gall bladder, and the walls of the duodenum itself. These juices turn the food from acid to alkaline, and complete the digestion process. Proteins become a mixture of ⋆amino acids. Carbohydrates form simple sugars such as glucose. Fats are first emulsified (turned into small droplets) by ⋆bile from the gall bladder, and then broken down into fatty acids and glycerol. **f** The amino acids, simple sugars, and minerals are absorbed into the bloodstream from the small intestine (Fig. V3). The fatty acids and glycerol are absorbed into the cells in the wall of the intestine where they recombine to form fats, and later enter the ⋆lymphatic system which eventually drains into the blood stream. **g** Undigested material then passes to the ⋆large intestine where moisture is absorbed and bacteria make further changes. **h** Finally, any remaining undigested material is eliminated through the ⋆anus (⋆egestion).

**digitigrade** Describes mammals which walk with their body weight supported by the digits (metacarpals or metatarsals). e.g. Dogs, horses. Compare plantigrade and unguligrade.

**dihybrid** An organism that is hybrid (⋆heterozygous) in at least two genetic features. e.g. TtRr. Compare monohybrid.

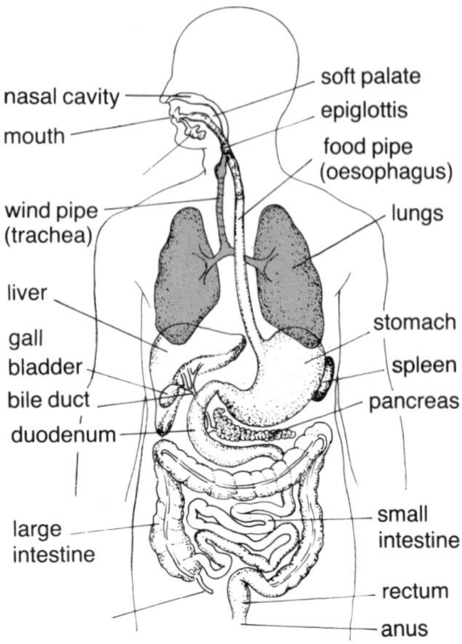


**Fig. D2 Digestive system**

**dilation** The act of making something wider or larger. e.g. When the pupil of an eye dilates, the hole in the ⋆iris increases in diameter allowing more light to reach the retina. Compare constriction. See also vasodilation.

***Dillwynia*** A genus of ⋆endemic Australian shrubs in the pea family Fabaceae. e.g. Parrot pea or 'eggs-and-bacon'. See Appendix 1.

**dimorphism** **1** Amongst animals, the occurrence of two different forms of the same species. e.g. Reef herons exist as either dark grey or white forms. South American Jaguars are either spotted or are black. **2** Amongst plants, the occurrence of two types of a particular organ such as leaves, flowers etc. on the same or on different plants of the same species. e.g. In conifers, male and female cones are different.

**dingo** Australian placental wild dog, *Canis familiaris dingo*.

**Dinoflagellates** Members of Phylum (Division) Pyrrhophyta. Most are single celled, aquatic algae with two ⋆flagella. Generally they are the dominant ⋆phytoplankton of tropical seas. See also Appendix 1.

**dinosaur** Any of a large variety of often gigantic fossil reptiles from the Mesozoic era. See *Tyrannosaurus*. Appendix 1 and 2.

**dioecious** Having separate male and female plants; male flowers with only stamens on one plant, and female flowers with only carpels on another plant of the same species. e.g. Holly, willows and hemp; some *Allocasuarina* in Australia. Compare monoecious.

**dioxin** A group of complex organic compounds, but especially 2,3,7,8-tetrachlorodibenzodioxin, a highly toxic impurity formed during the manufacture of the herbicide ⋆2,4,5-T. Causes birth defects.

**diphtheria** A highly ⋆contagious and serious infection caused by the bacterium *Corynebacterium diphtheriae*, once common in children. Most children are now immunised against this disease. See also triple vaccine.

**diploid** Having two of each type of chromosome (except for the ⋆sex chromosomes). Hence there must be two ⋆alleles present for each gene (e.g. TT, Tt or tt). In humans, the body cells are diploid. Only the sex cells (sperm and ova) differ, having only one of each type of chromosome (see haploid).

**Diplopoda** The class of arthropods that is also called ⋆millipedes. See Appendix 1.

**diplotene** A stage during prophase I of ⋆meiosis where ⋆crossing over (Fig. H3) occurs. See Fig. M1.

**diprotodont** 1 Diprotodonta, an order of marsupials having fewer than three upper incisor teeth on each side of the jaw. e.g. Kangaroos, wombats, and some giant marsupials only found as fossils. 2 *Diprotodon*, a genus of giant fossil marsupials.

**Diptera** One of the larger orders of insects, the 'true' flies. It includes the two-winged flies, blowflies, gnats, midges, and mosquitoes. The front wings are membranous; the rear wings are club-like 'balancers' called ⋆halteres. They have sucking or piercing/sucking mouthparts. See Appendix 1.

**disaccharide** A molecule formed when two simple sugars (monosaccharides) are chemically joined together. e.g. Sucrose (table sugar) is formed when a molecule of glucose is chemically bonded to a molecule of fructose. See Fig. C2.

**disc** 1 In plants, the flattened head of some daisy-like flowers. e.g. Sunflower. 2 In humans, the discs between the ⋆vertebrae (Fig. V2) made of cartilage, allowing some movement between vertebrae, and absorbing shock in the vertical direction. See also slipped disc.

**disease** 1 A sickness, illness, malady, ailment. 2 The result of the body not functioning correctly due to infection, injury or incorrect body metabolism.

**disinfectant** A chemical that kills bacteria and other disease-causing microbes. Disinfectants are often toxic to humans, and thus only applied to non-living structures. Compare antiseptic.

**dislocate** To put out of place. e.g. When a bone is dislodged from its socket at a ⋆joint.

**dispersal** To scatter, spread far and wide. e.g. Dispersal of ⋆seeds from a ripened fruit. Some seeds are very light, and may have hairs or 'wings' to assist their dispersal by the wind (dandelion). Other seeds have spines or hooks (bindii), or are sticky (paspalum), so that they can attach themselves to passing animals. Many seeds are found inside edible fruits, and pass through the digestive system of animals to leave with the faeces (mistletoe).

**display** Specific, and sometimes very elaborate, sets of behaviour used to communicate to members of the same species. e.g. ⋆Courtship display of the male Bower bird in its bower. Aggressive territorial display by nesting kookaburras. Display rarely leads to actual violence. See also agonistic.

**dissect** To carefully cut open and expose the interior parts of a plant or animal for study purposes. See also vivisection.

**distal** Situated away from the place of origin or point of attachment. Compare proximal.

**distend** To swell, become stretched or bloated.

**distribution** The natural range or area that a species or group of species inhabits. Compare abundance.

**diuretic** A substance that increases the volume of urine produced.

**diurnal** Occurring every day; active during the daytime. See also nocturnal, crepuscular, circadian rhythm.

**divergence** Moving or spreading apart as from a point. e.g. **Divergent evolution**: The many types of Darwin's finches on the Galapagos Islands have developed from perhaps one original pair.

**diversity** The many different types (variety) in living things.

**Division** In plants, a major classification group (taxon) made up of many classes, comparable to a Phylum in animals.

**DNA** Short for **deoxyribonucleic acid**. The 'molecule of life', mostly found in the nucleus of a cell. Contains a chemical sequence of organic bases that specify the ⋆genetic code. The genes that control what happens in a cell are thus arranged in sequence along this thin long molecule. (See Fig. B3 and D3.)

The two strands of the DNA molecule are wound around each other in the form of a ⋆double helix. The 'backbone' of each strand is on the outside, and is made from ⋆deoxyribose sugar and ⋆phosphate. Projecting towards the centre are the organic bases guanine, cytosine, thymine and adenine, which act as the letters in the genetic code. Each combination of an organic base, deoxyribose sugar, and phosphate is called a ⋆nucleotide. The string of nucleotides when joined together makes up the ⋆nucleic acid. DNA molecules produce copies of themselves during each ⋆mitosis and ⋆meiosis by a process called ⋆replication. The DNA is ⋆transcribed into messenger RNA during protein synthesis. See also base pairing, nucleosome. Compare RNA.

**DNA fingerprint** See fingerprint.

**DNA ligase** An enzyme that can join together the 'sticky ends' of segments of DNA inserted into a ⋆plasmid during ⋆genetic engineering. It is also part of the mechanism cells use to repair DNA. See Fig. G2. Compare DNA polymerase, and restriction enzymes.

**DNA packaging** See nucleosome.

**DNA polymerase** An enzyme that, during ⋆replication (which occurs during cell division), chemically links together the organic bases that form the ⋆genetic code of DNA. The phosphate part of each ⋆nucleotide is joined to the next deoxyribose sugar to form a DNA molecule that is a ⋆complementary copy of the ⋆template (original) DNA. See base pairing (Fig. B3) and double helix (Fig. D3).

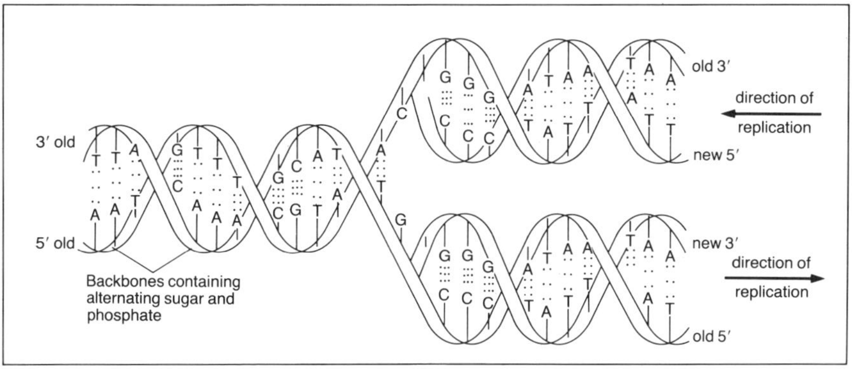


**Fig. D3 Double helix**

**dog** Any member of *Canis familiaris,* part of the placental mammal order Carnivora. See also dingo and Appendix 1.

**dominance** In genetics, the ability of one form of a gene (or ⋆allele) to determine the resulting character (⋆phenotype) in a hybrid (or ⋆heterozygous) organism. The other allele is said to be ⋆recessive. e.g. With Mendel's peas, T (tall) dominates t (short), resulting in the hybrid Tt growing tall.

**donkey** Any of several horse-like ass, especially the domestic donkey *Equus asinus,* order Perissodactyla, often used as a beast of burden. See Appendix 1.

**dopamine** A complex organic molecule that is used to make ⋆adrenaline and noradrenaline. It also acts as an inhibiting ⋆neurotransmitter in certain parts of the brain. See also Parkinson's disease, catecholamine, schizophrenia.

**dormant** A period during which growth ceases, and other activity is at a minimum, so that an organism can survive hard times. e.g. Animals during ⋆hibernation, ⋆aestivation and ⋆diapause. Plants can lie dormant in the soil as seeds, bulbs, tubers, etc.

**dorsal** Towards the back or upper surface of an animal. See Fig. S11.

**double helix** The 'spiral staircase' structure of the two strands of the DNA molecule. The DNA unwinds during ⋆replication so that ⋆complimentary copies can be made. See DNA above, and Fig. B3 and D3.

**Down's syndrome** A birth defect caused by there being three chromosome number 21s in the body cells, rather than the normal two. A number of deformities are visible (slanting eyes, broad face, stubby nose), and mental development is limited. See also chromosome mutation.

**dragon** ⋆Lizards in family Agamidae, including the bearded dragon, thorny devil, and frilled-neck lizards. See Appendix 1.

**dragonfly** Long thin insects in the order Odonata which differ from damselflies by holding their membranous wings horizontally from their body when they have landed. See Appendix 1

**drench** To administer medicine, often forcefully, to an animal through its mouth.

**drone** A male honey ⋆bee.

***Drosophila*** A genus of fruit-flies, especially *D. melanogaster* in Order Diptera, which is widely used in genetics experiment. See also mutation.

**drought** Lack of sufficient rain in an area for normal growth.

**drug** 1 A chemical substance that, when taken into the body, may modify one or more of the functions of the body. Usually used to treat sickness or pain. e.g. ⋆Aspirin, ⋆penicillin. 2 A chemical substance the taking of which may be habit forming. e.g. Alcohol, narcotics. See also dependence.

**drupe** A stone fruit such as the cherry, plum and peach. It contains a skin (⋆epicarp), succulent pulpy fruit (⋆mesocarp), and a hard woody inner shell or stone (⋆endocarp) that encloses the seed. Compare berry. See Fig. F4.

**duct** A tube or canal carrying body fluids. e.g. The pancreatic duct carries digestive juices from the pancreas to the small intestine (duodenum). See Fig. D2.

**ductless gland** A gland that secretes its products directly into the blood without use of a duct or tube. e.g. Thyroid, pituitary, adrenals. See also endocrine. Compare exocrine.

**dugong** A large aquatic herbivorous placental mammal (*Dugong dugon*) found in tropical waters around northern Australia and the Indian Ocean. Similar to manatees found in other parts of the world. See Appendix 1.

**dung** Manure. Animal excrement. Faeces.

**dung beetle** Any of the many scarab ⋆beetles that lay their eggs in and feed upon the dung of animals. e.g. The introduced *Sisyphus quadricollis* is able to remove and bury cow dung pats that native species are too small to 'handle'.

**dunnart** Any of the small, nocturnal, insect eating narrow-footed marsupial 'mice' of the genus *Smithopsis*, family ⋆Dasyuridae. See Appendix 1.

**duodenal ulcer** A type of ⋆peptic ulcer which occurs on the inside surface of the duodenum (first section of the small intestine).

**duodenum** The first part of the ⋆small intestine where ⋆digestion of the food received from the stomach is completed, and most of the digestion products are absorbed into the blood stream and lymph. See also jejunum, ileum, and Fig. D2.

**duplication** A type of ⋆chromosome mutation where a section of DNA is doubled during mitosis or meiosis.

**dwarf** An animal or plant that is much less than the normal size for that species.

**dysentery** A disease of the lower part of the intestine which is ulcerated and inflamed, causing severe ⋆diarrhoea. The disease is very painful, and the faeces often contain blood and mucus. The disease can be caused by amoeba or bacteria.

**dysfunction** Any impairment of normal functioning.

**dyspepsia** Disturbed digestion; indigestion.

# Ee

***E. coli*** Short for *Escherichia coli*. See coliform bacteria and intestinal flora.

**ear** The organ of hearing and equilibrium (balance).

The ear consists of three parts: **a** External ear collects the sound waves and channels them to the eardrum. **b** Middle ear where sound waves cause the eardrum to vibrate. These vibrations are sent to the inner ear through three small bones called *ossicles. These bones are called the malleus (hammer), incus (anvil) and stapes (stirrup). The ossicles act as a lever system, amplifying the vibration before it reaches the inner ear. **c** Inner ear structure that detects sound is the *cochlea. This coiled tube contains sensitive nerve endings projecting out into a fluid. When the fluid is set vibrating the nerves detect this, and send the signal produced to the brain (via the auditory nerve) where they are interpreted as sound. High frequency notes are detected at the start of the coil, and low frequency notes at the far end. On top of the cochlea are the organs of equilibrium and balance—see inner ear. See Fig. E1.

**eardrum** Also called the **tympanum**. The thin membrane separating the external from the middle ear. Sound waves cause this membrane to vibrate. The vibrations are then passed to the small bones (ossicles) of the middle ear. See Fig. E1.

**earths** Types of soil. **1 Black earths.** Mostly occur on the western side of the Queensland highlands, and in northern New South Wales where there are seasonal rains of 400–600 mm. They become very sticky when wet, and are dark due to their humus content. **2 Red earths.** Mostly occur on the inland slopes of south-eastern and south-western Australia where there are seasonal rains of 350–600 mm. Not as much humus, and less fertile than the black earths.

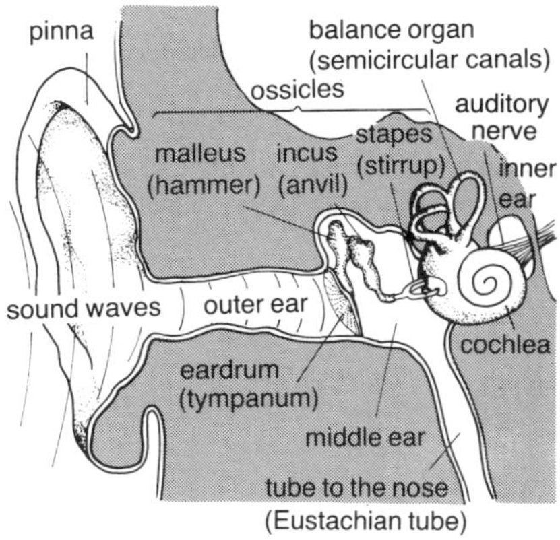


**Fig. E1 Ear**

**earthworm** A soil inhabiting segmented (*annelid) worm that feeds on decaying organic matter as it burrows through the soil. See also clitellum, hermaphrodite, and Appendix 1.

**earwig** An insect in order Dermaptera, having two forceps-like *cerci on the tip of the abdomen. See Appendix 1.

**Ebola virus** A *filovirus which causes a high fever and uncontrollable internal bleeding. It spreads through contact with body fluids.

**ecdysis** Also called ***moulting**.

**ecdysone** Also called ***moulting hormone**.

**ECG** Short for **electrocardiogram** or **electrocardiograph**. 1 Electrocardiogram, a graph or other record produced by an electrocardiograph. 2 Electrocardiograph, a device used to measure and record the small differences of electrical potential (voltage) that occur as the heart is pumping. Electrodes are attached to the skin, usually on the arms and legs, and the results can be used to help diagnose heart disease.

**Echidna** A spine-covered *monotreme mammal found in Australia and nearby islands. Eats ants and termites. Lays leathery shelled eggs from which the young are born. See Appendix 1.

**Echinodermata** A phylum of marine invertebrates that includes the *sea stars, *brittle stars, *sea urchins, *sea cucumbers, *crinoids, etc. The outer surface of their body is covered with limy plates, and often with spines. They also have an internal *water vascular system. See Appendix 1.

**Echinoidea** A class of echinoderms that includes the sand-dollars, and the spine-covered *sea-urchins. See Appendix 1.

**echolocation** The use of high frequency sound waves produced by bats, dolphins and whales to determine the location of objects around them, by detecting the reflections.

**ecology** The study of the interactions of organisms with their physical environment and with other organisms.

**ecosphere** Those parts of the planet where living things can be found. The world's life zone. See also biosphere.

**ecosystem** An ecological system. All the organisms present in a particular area together with the physical environment with which they interact. Compare biome and habitat. See Fig Z2 for Australian ecosystems.

**ecotourism** Travelling for recreation where the emphasis is on visiting areas of environmental interest, often guided by experts in *ecology.

**ectoderm** The outermost of the three cell layers (ectoderm, mesoderm, endoderm) that form the ball-shaped *gastrula stage of the *embryo. The ectoderm develops to become the skin, nails, hair, sense organs, brain, spinal cord, etc.

**ectomorphic** A lightly built body type. Compare endomorphic and mesomorphic.

**ectoparasite** A *parasite that lives on the surface of another organism. e.g. Tick, louse, flea.

**ectotherm** A *poikilothermic or 'cold blooded' animal. An animal whose body temperature is regulated by the temperature of the external environment. e.g. Reptiles and amphibians.

**edema** See oedema.

**Edentata** An order of placental mammals that includes the toothless, insect-eating sloths and armadillos. See Appendix 1.

**editing** The converting of *primary transcript RNA into *messenger RNA by the removal of non-coding *introns. See transcription, capping, tailing, and Fig. T2.

**EEG** Short for **electroencephalogram** or **electroencephalograph**. 1 Electroencephalogram, a graph or other record produced by an electroencephalograph. 2 Electroencephalograph, a device used to measure and record the rhythmic electrical activity of the brain. These signals are mainly produced by the cerebral cortex and can be measured by attaching electrodes to the skin covering the skull. The results of such measurements are useful in brain research (e.g. study of sleep), and in diagnosis of diseases such as epilepsy.

**effector** 1 In humans, a ⋆muscle or ⋆gland capable of producing a response to a ⋆stimulus. 2 Various other animals have additional effectors such as ⋆cilia and ⋆chromatophores.

**efferent** Travelling or leading away from a central point. Usually applied to nerves and blood vessels. Compare afferent. See Fig. R2.

**efflorescence** The period of flowering, or ⋆anthesis.

**egestion** To discharge from the body. e.g. Elimination or voiding of undigested materials (faeces) from the body. See digestive system.

**egg** The female reproductive cell or ⋆gamete. 1 In humans, the ⋆ovum. 2 In other animals, the egg may be enclosed in a shell and contain varying amounts of food reserves. The egg white is a mixture of ⋆glycoprotein and water, while the yolk is a mixture of protein and fat. e.g. Birds and reptiles. The eggs of amphibians, fish, etc. may have a jelly-like covering to help in their protection.

**egg tooth** A projection from the upper beak of a bird, or upper jaw of a reptile, that it uses to break the egg-shell from the inside on hatching.

**ejaculation** The forceful and rhythmic discharge (throwing out) of semen during sexual intercourse.

**El Niño event** The reversal of normal climate when warm water flows from Indonesia towards the Americas, taking tropical storms with it that cause floods in Peru but leave droughts in Eastern Australia. Normally, trade winds push water from east to west across the Pacific towards the 'warm pool' of ocean surrounding eastern Indonesia. When an El Niño event occurs, the surface temperature of the 'warm pool' drops so that trade winds weaken, allowing warmer water to flow eastwards towards the Americas.

**Elapidae** The family of venomous snakes with fixed fangs in the front of the jaw. It includes some of the most dangerous types in the world, including mambas, cobras, and kraits. Australian examples include the death-adder, taipan, tiger snake, king brown, etc. See Appendix 1.

**elastin** The main protein found in the elastic 'yellow fibres' of ⋆connective tissue. Elastin is common in the elastic walls of blood vessels and the ⋆alveoli of the lungs. Compare collagen.

**electric organ** A structure found in some fish that can produce voltages of up to 600 volts. Used to paralyse predator and prey.

**electrocardiogram** Called ⋆ECG for short.

**electrocardiograph** See ECG.

**electroencephalogram** Called ⋆EEG for short.

**electroencephalograph** See EEG above.

**electron** The negatively charged particle (mass 1/1836th of a ⋆proton) that orbits around the atomic ⋆nucleus. The chemical, and hence biological properties of chemicals, depends on the number and arrangement of these electrons.

**electron micrograph** A photograph taken using an ⋆electron microscope. (See inside front and back covers for examples.)

**electron microscope** A microscope that uses ⋆electrons instead of light rays to produce a magnified image of objects (⋆specimens). The image can be produced on a fluorescent screen (like a TV screen), or focused onto a photographic plate. Because the 'wavelength' of electrons is much shorter than light, much higher magnifications can be produced (more than 300 000X compared to 1500X with a ⋆light microscope). See Figs E2 and L2. There are two main types: **a** Transmission electron microscopes require ultra-thin specimens so that the electrons can pass through for the image to be formed. e.g. Images of the internal structure of cells. **b** Scanning electron microscopes produce images of the surface features of objects. e.g. Details of the eye of an insect (see inside the back cover). The specimen often has to be coated by an ultra-thin layer of metal atoms for the image to be produced.

**electron transfer** When a molecule loses or donates an electron in a chemical reaction, it is said to be ⋆reduced. If a molecule accepts an electron, it is said to be ⋆oxidised.

**electron transport chain** Part of the complex series of chemical reactions that occur during cellular ⋆respiration. Along the inner walls of ⋆mitochondria a series of enzymes are arranged. Chemicals such as NADH produced in ⋆glycolysis and the ⋆Krebs' (citric acid) cycle donate electrons to this chain of enzymes. As the electrons pass along this chain, energy is stored in ATP molecules, or released as heat. At the end of the electron transfer chain, the electrons, hydrogen ions and molecular oxygen combine to produce water molecules. See Fig. E3.

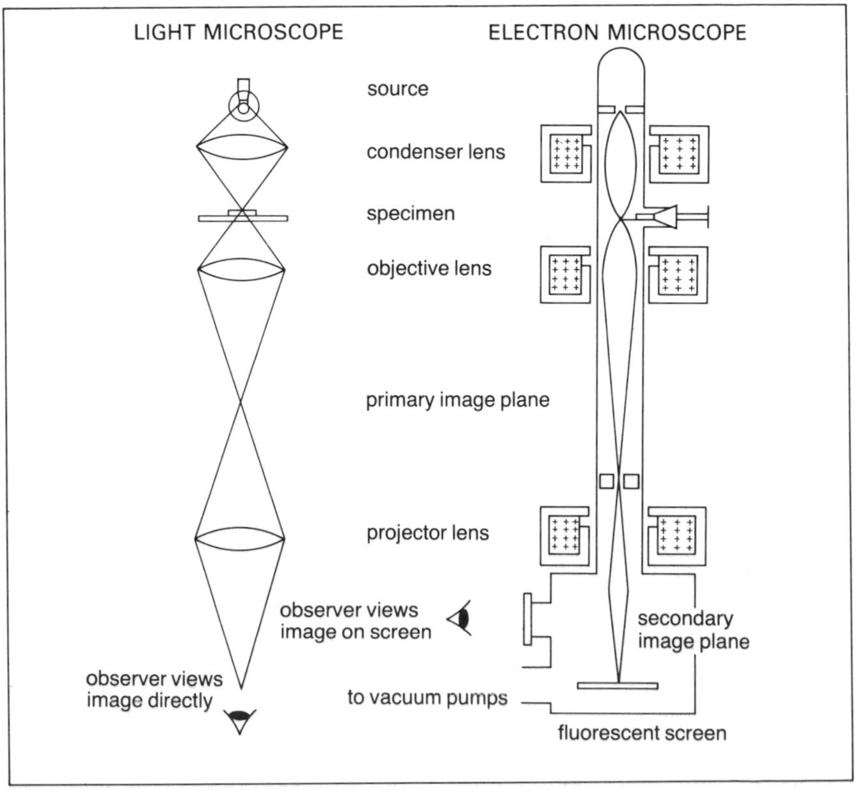


**Fig. E2 Electron microscope**

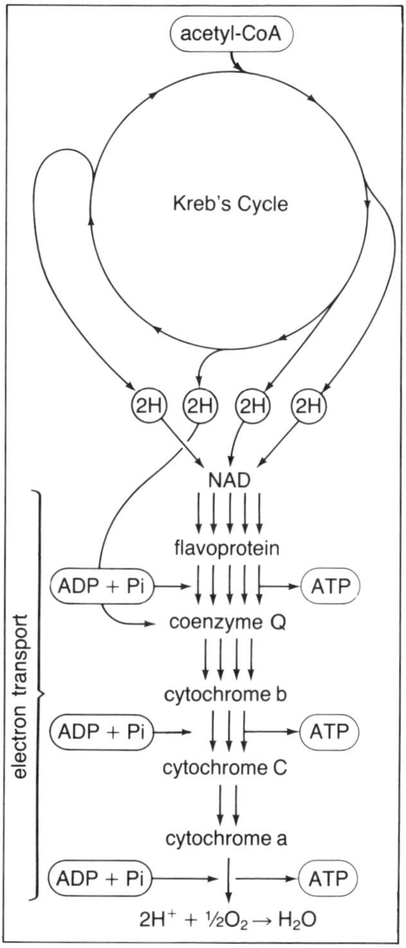


**Fig. E3 Electron transport chain (transpiration)**

**electronegative** Attracting electrons. Having an affinity for electrons.

**electrophoresis** A method used to separate charged molecules, especially proteins and small lengths of DNA. If each molecule in a mixture is given a charge, and then placed in a porous material (e.g. filter paper, ⋆gel) with an electric field, then the molecules will move at a speed depending on the size, shape and charge of the molecule, thus separating themselves. Used in DNA ⋆fingerprinting. Compare chromatography.

**element** A substance that cannot be chemically decomposed into a simpler substance. Each of the approximately 100 elements is made of only one type of ⋆atom. i.e. all the atoms of an element have the same number of electrons and protons.

**elephantiasis** A disease caused by an infection of the ⋆filaria worm (a nematode or roundworm) where the extremities of the body (e.g. arms and legs) become enormously enlarged.

**elimination** Egestion. Defecation. Muscular expulsion of undigested body wastes (faeces) from the end of the large intestine (the anus).

**elliptic** The shape of a leaf that is broadest across the middle, and becomes narrow towards the ends. See Fig. L1.

**emaciation** Becoming very thin due to lack of proper nutrition, or due to disease.

**emarginate** A leaf notched at the summit or apex. See Fig. L1.

**embolism** The blocking of a blood vessel by a blood clot (⋆thrombus), air bubble or tissue debris.

**embryo** The early stages of ⋆development of an organism that has been produced from a fertilised ovum (zygote). Compare foetus. **1** In humans, the embryo is the first three months of growth after fertilisation. **2** In animals in general, the growth stages before hatching or birth. **3** In plants, the growth before the young plant emerges from the seed. See Fig. E4.

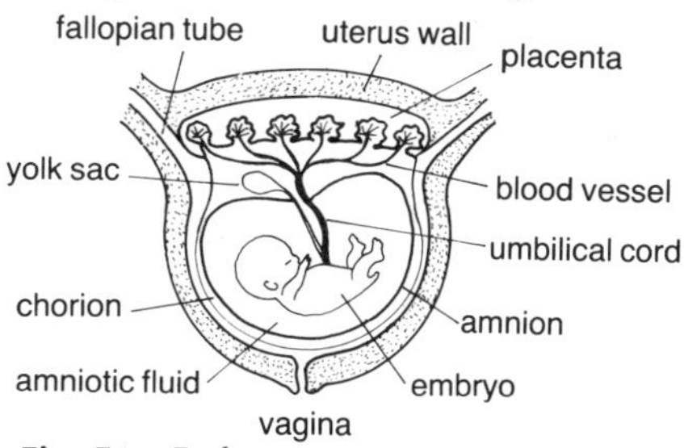


**Fig. E4 Embryo**

**embryology** The study of ⋆embryo development.

**emergent** Of trees. The tall trees of a forest that project above the ⋆canopy, and especially applied to rainforests.

**emetic** A medicinal substance to induce vomiting.

**emigration** Leaving an area of residence for some other place.

**emphysema** Damage to the ⋆lungs, usually caused by smoking cigarettes, where the walls of the ⋆alveoli in the lungs lose their elasticity and break (rupture). This reduces the amount of surface area available for exchange of oxygen and carbon dioxide. People with emphysema are thus often short of breath.

**empirical** Results and ideas obtained from experiments rather than from theory or deduction.

**emu** A large flightless bird *Dromatius novaehollandiae,* found throughout Australia. It is part of Australia's coat-of-arms.

**emulsion** A suspension of droplets of one liquid in another. e.g. Oil in water. It is a type of ⋆colloid.

**enamel** The very hard material (hardest in the body) that covers the ⋆crown of the teeth. Enamel is 97% inorganic (calcium salts) and 3% organic. See tooth.

**encephalitis** An inflammation in the brain.

**encephalon** The brain.

**endemic** 1 Restricted to a particular region or locality. 2 Native, in contrast to that which has been introduced from elsewhere. e.g. Native (endemic) animals such as the kangaroo compared to introduced animals such as rabbits.

**endergonic reactions** Also called ⋆endothermic.

**endocarp** The innermost of the three layers that make up the wall of a plants ovary (⋆pericarp) as it swells to form the ⋆fruit (from the outside, epicarp, mesocarp, endocarp). The endocarp may not be separable from the mesocarp (the thickest layer we eat) in berries, may be leathery as in citrus, or hard and stony as in peaches.

**endocrine gland** A ductless gland which produces a ⋆hormone (e.g. adrenaline, insulin, oestrogen) that is secreted directly into the blood stream. The hormone then acts on cells distant from the gland where it was produced.

**endocrine system** That system of ⋆endocrine glands that helps in the long term coordination of the body. In humans, this includes the ⋆pituitary, ⋆pineal, ⋆thyroid, ⋆parathyroid, ⋆adrenal glands, the ⋆gonads, ⋆placenta, ⋆Islets of Langerhans (in the ⋆pancreas).

**endocytosis** The process by which all materials enter a cell across the cell membrane, except for those substances small enough (e.g. ions, small molecules) to pass through the pores. The membrane either folds inwards to form a small sack (⋆vacuole), or if the particle is larger and solid, extends outwards to surround and ⋆engulf it (phagocytosis). Compare exocytosis. See Fig. E5.

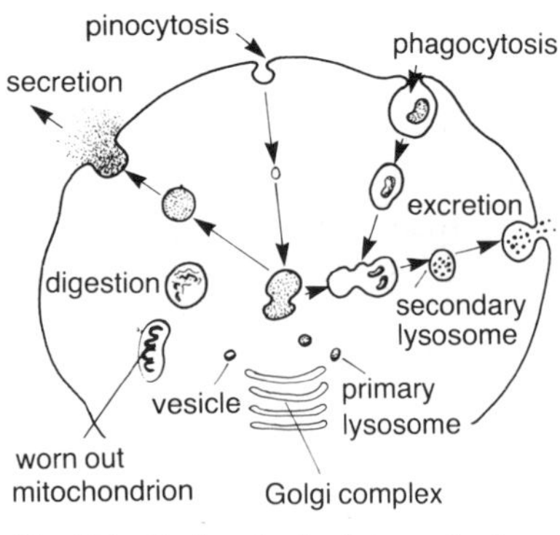


**Fig. E5 Endocytosis (pinocytosis and phagocytosis) and exocytosis (secretion and excretion)**

**endoderm** The innermost of the three layers (ectoderm, mesoderm, endoderm) that form the ball-shaped ⋆gastrula stage of the ⋆embryo. The endoderm develops into the lining of the alimentary canal (digestive tract), most of the respiratory tract, and the urinary bladder, liver, pancreas, and some of the endocrine glands. See also development.

**endodermis** The layer of ⋆parenchyma cells surrounding the ⋆vascular tissue of plants. The endodermis is most obvious in roots where it includes the ⋆Casparian strip. The Casparian strip forces water and minerals absorbed from the soil to pass through the cells. In this way the root cells exercise control over which chemicals can enter the plant's vascular system. See also roots and stele.

**endometrium** The glandular mucous (slimy, slippery) membrane that lines the uterus (womb) of mammals. In women, the lining thickens as the ⋆menstrual cycle progresses, under the influence of oestrogens and progesterones. If the egg is fertilised, the developing embryo may settle into the thickened walls of the uterus, where the ⋆placenta grows. If fertilisation does not occur, the thick wall breaks down during the menstrual flow.

**endomorphic** Having a heavily built body type. Compare ectomorphic and mesomorphic.

**endoparasite** A parasite that lives within its host. e.g. Tapeworm.

**endoplasmic reticulum** Called ER for short. An extensive system of membranes, present in most eukaryotic cells (those with a separate nucleus), that divides the ⋆cytoplasm into compartments and channels. See Fig. C4. The surface of ER may in parts be covered with ⋆ribosomes where protein synthesis takes place. The ⋆Golgi bodies are specialised parts of the ER. In plant cells, the ER often extends through pores into neighbouring cells. Compare cytoskeleton.

**endorphin** A substance produced in the brain that helps the body endure pain. ⋆Opiates such as heroin have a similar chemical structure and influence the same receptor sites in the brain. See also enkephalin.

**endoscope** An instrument that uses fibre optics for looking at the interior of a hollow organ such as the ⋆rectum or ⋆urethra.

**endoskeleton** An internal skeleton. e.g. The bony skeleton of vertebrates.

**endosperm** The food reserves that surround the embryo of a seed in flowering plants. In some plants (pea, bean), the endosperm is completely absorbed by the embryo when the seed is fully developed. More food reserves are in the ⋆cotyledons. In other seeds (wheat, caster oil), part of the endosperm remains after the seed has developed, and is not absorbed and used until the seed germinates.

**endothelium** The single layer of smooth flattened cells lining the insides of the heart, blood vessels, and lymph vessels, in vertebrates. Is much the same as ⋆epithelium in structure.

**endotherm** A ⋆homoiothermic or 'warmblooded' animal. An animal that regulates its own internal temperature through metabolic processes. e.g. Mammals and birds.

**endothermic** Also called **endergonic**. Chemical reactions where energy must be supplied from elsewhere. Thus the energy of the products is greater than that of the reactants.

**energy** The capacity for doing work.

**engineering, genetic** See genetic engineering.

**engulf** To swallow up. Single celled organisms (e.g. *Amoeba*) feed by extending their cell body (★pseudopodia) around their food and engulfing (absorbing, swallowing) it. See also endocytosis, phagocytes, and Fig. E6.

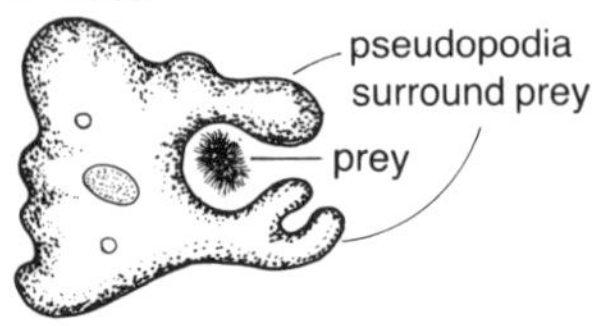


**Fig. E6 Engulfing**

**enhancer** In eukaryotic cells, a sequence of DNA that increases the activity of the ★promoter. See also operon.

**enkephalin** A substance produced in the brain that helps the body endure pain. ★Morphine has a similar chemical structure and influences the same receptor sites in the brain. See also endorphins.

**enteritis** Inflammation of the intestine, especially the small intestine.

**enthalpy** The heat content of a chemical.

**entire** Used to describe smooth-edged leaves without toothing, lobing, or division.

**entomology** The study of insects.

**entrails** The internal parts of the trunk of the body, especially the intestines.

**entropy** A measure of the randomness or disorder of a system.

**environment** An organism's physical and biological surroundings. The conditions under which an organism lives. Compare habitat.

**environmental impact study** A scientific study to help predict the effect on the surrounding environment of some change, such as new buildings, roads, airports, and other human activities.

**enzyme** A biological ★catalyst that speeds up chemical reactions. Enzymes are proteins, and each chemical reaction in a cell is controlled by its own enzyme. Thus a cell nucleus can control which chemical reactions are to take place by controlling the type and amount of enzymes produced.

Each enzyme is a protein with a unique three-dimensional shape. On the surface of the enzyme is an ★active site where another molecule (called the ★substrate) can be held by chemical bonds (often ★hydrogen bonds). The enzyme binds the molecule so that the bond that is to undergo the chemical reaction is exposed (Fig. E7). The enzyme may also weaken this chemical bond. After the reaction is complete, the changed molecule (product) is released leaving the enzyme unchanged. In this way less energy (★activation energy) is needed to start a particular reaction. This speeds up the chemical reaction, often by more than a million times. No artificial catalyst is so effective.

**enzyme substrate** See substrate.

**Eocene** The second epoch of the Tertiary period when modern mammals are believed to have mainly evolved. See Appendix 2.

**Epacridaceae** A family of dicot flowering plants often called native heath, and including ★*Epacris* and *Leucopogon*. See Appendix 1.

***Epacris*** A genus of heath-like shrubs in family Epacridaceae, with often spectacular masses of red, white or pink flowers. e.g. *Epacris impressa*, the floral emblem of Victoria. See Appendix 1.

**ephemeral** 1 Existing for only a short time; short lived. 2 A plant that completes its lifecycle—from seed

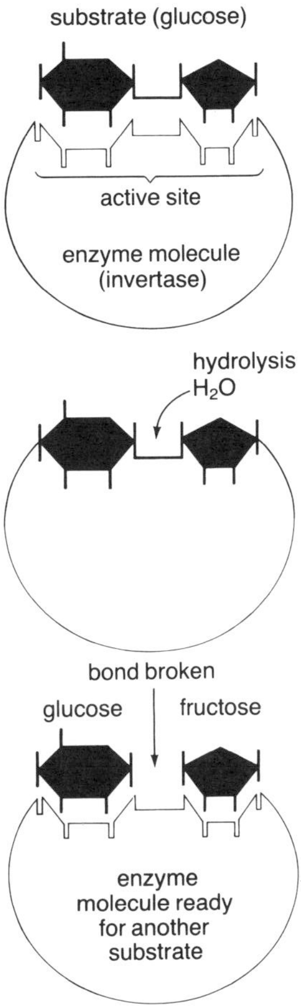


**Fig. E7 Enzyme**

germination to seed dispersal—in a relatively short period of time. Such life cycles are very short, with germination and growth being very rapid. e.g. Many desert flowers respond to rare rains in this manner. Compare annual, biennial, perennial.

**Ephemeroptera** An order of insects that includes the ⋆mayflies. See Appendix 1.

**epicormic growth** A shoot or branch growing from a ⋆dormant bud on the trunk of a tree, usually as a result of damage to the tree. e.g. Fire.

**epidemic** A temporary dramatic increase in numbers of living things suffering from a disease that is not normally present in an area. Such diseases are often contagious (readily spread from organism to organism) e.g. Cholera in humans.

**epidemiology** The branch of medicine that normally studies disease epidemics.

**epidermis** The outermost layer of cells covering a plant or animal. **1** Invertebrates usually have a covering of a single layer of cells. This is often covered by a layer of wax (⋆cuticle) that helps reduce water loss. 2 Vertebrates have an epidermis several layers thick, the outer–surface of which is made of dead, horny (⋆keratinised) cells. The outer surface is continually being rubbed off and replaced by new cells from beneath. 3 The outer surface of plants is covered by a 'pavement' of interlocking cells. These cells often produce a protective waxy layer (cuticle) to help reduce loss of water.

**epididymis** A mass of small tubes on the surface of the ⋆testes. Sperm produced in the testes are stored in the epididymis before passing along the sperm duct (vas deferens) to the penis. See Fig. R3.

**epiglottis** A 'hinged' flap of cartilage and membrane at the back of the mouth (pharynx) which closes over the opening (glottis) to the wind pipe (tracheae) when we swallow. This stops food and liquid passing along the wind pipe to the lungs. See also Adam's apple and Fig. D2.

**epigyny** Refers to the arrangement in flowers where the sepals, petals and stamens are attached above the ovary. The ovary is then called inferior. Compare hypogyny. See Fig. F2.

**epilepsy** A nervous disorder that usually involves convulsions and a loss of consciousness.

**epinephrine** Also called ★**adrenaline**.

**epiphyte** A plant that grows on another plant, but only to help gain an advantageous position or support. An epiphyte does not depend on its supporting plant for nutrients. e.g. Many orchids, ferns, moss and lichens.

**episome** A genetic unit (segment of DNA) found in bacteria and some other organisms that can exist and replicate itself either as part of a chromosome, or else independently. e.g. Some ★bacteriophages (a virus), the ★F factor transferred between bacteria during sexual reproduction.

**epithelial tissue** In animals, a type of tissue that covers a body or structure, or which lines a cavity. Epithelial cells form one (simple) or more (stratified) regular layers with little space or material between the cells. The cells are often classified by shape: squamous (flat), cuboidal (cube like) or columnar (column like).

**epoch** In geological time, a subdivision of a period. See Appendix 2.

**Epstein–Barr virus** Called **EBV** for short. A herpes virus that causes ★glandular fever and is involved in other diseases such as Burkitt's lymphoma, a relatively common cancer of children in Africa.

**equatorial plate** The plane at roughly right angles to the ★spindle, along which the chromosomes are arranged during the metaphase stage of cell division. In ★mitosis, the ★centromeres of the chromosomes are located along the equatorial plate. In ★meiosis, the centromeres of homologous chromosomes come to lie a little distance from the plate, on opposite sides.

**equilibrium** The state of balance between opposing actions. **1 Ecological equilibrium** occurs when there appears to be no change in an environment. e.g. For a particular species, death rates equal birth rates. **2 Chemical equilibrium** occurs when the rate of forward reaction equals the rate of reverse reaction.

**equine morbillivirus** A ★morbillivirus lethal to horses, but has been known to be transferred to humans where it can also cause death.

**era** A major division of geological time. See appendix 1.

**erect** Upright; not prostrate or leaning.

**erectile tissue** Spongy tissue surrounding the male reproductive organ, the ★penis. The spaces in this tissue fill with blood during sexual stimulation causing the penis to become erect.

**erection** Becoming swollen and rigid. e.g. The penis when filled with blood during sexual arousal as preparation for sexual intercourse.

**ericoid** Of leaves, small, narrow, stiff, and sharp-pointed as those of *Eric* (Heather). Applied to leaves of many species of heath.

***Eriostemon*** A genus of dicot flowering plants in the family Rutaceae, often called Wax-flowers. See Appendix 1.

**erythrocyte** Also called ★red blood cells or red corpuscles.

**essential** Absolutely necessary; indispensable. e.g. **a** Essential amino acids and essential fatty acids cannot be produced by the body and must be supplied in the diet. **b** Essential elements: See macronutrients and micronutrients.

**estivation** Also spelt ★aestivation.

**estrogen** Also spelt ★**oestrogen**.

**estrus** Also spelt ★**oestrus**.

**estuary** The mouth and lower portions of a river where ocean tides meet river flow, and thus where salt and fresh water mix.

**ethanol** Alcohol, ($C_2H_5OH$) produced by fermentation where yeasts undergo anaerobic (without oxygen) respiration and convert sugars (e.g. found in wine grapes) into alcohol.

**Ethiopian realm** A biogeographic region comprising Africa south of the Sahara desert. Typical mammals of the region include the chimpanzee, gorilla, giraffe, hippopotamus, and aardvark. See Fig. B5.

**ethnography** The scientific study and classification of the various racial and cultural groups of mankind.

**ethology** The scientific study of animal behaviour.

**ethyl alcohol** See ethanol above.

**ethylene** Also called **ethene** ($C_2H_4$). Plays a role in the ripening of fruit. See plant growth substances.

**etiolation** The appearance of plants grown in the dark: white or pale due to less chlorophyll, and with a spindly appearance due to elongated ⋆internodes and reduced leaves.

***Eucalyptus*** A genus of trees (Family Myrtaceae) native to and characteristic of the Australian region. The main distinguishing feature is the cap covering the flower buds, which gives the genus its name. The main groups of eucalypts are distinguished by their bark. e.g. ⋆Bloodwoods, ⋆ironbark, ⋆stringy bark, ⋆gums, ⋆peppermints, ⋆mahogany, etc. See also gum nuts, Appendix 1, and Fig. G5.

**eucaryotic** Also spelt ⋆**eukaryotic**.

**eugenics** The application of genetics in an attempt to 'improve' the heredity qualities of humans.

**Euglenophyta** A Phylum (Division) of mostly single-celled algae with flagella, including *Euglena*. See Appendix 1.

**eukaryote** Also spelt **eucaryote**. Cells in which the chromosomes are found in a nucleus separated from the cytoplasm by the ⋆nuclear membrane. The cells of most organisms are eukaryotic, except bacteria and blue-green bacteria. Compare prokaryotic.

**Eumycota** The phylum also called fungi, with cells enclosed in cell walls like a plant, but lacking chlorophyll. e.g. Mushrooms, moulds, yeasts. See Appendix 1.

**euphoriant** A substance that produces a feeling of well being, especially the exaggerated elation associated with some drugs.

**euro** See wallaroo.

**Eustachian tube** The tube that connects the ⋆middle ear to the back of the throat (pharynx). This tube allows movement of air to ensure that air pressure is equal on both sides of the eardrum. See Fig. E1.

**euthanasia** Assisting the death of someone who is suffering a painful and incurable disease, or allowing them to die by removing their life support apparatus.

**eutherian** Another name for the placental ⋆mammals, where the young are fed before birth through a ⋆placenta. See also metatherian and prototherian.

**eutrophic** Used to describe bodies of water with adequate nutrition available to its inhabitants.

**eutrophication** The process by which bodies of water such as lakes become enriched with dissolved nutrients. However this may ultimately result in much growth of algae and other organisms with an equivalent drop in dissolved oxygen. Humans can speed the process by allowing sewage or excess fertiliser to enrich the water. In extreme cases oxygen levels fall so low that only bacteria survive, and the lake is almost 'dead'. See also pollution.

**evaporation** The gradual loss of moisture to the air from an uncovered water surface. This results in a drop of temperature. ⋆Perspiration evaporates from the body thus cooling it. See also latent heat.

**evergreen** A tree or shrub that retains leaves all the year. e.g. Eucalypts. The last season's leaves are only gradually shed as new ones grow. Compare deciduous.

**everlasting** Applied to flowers that retain their colour and shape even after they have been dried, especially some daisies with coloured ⋆bracts.

**eviscerate** To disembowel; remove the intestines and other internal organs.

**evolution** The gradual development of new varieties of organisms from pre-existing organisms over millions of years.

The modern theory of evolution has developed from the ideas of Charles Darwin published in his book *The Origin of Species* in 1859. According to the modern theory (called neo-Darwinism), changes occur in organisms by ⋆mutations of genes. This leads to the existence of ⋆variation amongst individuals. Some of these individuals may survive more successfully than others (called ⋆natural selection), thus producing more offspring with their new features. Gradually these new features will extend throughout the ⋆population. If, however, the population is ⋆isolated from others, differences cannot spread, and over a period of time two varieties come to exist. Only small changes to organisms have been actually observed to occur by this mechanism. e.g. Industrial melanism, resistance to antibiotics and insecticides. Evidence for larger changes must be deduced from the fossil record.

**excise** To cut out. To cut off.

**excitation** To increase the activity of. To stimulate.

**excrement** Solid waste material from the body. Faeces.

**excretion** The removal of wastes produced by ⋆metabolism. e.g. Carbon dioxide from the lungs; ⋆perspiration from the skin; filtration of wastes from the blood, and their removal from the body as ⋆urine. See also excretory system.

**excretory system** The system of organs that help filter waste products from the blood, and remove it from the body. In humans, the excretory system consists of the ⋆kidneys, bladder and connecting tubes. Inside the kidneys, blood is filtered by the ⋆nephrons. The urine produced passes along the ⋆ureter to the bladder where it is stored until convenient for release. Finally, muscles relax at the start of the ⋆urethra, and the urine passes along this tube and leaves the body. See Figs E8 and K1.

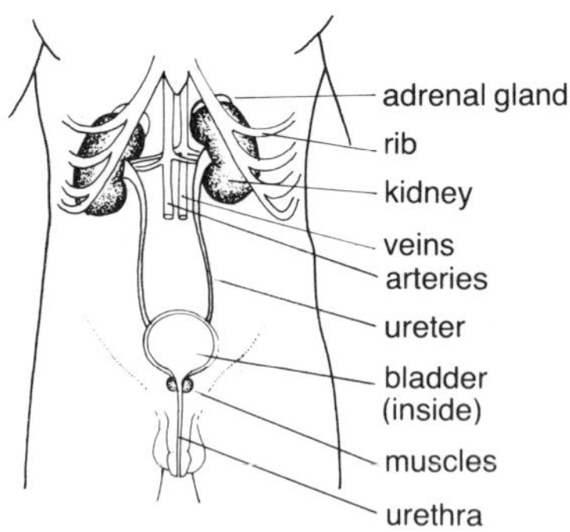


**Fig. E8 Excretory system**

**exergonic** Also called ⋆**exothermic**.

**exfoliate** The shedding in flakes or thin layers. e.g. The shedding of the skin or epidermis of an animal. The shedding of bark by certain trees, or scales from buds.

**exhalation** The act of breathing out. Air passing out from the lungs.

**exocrine** Secreting externally. e.g. An exocrine gland delivers its secretion through a duct to an epithelial surface (sweat glands, pancreas).

**exocytosis** The carrying of waste products or secretions in a ⋆vacuole to the surface of a cell where they are released through the membrane. Compare endocytosis. See Fig. E5.

**exogamy** Outbreeding as distinct from inbreeding. The gametes (sperm and ova) come from parents that are of the same species, but not closely related. See also cross-pollination, cross-breeding, cross-fertilisation.

**exon** The part of a DNA molecule that codes for a protein. Located between ⋆introns which do not appear to code for anything. See also transcription.

**exoskeleton** A skeleton that lies outside the body tissues of an animal. **1** The protective external skeleton of most arthropods. It supports some organs, and provides support for muscles. e.g. The ⋆cuticle of insects; lime-strengthened cuticle of crabs; shell of molluscs. **2** In many vertebrates, the exoskeleton is bony plates formed in the skin e.g. armadillo, tortoise.

**exothermic** Also called **exergonic**. Chemical reactions that produce and liberate heat energy. e.g. ⋆Respiration releases energy from sugars.

**exotic** A plant or animal that is not ⋆native, but has been introduced from abroad. Mainly applied to plants.

**exploratory** Applied to surgery where the aim is to investigate the source of a problem that could not been able to be ⋆diagnosed by external examination, X-rays, etc.

**exponential growth** A very rapid increase in numbers described by an equation such as $y=a^x$. Human population has been increasing exponentially, producing a graph that looks like the letter J.

**express** **1** To force out as the result of pressure. **2** A gene is said to be expressed when it is turned on and able to produce its effects.

**exsert** Projecting or protruding beyond surrounding parts, as with valves on a ⋆capsule, or stamens from a flower.

**extensor** A muscle or tendon that straightens a limb or joint. e.g. ⋆Tricep (Fig. B4). Compare flexor.

**exterminate** To totally destroy or rid.

**external ear** Also called a ⋆pinna.

**external fertilisation** During external fertilisation, the gametes (sperm and ova) are released by the parents and meet more or less by chance. e.g. Oysters and other molluscs release large numbers of gametes into the water thus increasing the chance of them meeting. A male frog climbs onto the back of the female and secretes sperm over the eggs as she lays them.

**extinct** No longer in existence. No living representatives. See biodiversity.

**extracellular fluid** Fluid surrounding and bathing a cell. In mammals, this fluid is usually part of the ⋆lymph.

**extrapolate** To estimate a value that is bigger or smaller than those that have been measured. e.g. With a graph, extending the graph beyond the points measured by experiment. Compare interpolate.

**exude** To discharge slowly through small pores. To ooze out.

**eye** A sense organ that responds to light. **1** A simple eye including ⋆eye-spots (e.g. protozoan), and ⋆ocelli (found in many invertebrates). **2** Eyes of insects and crustaceans are ⋆compound eyes, made up of many light-sensitive ⋆ommatidia. These eyes produce only an outline (shadow-like) image, but are excellent for detecting movement. The shadow only has to move slightly and the next ommatidium will be stimulated. See Fig. O3. **3** The vertebrate eye found in mammals, birds and reptiles is also found in certain ⋆cephalopod molluscs (e.g. octopus, squid). The lens can change its shape to focus images from different distances onto the light-sensitive ⋆retina (see accommodation). The ⋆iris opens and closes the ⋆pupil to keep the intensity of light reaching the retina as constant as possible. The light reaching the retina stimulates the light-sensitive ⋆rods (which detect brightness) and ⋆cones (which detect colour and fine details) which send their impulses along the optic nerves to the brain. See also cornea, vitreous humour, aqueous humour, and Fig. E9.

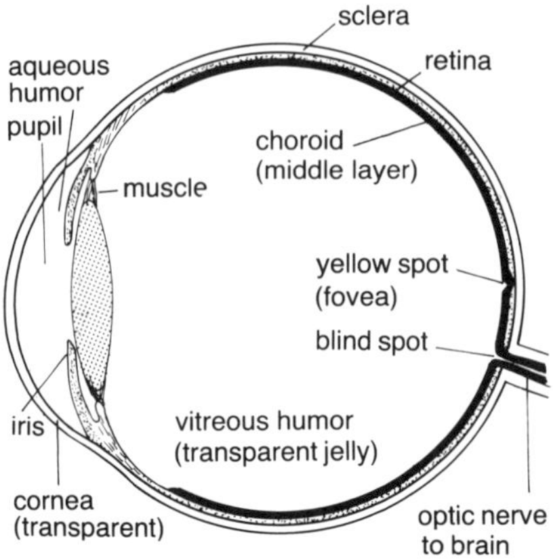


**Fig. E9 Eye**

**eye-spot** Also called a **stigma**. A light-sensitive spot containing various pigments. Found in some protozoa, jellyfish, flatworms, flagellate algae.

# Ff

**F factor** Short for **fertility factor**. A piece of DNA that is transferred during sexual reproduction in bacteria (*conjugation). A copy of the F factor is transferred from one cell to the other through a tube that temporarily connects the two bacteria. See also pili.

**$f_1$** Short for **first filial generation**. The offspring that result when plants or animals from the parent generation are crossed (allowed to breed).

**$f_2$** Short for **second filial generation**. The offspring that result when plants or animals from the F1 generation are crossed (allowed to breed) among themselves.

**Fabaceae** Also called **Papilionaceae** or the Pea family of dicot flowering plants. See Desert Pea and Appendix 1.

**face** The front (anterior) part of the head where the eyes, nose and mouth are usually located.

**facilitation** 1 An important process in nervous *integration. As several impulses arrive at a synapse, they make it more and more sensitive until one final impulse produces a new impulse in the next (postsynaptic) neurone. Compare summation. 2 Social facilitation is the effect of the behaviour of one or more organisms on the behaviour of others by making it easier for them to conform. e.g. Some birds need crowded conditions of a rookery, where they are surrounded by many other birds, before they will permit mating.

**factor** 1 An historic term that was replaced by the term gene. 2 See clotting factor.

**facultative** An organism that has the ability to live under varying conditions. Compare obligate.

**FAD** Short for **flavin adenine dinucleotide**. A *co-enzyme derived from the vitamin riboflavin (B2). Important in many *oxidation-reduction reactions. e.g. Glycolysis.

**faeces** Solid wastes produced in the digestive system (intestines), and expelled through the anus.

**fainting** The temporary loss of consciousness due to reduced supply of blood to the brain. The head is thus effectively lowered, restoring the blood flow.

**falcate** Used to describe leaves which are sickle-shaped. e.g. Many eucalypt leaves. See Fig. L1.

**fallen arch** Insufficient arching of bones under the middle part of the foot, resulting in flat feet.

**fallopian tube** Also called **oviduct** or just '**tubes**'. In female mammals, a tube with a funnel-shaped opening next to the ovary, which then leads to the womb (uterus). See Fig. R4.

There is a fallopian tube on each side of the body. The combined action of muscular contraction and beating cilia help move the egg towards the uterus, as sperm move from the uterus up towards the descending egg. It is the site of *fertilisation, and in some mammals, the site of *implantation. See also Müllerian ducts.

**false fruit** Also called ***pseudocarp**.

**false ribs** The five lower pairs of ribs that are not directly attached to the breastbone (sternum). The top three pairs of false ribs are attached to each other and then to the true ribs above. The bottom two pairs are called floating ribs. See Fig. S3.

**Family** A major division of classification. One or more genera make up a family. Like Families are grouped into Orders. Family names end in -aceae in botany, and -ideae in zoology. See Appendix 1.

**fang** A long, sharp, pointed tooth or toothlike structure. e.g. The canine tooth of a carnivore; the long, hollow, curved tooth of a venomous snake, down which venom is passed; the venom carrying ⋆chelicera of a spider.

**far-red light** Red light of the longest wavelengths (about 740 nm wavelength), at the far end of the visible ⋆spectrum.

**fat** 1 A usually soft, greasy substance present in ⋆fat (adipose) tissue, and in certain seeds such as those of cotton, soy beans, and corn. 2 A ⋆lipid, containing the elements carbon, hydrogen and oxygen, and soluble in alcohol and ether, but not in water. When reacted (boiled) with alkali, fats produce glycerol and soap (saponification). 3 A compound (ester) of glycerol and fatty acid called triglycerides. 4 Fats act as a long term energy reserve (compare ⋆glycogen found in the liver).

**fat tissue** Also called **adipose tissue**. Tissue containing large numbers of fat storage cells, each cell with only a thin layer of cytoplasm surrounding a large droplet of fat. Fat tissue acts as an energy reserve, and in some organisms as insulation against cold. Most fat in fat tissue is white fat, but in many animals there are deposits of special brown fat. By comparison, brown fat stores more energy, and this energy is more quickly available than the energy in white fat. Fat tissue occurs under the skin and around the internal organs such as the kidneys.

**fatty acid** 1 A group of organic acids such as palmitic, stearic and oleic acid that occur widely, and are chemically combined with ⋆glycerol to form ⋆fats. 2 Fatty acids are carboxylic acids with long hydrocarbon chains. The most common fatty acids have even numbers of carbon atoms, ranging from $C_{14}$ to $C_{24}$. e.g. Stearic acid $CH_3(CH_2)_{14}COOH$.

**fauna** 1 The animal life that exists in a particular place or locality. 2 The animal life that existed during a particular period of geological time. Compare flora.

**feather** A surface (epidermal) structure that provides the body covering of birds. Most birds have three types: **a** Contour feathers help give the bird shape and help it fly, tiny ⋆barbules (hooks) holding the thin ⋆barbs in place. **b** Down feathers exist under the contour feathers to help insulate the bird. **c** Filoplumes are small hair-like feathers scattered over the body between the contour feathers. See Fig. B2.

**feather stars** A common name for crinoids, a class of ⋆echinoderms having a cup-shaped body with feathery arms. The crinoids are able to move about. Compare sea lilies.

**feathertailed glider** See pygmy glider.

**fecund** Able to produce abundant fruit or numerous offspring.

**feedback** The return of a part of the output of a system or process, as input, to help correct and control the system. 1 **Negative feedback**: An increase in some product of a reaction leads to an inhibition or reduction in the reaction that produces that substance. e.g. An increase in the amount of glucose sugar in the blood decreases the production of the hormone glucagon (by the liver) needed to convert stored glycogen into glucose. An increase in the amount of the male sex hormone testosterone in the blood is detected by the hypothalamus, which decreases the production of the hormone LH. Since LH directs the production of testosterone by the testes, less LH means that less testosterone is produced. See Fig. F1. **2 Positive feedback**: An increase in some product of a reaction leads to a further increase in the reaction that produced the substance. e.g. Release of ⋆oxytocin by the pituitary in response to nerve impulses from the hypothalamus. Most feedback in the body is negative feedback.

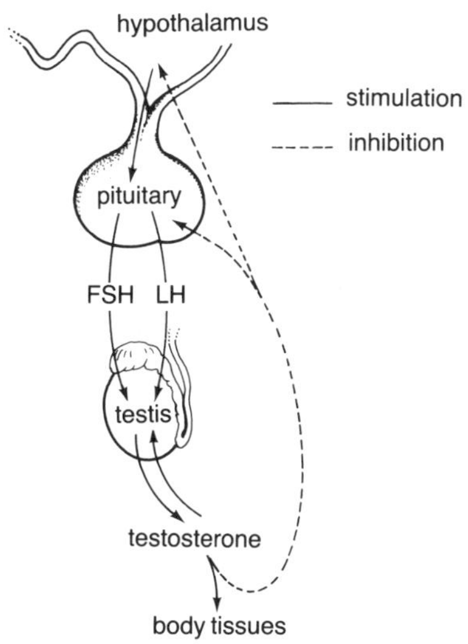


**Fig. F1 Feedback**

**Fehling's test** A mixture of two solutions used to detect the presence of ⋆reducing sugars such as glucose. Fehling's A is copper (II) sulfate, and Fehling's B is a mixture of sodium potassium tartrate and sodium hydroxide solutions. If mixed with glucose solution, and boiled, a brick-red precipitate of copper (I) oxide is produced. Compare Benedict's test.

**female** An organism that produces the ovum or its equivalent (female gamete). After fertilisation, the female usually nurtures the developing embryo.

**femur** 1 The thighbone. The long upper bone of the leg (rear leg in four-legged animals). It is connected by a ball-and-socket joint to the pelvis. See Fig. S3. 2 In insects, the section of the limb nearest the body.

**feral** Domesticated animals that have gone back to the wild state. e.g. Feral cats.

**fermentation** 1 The action of yeast cells on the sugar in fruit juices (e.g. grape) to produce alcohol. 2 A form of ⋆anaerobic respiration where carbohydrates such as sugar are oxidised to produce energy (ATP molecules) without the presence of oxygen. In plants, the by-product is alcohol. In animals, the by-product is lactic acid (causing muscular cramp). Neither of these reactions produces as much ATP as does normal respiration using oxygen.

**fern** A group of vascular plants, Class Filicopsida, with an underground stem or rhizome from which fronds (large divided leaf) and roots develop. Under the fronds of this ⋆sporophyte plant the reproductive organs called ⋆sporangia develop. These produce spores that germinate to grow into a small heart-shaped gamete plant (⋆gametophyte). This produces sperm (in ⋆antheridia) and ova (in ⋆archegonia). The fertilised egg grows into the much larger spore plant (sporophyte). See life cycles, Appendix 1.

**ferredoxin** A group of red-brown coloured proteins containing oxygen and sulfur. They are an important part of the ⋆electron transport chain in photosynthesis.

**ferruginous** 1 Containing iron. 2 Having the appearance of rusty iron.

**fertile** 1 Soil (land) that is capable of bearing abundant crops, vegetation, etc. 2 A plant or animal capable of producing offspring. Often implies the ability to produce many offspring.

**fertilisation** The union or fusion of a female gamete (e.g ovum) and a male gamete (e.g. sperm) to produce a ⋆zygote. In animals, both ⋆internal and ⋆external fertilisation occur. In flowering plants, the male gamete passes down the ⋆pollen tube to the ovary. In ferns, moss, etc., the male gamete needs droplets of water to swim through to reach the female gamete. Compare pollination, which is different.

**fertiliser** Any substance used to add minerals and other nutrients to the soil. e.g. Compost and manure are natural fertilisers. Superphosphate and phosphate of ammonia are manufactured fertilisers.

**fester** To produce pus such as in a small sore. To cause ulceration. To putrefy or rot.

**fetid** To have a strong smell that is offensive to people. Stinking.

**fetus** Also spelt ⋆**foetus**.

**fever** Abnormally high body temperature due to illness.

**fibre** 1 A fine thread-like cell, such as ⋆nerve or ⋆muscle fibres in animals. 2 An elongated thick-walled plant cell forming part of the plant strengthening ⋆sclerenchyma. These cells have tapered ends and simple ⋆pits in their walls. 3 A ⋆collagen or ⋆elastin (types of protein) fibre found in ⋆connective tissue.

**fibril** 1 Any minute, thread-like organelle (structure) within a cell. 2 A minute thread-like component of a fibre.

**fibrillation** Occurs when individual muscle fibres of the heart act in an uncoordinated manner causing rapid ineffective heartbeats. Electric shock can be applied to help restore correct heartbeat.

**fibrin** A fibrous protein made from ⋆fibrinogen during the formation of a blood clot. See blood clotting.

**fibrinogen** A soluble protein present in normal blood plasma, which can be converted, by the enzyme ⋆thrombin, into the fibrous protein called ⋆fibrin during the formation of a blood clot. See blood clotting.

**fibrositis** Inflammation of ⋆connective tissue.

**fibrous protein** Fibrous proteins are generally in the form of long coiled strands or flat sheets. e.g. ⋆Keratin, ⋆collagen, ⋆actin, ⋆myosin, ⋆fibrin. Compare globular proteins.

**fibula** The smaller of the two bones in the hind (rear) or lower limbs, connecting the knee to the ankle. It lies next to the tibia. See Fig. S3.

**filament** Any slender thread-like structure. 1 A chain or string of cells. e.g. Algae; bacteria; hyphae of fungi. 2 In flowers, the ⋆stalk carrying the ⋆stamens.

**File snakes** A small family (Acrochordidae) of aquatic snakes living in northern Australia. See Appendix 1.

**filial** Relating to the son or daughter. e.g. In genetics, the sequence of generations from parents onwards. See $f_1$ and $f_2$ above.

**Filicopsida** The Class of vascular plants generally called ⋆ferns. See also life cycles and Appendix 1.

**filovirus** A type of virus with one strand of RNA in the form of a helix. e.g. ⋆Ebola and Marburg viruses.

**filter feeder** An organism that obtains food by passing water through a filtering mechanism. e.g. Marine worms (such as *Galeolaria*) project feather-shaped filters into the water to catch small prey. ⋆Barnacles feed in a similar manner.

**fin** A thin wing-like or paddle-like structure attached to the bodies of fish and certain other aquatic animals by which they swim, balance, and brake.

**finger** A thin projection (digit) on the end of the forearm, other than the thumb.

**fingerling** A small young fish, usually from the time that the yolk-sac disappears to the end of its first year of life.

**fingerprint** 1 The pattern of ridges produced by an ink-covered human finger when pressed against an object. It is believed that each person's fingerprints are unique. 2 **DNA fingerprints** The patterns produced when certain highly variable sections of human DNA are separated using ⋆gel ⋆electrophoresis. Most DNA in humans is very similar, but certain sections are highly variable and believed to be unique to each individual. Used in criminal investigations and paternity disputes.

**fish** A general term used to describe three main groups of vertebrates: the jawless fish (Agnatha) such as lampreys; the cartilaginous fish (Chondrichthyes) such as the sharks; the bony fish (Osteichthyes) such as snapper, bream and tuna. See Appendix 1.

**fissile** Easily split or separated into two parts.

**fission** Asexual reproduction where the body or single-celled organism divides into two or more equal parts. 1 Binary fission is division of a single-celled organism into two cells. e.g. *Paramecium, Amoeba*. 2 Multiple fission (sporulation) in certain protozoa results in the formation of many spores.

**fissure** A deep groove or cleft.

**fit** 1 In good health. 2 Capable of strenuous work or exercise. 3 A sudden convulsion or loss of control e.g. Epileptic fit. 4 See also fitness.

**fitness** In genetics, a measure of the ability of an organism to survive ⋆natural selection (i.e. produce surviving offspring).

**fixation** 1 To kill, and preserve. 2 The first stage in the preparation of a microscope slide. The fixing agent (e.g. alcohol, formalin-alcohol) kills the tissue, helps ensure that the original shape is maintained, prepares it for staining and helps it to harden so that thin sections can be cut. 3 The converting of a substance from a gaseous to a more useable form. e.g. Nitrogen fixation converts atmospheric nitrogen into soluble nitrates, etc.

**flaccid** Soft and drooping; limp; flabby. e.g. Wilted leaves. Used to describe a cell when its ⋆vacuoles have lost water and its cell membrane has pulled away from the cell wall.

**Flagellata** A protozoan of the class Mastigophora (Flagellata) able to swim by the movement of ⋆flagella. e.g. *Euglena*. See Appendix 1.

**flagellate** 1 An organism with one or more flagella. 2 See Flagellata above.

**flagellum** Plural is **flagella**. Long, thin, thread-like ⋆organelles projecting from a cell. Found in certain protozoa (flagellates), algae, and gametes (sperm) where its movement propels the organism along. Certain cells (e.g. in sponges) have flagella to produce moving currents of water. See also cilia.

**flannel flower** A dicot flowering plant, *Actinotus helianthi*, in family Apiaceae, common in the Sydney region. See Appendix 1.

**flatulent** The often uncomfortable accumulation of gas in the alimentary canal.

**flatworm** Any member of the phylum ⋆Platyhelminthes e.g. ⋆Planarians, ⋆flukes, ⋆tapeworms, marine flatworms. See Appendix 1.

**flea** A group of insects in order ⋆Siphonaptera, especially human fleas, *Pulex irritans*, notorious for transmitting bubonic plague and a form of typhus. See Appendix 1.

**fledgling** A young bird that has just grown its feathers.

**flesh** The muscle and fat of an animal body.

**flesh-eating bacteria** See necrotizing fasciitus.

**flexor** A ⋆muscle or ⋆tendon that bends a part of the body. e.g. The ⋆bicep muscle at the arm joint. See Fig. B4. Compare extensor.

**flight feathers** The large stiff feathers that make up most of a bird's wing, and which are essential for flight. See Fig. B2.

**flipper** A broad flat limb used by penguins, whales, seals, etc. for swimming.

**floating rib** One of the two lowest pairs of ribs in humans which are not attached at the front to the breast bone (sternum), nor to the cartilages of other ribs. They are attached only to the spine at the rear. Compare false ribs. See Fig. S3.

**floccose** Having tufts of soft, woolly hairs.

**flora** 1 The plant population in a particular area or locality, or in a particular geological period or era. 2 A list of plants, with descriptions, for a particular area, and with a key for their identification. Compare fauna. See also normal flora and intestinal flora.

**floral** Relating to or consisting of flowers.

**floral diagram** Shows the relative position and number of the parts of a flower.

**floret** 1 A small flower, especially one of the many small flowers that make up the head of a daisy or similar flower. e.g. Sunflower. See capitulum. 2 The flower and enclosing bracts (small leaves) of grasses.

**florigen** Also called **flowering hormone**. This plant ⋆growth substance (hormone) has been postulated to transmit the flowering stimulus from the plant leaves to the buds on the ends of the stem and branches. It has not been identified chemically.

**flower** The ⋆blossom of a plant used for sexual reproduction. See Fig. F2.

The female gametes (ova) are produced within the ⋆ovary. The ovary contains one or more ⋆ovules inside of which are the ⋆ova (eggs). The male gametes are produced within the ⋆anthers on the ends of a stalk. Anther and stalk are together called the stamen. The male gametes (actually nuclei) are contained within the ⋆pollen grains. Insects, birds, wind and occasionally mammals, transfer the pollen from one flower to another (⋆pollination). Brightly coloured and marked ⋆petals help attract such pollinators, as does the sweet ⋆nectar produced at the base of the petals. The pollen sticks to the sticky end of the female part of the flower, called the ⋆stigma. The stigma, style and ovary are together called the ⋆carpel or pistil. A pollen tube then grows from the pollen on the stigma, down the style, and finally into the ovary. The male nuclei pass along this pollen tube to reach the ova in the ovary, where fertilisation takes place. After fertilisation, the zygote thus formed develops into the ⋆embryo and the ovules into ⋆seeds. The ovary wall develops into the fruit.

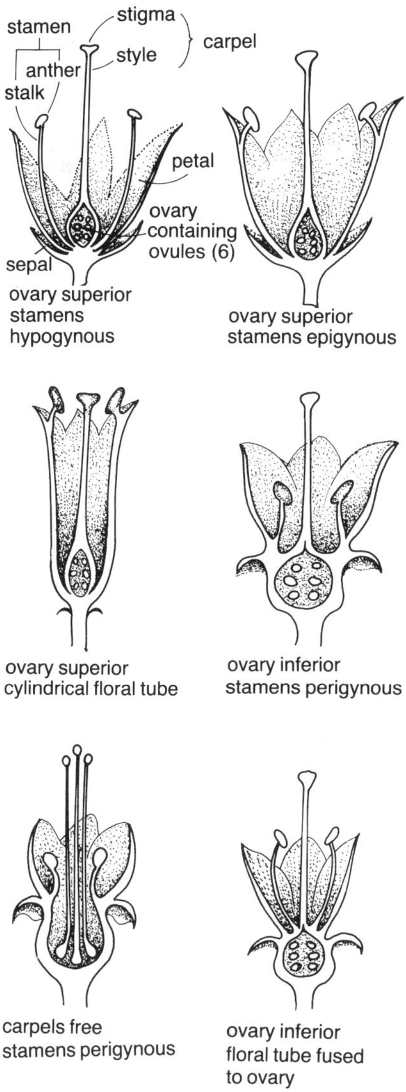


**Fig. F2 Flower structure**

**flowering plant** Any member of the plant Class called Angiospermae (or angiosperms). All such plants bear flowers, and produce seeds contained within fruit. Not all such fruits are classed as fruit by the general public. e.g. Gum nuts and tomatoes are classed as fruit by botanists. See also monocotyledon, dicotyledon and Appendix 1.

**flu** Short for ⋆**influenza**.

**fluid** A substance that can flow and offers no major resistance to a change in shape. e.g. A liquid or a gas.

**flukes** Any member of the class of flatworms (⋆Platyhelminthes) called Trematoda. e.g. ⋆Liver fluke. See Appendix 1.

**fluorescence** The property of certain substances of absorbing light of a particular wavelength, and then emitting light of a different wavelength (and hence colour). See also luminescence.

**fly** Any member of the insect order ⋆Diptera, which have two pairs of wings. e.g. *Musca domestica* or the house fly. See Appendix 1.

**flying fox** Large fruit-eating ⋆bats of the order Chiroptera and genus *Pteropus*. See Appendix 1.

**foetal graft** A controversial treatment for ⋆Parkinson's disease where tissue from a foetus is ⋆grafted into the brain. The cells in this tissue produce ⋆dopamine to replace that no longer being made by the person with the disease.

**foetus** Also spelt **fetus**. In mammals, the embryo in later stages of development when features such as legs, digits etc. are clearly visible. In humans, the period after about 3 months when arms, fingers, etc. are visible. Compare embryo and Fig. E4.

**foliage** All the leaves of a plant.

**foliate** Carrying leaflets. e.g. A trifoliate leaf has three leaflets.

**folic acid** One of the B group of ⋆vitamins. Folic acid is used by the body to make several different ★co-enzymes, and is important in cell division.

**foliose** A growth habit of ★lichens where they are loosely attached to a substrate and have a lobed appearance. Compare crustose and fruticose.

**follicle** 1 A small sack, cavity or gland. e.g. **a** ★Graafian (ovarian) follicle in the female ★ovary. **b** Hair follicles contain the roots of the growing hair. 2 A simple, dry fruit containing one carpel, which splits along one seam. e.g. Milkweed.

**follicle stimulating hormone** Called ★**FSH** for short.

**food** That which is taken into the body through the mouth for nourishment.

**food chain** A food relationship in an ★ecosystem in which energy and other nutrients are transferred from plants, to plant eaters (herbivores), to animal eaters (carnivores).

The plants obtain their energy from the Sun by ★photosynthesis, and are called ★producers. The herbivores and carnivores that follow are called ★consumers as they rely on other living things for their energy and other nutrients. There may be several carnivores in a food chain, but rarely more than five or six living things altogether. All food chains start with a producer (usually a plant), and the arrows are used to show the direction in which energy and nutrients flow, NOT which eats what. e.g. grass → grasshopper → fieldmouse → hawk. Generally there are larger numbers of organisms early in the food chain, and certainly the total weight of living matter (★biomass) of each successive stage decreases when considered over a period of time. By convention, the ★decomposers (e.g. bacteria acting on a dead hawk) are left out of a food chain. See also pyramid, amplification.

**food pipe** Also known as the ★**oesophagus**.

**food poisoning** An internal infection caused by bacteria or their toxins in foods. Commonly the infection is due to *Salmonella* or *Staphylococci* bacteria. See also botulism.

**food pyramid** See pyramid.

**food web** The many interacting and interconnecting ★food chains in an ecosystem. See Fig. F3.

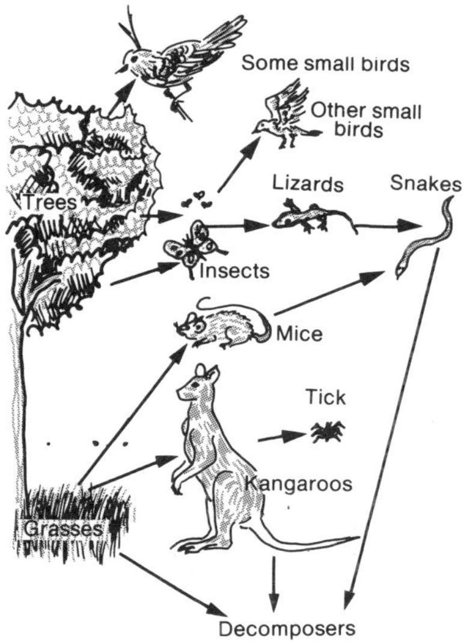


**Fig. F3 Food web**

**foot** 1 In animals, the end part of a limb upon which the body rests when standing. e.g. Tube feet of echinoderms, muscular foot of shellfish and other molluscs, ★tarsus of insects, foot of vertebrates. See pentadactyl limb. 2 In liverworts, mosses, club mosses, and ferns, the bottom part of the developing spore plant (sporophyte) where it is attached to the gamete plant (gametophyte).

**foot-and-mouth disease** A contagious (easily spread) virus disease of cattle and other cloven-hoofed animals, with pus-filled sacks found around the mouth and hooves.

**forage** 1 Food or fodder for horses and cattle. 2 Seeking or searching out food.

**foram** Short for **Foraminifera**. See below.

**foramen magnum** The hole in the base of the skull through which the spinal cord passes to reach the brain inside. See also occipital bone. In quadrupeds this hole is towards the back of the skull, whereas in bipeds it is located more towards the centre of the base of the skull. The position of the foramen magnum on the skull is an indicator of the possible degree to which an organism walked upright.

**Foraminifera** Called **forams** for short. An order of the protozoan class Sarcodina, having a calcium carbonate-rich shell usually perforated with small holes or pores through which project ⋆pseudopodia used to help capture food. Foraminifera shells form an important part of the rock called chalk, and of many deep-sea ⋆oozes. e.g. *Globigerina*. Their fossils are important in dating many rocks, especially those associated with the search for oil. See Appendix 1.

**forelimb** The front (anterior) pair of limbs on an animal with four limbs (tetrapod).

**forensic** The application of scientific knowledge to legal problems. e.g. **Forensic medicine** is the scientific use of medical, dental, psychological, biological, chemical, and mechanical techniques, in investigating the causes of a person's death, disability, injury, or disease.

**foreskin** Also called the **prepuce**. The soft fold of skin that covers the head of the penis. Removed during ⋆circumcision.

**forest** A large area of land covered with trees as the dominant plant. In Australia, forests are roughly divided into tropical ⋆rain forest, temperate rain forest, ⋆tall open (wet sclerophyll) forest, and ⋆open (dry sclerophyll) forest. Compare woodland.

**forester** A common name for the great grey kangaroo (*Macropus giganteus*).

**fossil** The remains of an organism, or direct evidence of its presence (e.g. tracks), as preserved in rocks. Fossils may be unaltered hard parts (tooth, bone, shell), partially altered (a frozen mammoth, insects in amber), a mould (hollow cavity) in a rock, ⋆petrified (wood, bone), etc.

**founder effect** The idea that a small population of organisms, genetically separated from the main population, has only a limited variety of genes. This may speed development of new varieties if the genes are distinct or unusual. Such populations are also vulnerable to environmental change due to the limited variety of genes, thus restricting their ability to adapt quickly. Compare with genetic bottleneck.

**fovea** A shallow pit in the ⋆retina of some vertebrate ⋆eyes, including mankind. There are no ⋆rods or blood vessels, but many close-packed ⋆cones giving the area the ability to see great detail. Located where images most readily fall on the retina. See Fig. E9.

**fox** A placental in order ⋆Carnivora, *Vulpes vulpes*, introduced into Australia in the 1860s. See Appendix 1.

**fracture** The cracking or breaking of a bone. Called a simple fracture if the skin is not broken, and called compound if the bone breaks through or lacerates the skin.

**fraternal twins** See twins.

**free radicle** A very reactive atom or group of atoms that exist for very short periods of time during chemical reactions.

**freeze drying** A method of removing water without the need to heat delicate chemicals or tissues. In a vacuum, ice will change directly to water vapour (called sublimation), thus drying the specimen. Used to prepare specimens for electron microscopy.

**freeze fracturing** A technique used in examining specimens with an electron microscope. The specimen is quickly frozen to very low temperatures, and then fractured (broken) along lines of weakness while still in a vacuum. This usually reveals various cell membranes and organelles. These are then coated with a very thin layer of carbon or platinum. This thin layer is then separated, being a replica of the original surface which can be examined under the electron microscope.

**freshwater** Water that is low in dissolved minerals and is suitable for drinking; water that is not salt water (i.e. marine).

**frog** Any of the tailless animals of class Amphibia, order Anura (Salientia). Especially those of genus *Rana* and similar animals. All frogs start life hatching as ⋆tadpoles from jelly covered eggs, before undergoing ⋆metamorphosis to become adult frogs. See Appendix 1.

**frond** 1 A large, often divided leaf of a fern, palm, or cycad. 2 A broad flattened leaf-like structure (thallus) of certain seaweeds and lichens.

**fructose** A hexose or six-carbon sugar (monosaccharide) which in combination with glucose forms the disaccharide sucrose (table sugar). See Fig. C2.

**fruit** The structure that develops from the ovary wall, and sometimes nearby structures, which encloses the ⋆seeds and aids in their ⋆dispersal. See Fig. F4.

Fruits can be classified in a number of ways. If the middle layer of the ⋆pericarp (called mesocarp) is soft and fleshy the fruit is called succulent; otherwise it is termed dry. It is called ⋆dehiscent if the fruit splits open to release the seeds (otherwise indehiscent). Fruits are called simple (or true) if they develop from one ovary, compound (or aggregate) if they develop from an ⋆inflorescence, or false fruits (⋆pseudocarps) if they form from other structures (e.g. receptacle) as well as an ovary. After ⋆pollination, the fertilised ovum develops into the embryo, and the ovule into the seed. The type of fruit that develops from the ovary wall (and perhaps nearby structures) helps determine the method of seed dispersal. e.g. Succulent fruits are usually eaten. Dry fruits are often dispersed by the wind, especially dehiscent types.

**Fig. F4 Fruit (a) Pseudocarps (b) True fruits**

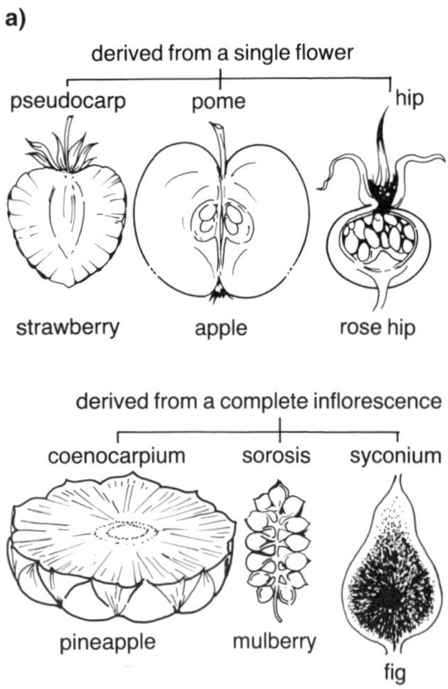

b)

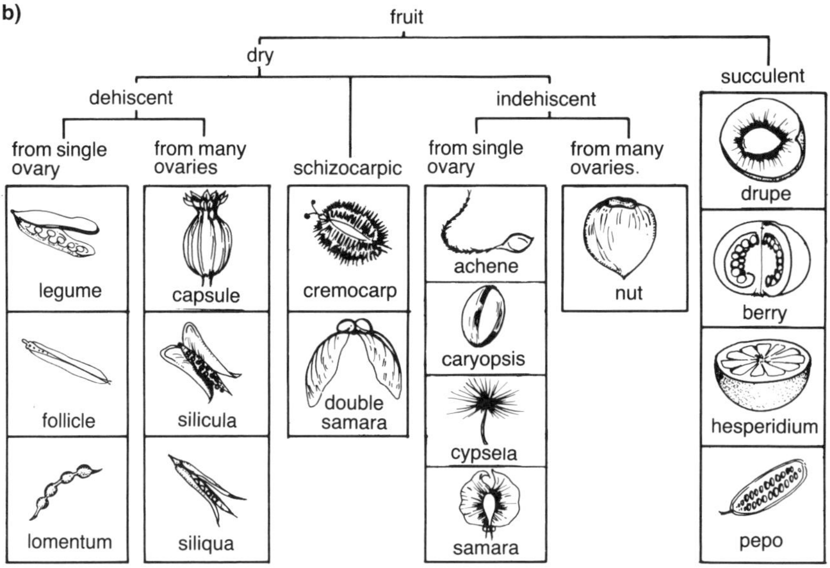


**fruiting body** A reproductive structure, especially the often visible part of a fungus (e.g. mushroom, toadstool, puffball) that produces ⋆spores.

**fruticose** Describing the growth habit of ⋆lichens which are either erect and bushy, or hanging and tassel-like. Compare crustose and foliose.

**FSH** Short for **follicle stimulating hormone**. A hormone secreted by the anterior (front) lobe of the ⋆pituitary gland which promotes the growth and maturing of ovarian ⋆follicles inside which the ovum develops. In males, FSH stimulated ⋆sperm production in the testes.

**function** The normal actions or activities of an organ or structure.

**functional group** The group of atoms responsible for the characteristic or special chemical reactions of a compound. e.g. Alcohols (–OH); organic acids (–COOH); amines ($-NH_2$).

**fungi** Any member of Phylum (Division) ⋆Eumycota having plant-like cells with cell walls, but lacking chlorophyll. Hence they are ⋆heterotrophs. e.g. Mushrooms, yeasts, mould. See hyphae, fruiting bodies, and Appendix 1.

**fungicide** A chemical that kills or inhibits the growth of fungi.

**funicle** A slender cord or stalk, especially that connecting an ⋆ovule to the inside surface of the ovary.

**funnelweb** Spiders in the genus *Atrax*, especially *A. robustus*, or the Sydney funnelweb which is notorious for its deadly venom. However, there are many species of *Atrax* extending over most of eastern Australia.

**funny bone** The ⋆humerus (Fig. S3), especially near the elbow where a nerve passes, which, when struck, produces a tingling sensation in the arm and hand.

**fur** The soft fine ⋆hair covering the skin of certain mammals.

**fused** Joined and growing together.

# Gg

**galactose** A ★monosaccharide (single sugar with formula $C_6H_{12}O_6$) which is a part of the ★disaccharide (double sugar) called ★lactose (milk sugar) as well as more complex molecules in plants (e.g. gums, ★pectins) and animals (★glycoproteins).

**galah** A pink and grey cockatoo, *Cacatua roseicapilla*, in bird family Cacatuidae.

**gall** A lumpy malformed growth on plants, caused by an infestation by various parasitic organisms such as bacteria, fungi, mites, ★roundworms, and gall insects.

**gall bladder** A small sac that opens into the ★bile duct near the liver. See bile and Fig. D2.

**gall stones** Stones found in the gall bladder. Gall stones form in people who are prone to making extra ★cholesterol. The excess cholesterol is excreted in the bile, and if it becomes overconcentrated, gall stones may be produced.

**gamete** A mature reproductive cell. When the gametes can be told apart, the male gamete is called a ★sperm, and the female gamete is called an ★ovum. The gametes contain only one of each type of chromosome (haploid). When two sex cells join together at ★fertilisation, the pair of each chromosome type is restored (diploid). The resulting cell is called a ★zygote which then develops into an ★embryo, and finally into a new individual. See also oogamy, spermatozoa.

**gametophyte** A plant that produces gametes ('sperm', ova). Gametophytes occur most obviously in plants that show ★alternation of generations. In moss and liverworts the gametophyte is the most prominent plant. In ferns, it forms a small heart-shaped ★prothallus. In the above plants, the sperm need water to swim through to reach the ova, so these gametophytes tend to grow mainly in moist environments. See also life cycles.

**gamma globulin** A group of globular proteins found in the blood serum that includes many ★immunoglobulins with ★antibody activity. See globulin and Fig. A7.

**gamma rays** A type of ★radioactivity made up of very short wavelength ★electromagnetic radiation. Because gamma rays are very penetrating, they are quite dangerous. However, gamma rays from cobalt-60 are used to treat cancer.

**gamophyllous** With leaves or leaf-like parts united along the edge.

**ganglion** Plural is **ganglia**. A small mass of nerve tissue containing the cell bodies of the ★neurones. In vertebrates, ganglia are found associated with each of the vertebrae. A swelling on the dorsal ★root contains a ganglion for nerves carrying impulses to the spinal cord (afferent nerves). Other ganglia are found elsewhere in the ★peripheral nervous system. In invertebrates, the nervous system consists largely of ganglia connected by nerves, the main ganglion being usually found in the head. See also brain, basal ganglion and Fig. R2.

**gap junction** The gap (synaptic cleft) in a ★synapse across which the

★neurotransmitter passes to reach the next neurone. See Fig. S12.

**gastric juice** A mixture of chemicals secreted by glands in the walls of the ★stomach to help digest food. Hydrochloric acid produces acid conditions so that enzymes such as pepsin can help break down proteins into short lengths of amino acids called ★peptides. A ★mucus is also produced which coats the walls of the stomach to prevent it being digested. See digestion.

**gastric ulcer** A type of ★peptic ulcer which occurs on the inside surface of the stomach.

**gastrocnemius** The calf muscles, connecting femur and the heel-bone through the Achilles tendon. Used to move the foot.

**gastrocoel** The cavity within a ★gastrula.

**gastroenteritis** Inflammation of the stomach and intestines, resulting in symptoms such as diarrhoea, abdominal pain, nausea, vomiting, fever. There are many causes, including infections and allergies.

**Gastropoda** A class of molluscs, also called **univalves**, all of which have but one shell, a distinct head with eyes and tentacles, and a large flattened foot. e.g. Snails, slugs, limpets, periwinkles, and whelks. Compare bivalves. See Appendix 1.

**gastrula** An early stage in the development of an animal ★embryo which follows the single-layered hollow ★blastula. During this stage the usually ball-shaped mass of cells develops the two or three germ layers: ★ectoderm, ★mesoderm (absent in some animals), and ★endoderm. Also, a hole called the blastopore develops in the side of the ball of cells. Cells from the surface gradually move inside the ball through this hole. See also development.

**Gause's principle** The hypothesis that two species with identical ecological requirements cannot coexist for extended periods in the same environment. The species that is better adapted to finding food and using the other biological and non-biological resources will eventually exclude the other. See also natural selection.

**Gecko** Any of a group of small ★lizards, family Gekkonidae, mostly nocturnal, many with adhesive pads on the toes. See Appendix 1.

**geebung** Any shrub or tree of the genus *Persoonia*, Family Proteaceae. Also used to refer to the small succulent fruit from this tree.

**gel** A jelly-like ★colloidal substance where a fine network of threads holds a solution between them. e.g. Gelatin and silica gel. See electrophoresis.

**gelatinous** Having the nature of jelly. Jelly-like. e.g. ★Agar. See also colloid.

**gemmule** 1 An asexual reproductive body produced inside a sponge. 2 The many spine-like projections from the dendrites of a ★neurone.

**gene** 1 The unit of ★heredity, which is part of a ★chromosome, and can be passed from one generation to the next. See also allele, dominant, recessive. 2 The sequence of organic ★bases that form part of a ★DNA molecule that codes for the production of an enzyme or structural protein. See also codon, transcription, translation.

**gene, dominant** See dominance.

**gene flow** The movement of a ★gene from one ★population of organisms into another population, due to interbreeding.

**gene frequency** The proportion of a given population that has a particular form of a ★gene (★allele). Usually expressed as a percentage.

**gene gun** The firing of tiny, high-velocity particles of tungsten coated with DNA into plant cells. See vectors and genetic engineering.

**gene locus** The location of a particular ⋆gene (or one of its ⋆alleles) along a ⋆chromosome.

**gene mutation** Also called a **point mutation**. A change in a gene that can be inherited. Caused by a change in the sequence of bases that make up the DNA.
Most such mutations are harmful to the organism. Base substitutions occur when one base is replaced by a different one, which may then cause the wrong amino acid to become part of a protein. When an extra base is added, or a base is left out, then a **frameshift mutation** occurs. This completely changes the groupings of the three bases (⋆triplets) and has drastic effects on gene function. If these errors only occur in one strand of the DNA, then the cell's DNA repair machinery can often repair the errors. Compare chromosome mutation. See also mutation.

**gene pool** All the genes in a given population of a species. If the gene pool becomes limited (e.g. ⋆genetic bottleneck; ⋆founder effect) then variation in the population is reduced and the species becomes less able to adapt in a changing environment.

**gene probe** A method of finding whether a known gene is present in a section of DNA, especially that which has been inserted into a bacterial ⋆plasmid during ⋆genetic engineering. A radioactively labelled piece of DNA or RNA (the probe) that is known to be ⋆complementary to the desired gene is added to the original DNA, and if it 'sticks', then the gene must be present in the bacteria concerned.

**gene, recessive** See recessive.

**gene splicing** A technique used in ⋆genetic engineering where a gene is cut from the DNA of one organism, and then inserted (spliced) into the DNA of another organism. An appropriate enzyme is needed for each stage of this process; one to cut it out of DNA (see restriction enzyme), and another to splice it into new DNA (see DNA ligase and 'sticky' ends). See Fig. G2. Since it is not possible to see which organism (e.g. bacterium) has successfully received the new gene, a ⋆gene probe is needed to help in identification. See also genetic engineering.

**gene therapy** The insertion of normal or genetically altered ⋆genes into cells, usually to replace defective genes, especially in the treatment of genetic disorders. e.g. Cells are taken from a person's body and a virus used to insert a gene that may be missing. The cells are then reintroduced back into the body.

**generalised** To devise a rule or principle that is not specific, but applies to many cases.

**generation time** As applied to populations of cells, the average period of time between the ⋆prophase of one cell division to the prophase in the next cell division.

**generator potential** See receptor potential.

**generic** Relating to a ⋆genus.

**genes, linked** See linked genes.

**genetic bottleneck** The effect on the ⋆gene pool where a population is reduced to a small number due to a chance disaster of some kind. This means that only a few organisms will give rise to a new population that could have a different gene pool from that of most of the original population. e.g. Cheetahs have a very limited variety of genes and it is thought this is due to a genetic bottleneck when its population crashed in the past.

Compare founder effect.

**genetic code** The sequence of three ⋆bases (called a triplet or ⋆codon) along the DNA or RNA that specifies the next ⋆amino acids in the protein. There are $4^3$= 64 different ways to arrange the four bases in groups of three (in DNA: C = cytosine, T= thymine, G= guanine, A= adenine; in RNA: thymine is replaced by U = uracil). As there is only 20 naturally occurring amino acids, each amino acid may be specified by more than one codon. e.g. CGA and CGG both specify arginine. See Fig. G1.

**genetic counselling** Counselling provided to people who have had a child with a genetic disease or who come from a family with a history of genetic diseases.

**genetic drift** A change in the frequency of genes that is due to chance rather than to ⋆natural selection. Genetic drift is thought to be more likely in small populations where mating is more restricted. See also the founder effect and genetic bottleneck.

**genetic engineering** The technology involved in altering the ⋆genes present in organisms, using artificial means. **1** The alteration of the chromosome structure in cells from which breeding takes place, in an attempt to control the characteristics of the offspring. e.g. ⋆hybridising cells; inducing ⋆polyploidy in plants; ⋆irradiation of flowers, fruitfly etc. **2** Especially applied to ⋆recombinant DNA technology. The separation of useful genes from one organism, and their insertion (see gene splicing) into another organism that does not normally have that gene (Fig. G2). e.g. The gene for production of human insulin (for treating ⋆diabetes) has been ⋆spliced into the bacterium *E. coli*. Although each bacterium produces only a little insulin, billions of bacteria can be grown together in tanks, and the

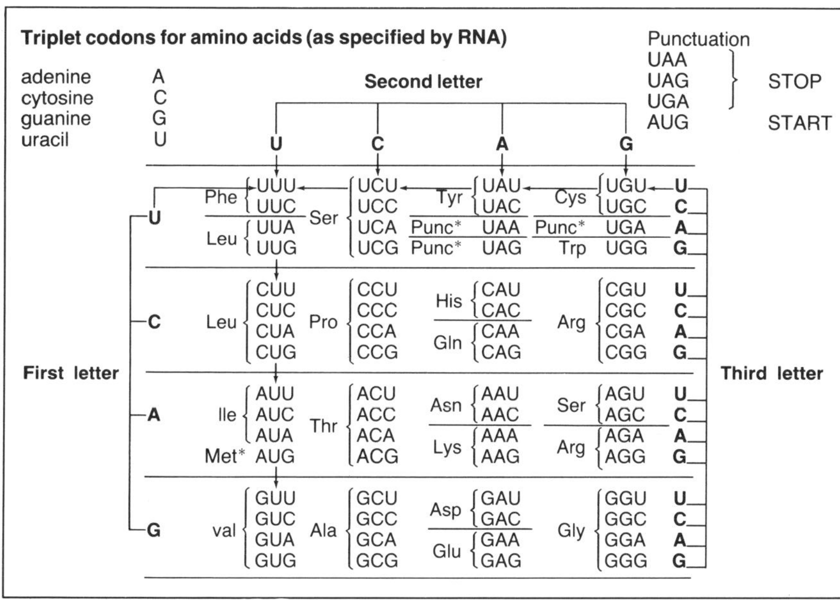


**Fig. G1 Genetic code**

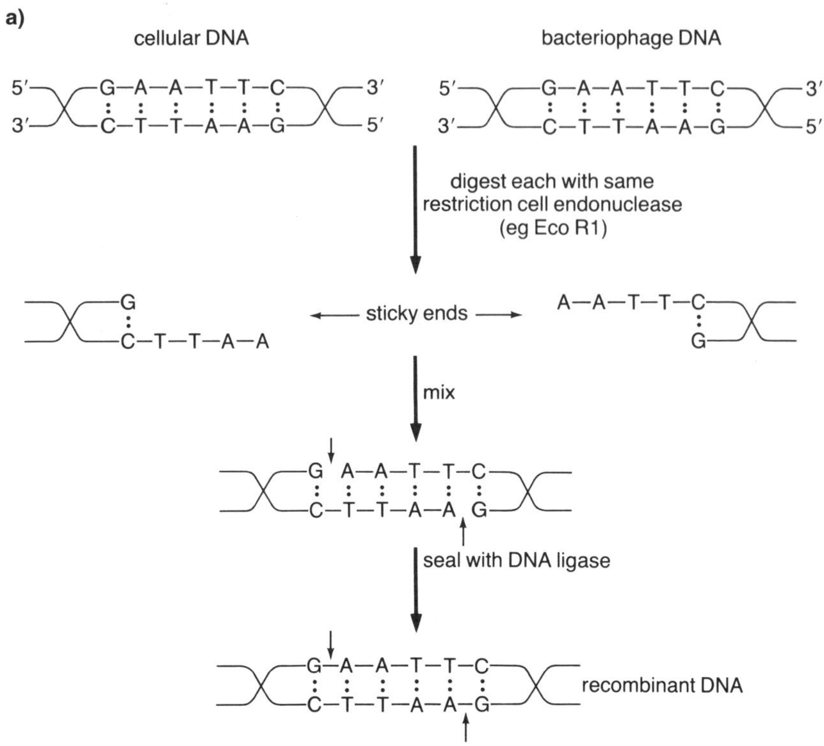


**Fig. G2 Genetic engineering (a) Sticky ends (b) Gene splicing**

larger amounts of insulin produced can then be separated. See also gene probe.

**genetic fingerprint** See fingerprint.

**genetic isolation** See isolating mechanisms.

**genetic marker** A usually *dominant gene or trait that helps to identify other genes or traits *linked with it when on the same *chromosome.

**genetic screening** The testing of people to see if they carry a defective gene, even if there are no symptoms. This is especially useful if a family has a history of a particular disease. e.g. Couples planning to have children to see if they have genes that would produce a disease such as Tay–Sachs in their children.

**genetics** The science that deals with *heredity and *variation.

**genital herpes** See herpes

**genital warts** See warts.

**genitals** The sex organs; the organs of reproduction. Especially applied to the external organs. e.g. Penis, testes, vulva.

**genome** 1 A complete (haploid) set of *chromosomes, and their associated *genes. 2 The project to find out where all the genes are located on human chromosomes.

**genotype** All the genes present in a cell, whether they are expressed or not. Compare phenotype.

**genus** A subdivision of a family. Each genus is made up of many

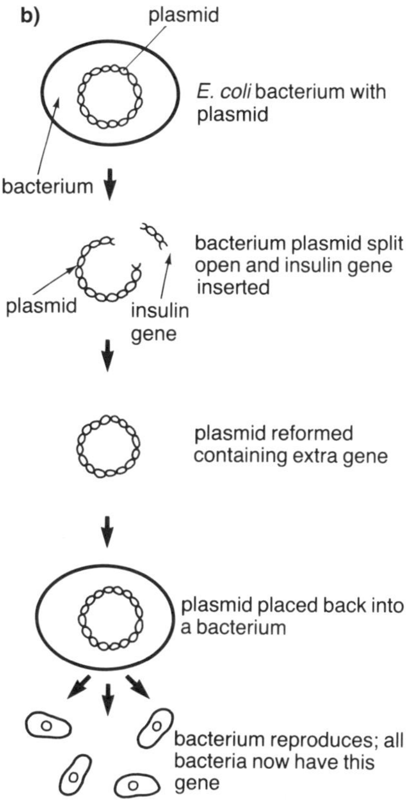


species. e.g. *Panthera leo* the lion, *Panthera tigris* the tiger, *P. onca* the jaguar, *P. pardus* the leopard. Notice that the first name (the genus) is the same in each case, and starts with a capital letter. The second name is the species name and starts with a small letter. Both names are in italics, or may be underlined (e.g. Panthera onca). If used often, the genus name may be reduced to one capital letter after it has been introduced. e.g. *P. onca*. Sometimes the species may not be known, and its name is written *Panthera* sp.

**geochronology** Determining the history of life on Earth based on a time scale derived from geological evidence. See Appendix 2.

**geographical distribution** The distribution of life over the surface of the Earth. The life found in the different regions tends to be characteristic, with often similar looking organisms occupying similar ⋆niches. e.g. Wombat in the Australasian realm, and the Marmots from Nearctic (North America) realm. See biogeographic realms, convergent evolution.

**geological time scale** The method of dividing the history of the Earth into ages based on fossil and other geological evidence. ⋆Radioactive dating is also used where possible. The main divisions are known as eras, which are further divided into ⋆periods and ⋆epochs. See Appendix 2.

**geometric increase** An increase in population size at a constant rate. e.g. When the numbers are doubled each generation, as when cells are dividing by ⋆mitosis. Compare exponential increase.

**geotaxis** An organism's response to the force of gravity. e.g. ⋆Geotropism in plants. See also taxis.

**geotropism** Also called **gravitropism**. The response of a plant to the force of gravity. The main stems of a plant are negatively geotropic, growing vertically upwards. The main roots of a plant are positively geotropic, growing vertically downwards. The change in growth direction is believed to be due to plant ⋆growth substances such as the ⋆auxins. These substances cause cell elongation to occur on the bottom side of a horizontal stem, causing it to bend upwards; they cause cell elongation on the top side of a horizontal root causing it to bend downwards. See also tropisms.

**Geraniaceae** The geranium family of dicot flowering plants. See Appendix 1.

**germ** **1** Any minute disease-causing organism, especially bacteria. **2** The embryo of a seed. e.g. Wheat germ.

**germ cell** Also called a ⋆**gamete**. A reproductive cell e.g. sperm, ovum.

**germ layer** One of the three layers of cells found in the early ⋆embryo stage called a ⋆gastrula. See also development.

**germ plasm** The reproductive cells of the body (e.g. sperm, ova) as distinguished from the ⋆soma (non-reproductive cells).

**German measles** A ⋆contagious virus infection (*Rubella*) with only mild symptoms in most people (e.g. pink rash, mild fever, etc.). However, in women in their first three months of pregnancy an infection can cause severe damage to the ⋆embryo.

**germination** The start of growth of a ⋆seed, ⋆spore, or other reproductive cell after a period of ⋆dormancy. Germination does not begin without the correct internal and external conditions. e.g. In warm conditions, a dormant wheat seed will absorb water, respiration starts, and protein synthesis begins. Germination is considered successful when the first root/shoot appears from the seed. Some seeds will not germinate without exposure to special conditions. e.g. Fire, freezing.

**gestation** The period of time between ⋆conception (⋆fertilisation) and birth (⋆parturition). Pregnancy. The carrying of young within the womb (uterus).

**giant axon** Relatively large ⋆axon (nerve fibre) found in various invertebrates (e.g. annelids, crustaceans, molluscs) and some vertebrates (e.g. fish, amphibians). Such axons are widely used in researching how the nervous system functions.

**giant fibre** See giant axon above.

**gibberellic acid** The first of the ⋆gibberellins to be isolated, being found in the fungus *Gibberella fujikuroi*. When this fungus infects rice, it causes excessive lengthening of the shoots and leaves. See also growth substance and Fig. G3.

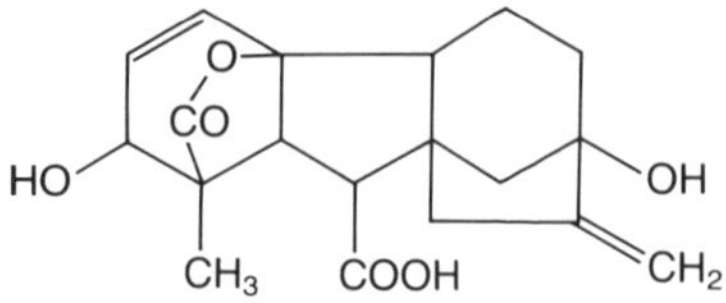


**Fig. G3 Gibberellic acid**

**gibberellin** A class of ⋆growth substances produced by plants (see gibberellic acid above). Gibberellins often only work when associated with other growth substances (e.g. ⋆auxins, ⋆abscisic acid). Thus gibberellins work with auxins in promoting the lengthening of cells. Gibberellins help stimulate flowering in certain day-long plants (e.g. cabbage, spinach). They also are involved in stimulating leaf and fruit growth, and the breaking of dormancy in some seeds and tubers. Gibberellins are widely used in agriculture and horticulture.

**Gibbon** The smallest and most slender of the ⋆anthropoid apes, genus *Hylobates*, living in the trees of south-east Asia. See Appendix 1.

**gidgee** A number of small wattle trees, including *Acacia cambagei*, or stinking wattle.

**gill** 1 In aquatic animals, the thin walled structures through which oxygen passes (⋆diffuses) from water to the blood stream, and through which carbon dioxide passes from the blood back to the surrounding water. 2 In fungi such as mushrooms and toadstools, the radially arranged plate-like structures (⋆lamellae) that hang under the cap, and on which the reproductive ⋆spores are produced.

**gill slits** In ⋆chordates, a series of openings connecting the back of the throat (⋆pharynx) to the exterior. In aquatic vertebrates (e.g. fish, amphibians), they are separated by arches which support the ⋆gills used to extract oxygen from water. In terrestrial vertebrates, similar structures only exist temporarily during growth of the ⋆embryo.

**ginkgo** The maidenhair tree (*Ginkgo biloba*) is often grown as an ornamental tree, and is similar to the gymnosperms. This tree is often considered a 'living fossil' as it is the only living representative of the many types of related trees that existed in the Mesozoic. See Ginkgopsida and Appendix 1.

**Ginkgopsida** A class of vascular plants, also called ⋆ginkgo. See Appendix 1.

**gizzard** Part of the gut in many animals in which the food is ground up into small particles, thus speeding digestion. e.g. Between the ⋆crop and the intestine in many worms and ⋆arthropods; the second part of the stomach in birds. Seed eating birds often swallow small stones that line the inside of the thick muscular walls of the gizzard, thus producing a hard grinding surface.

**glabrous** Used to refer to surfaces, especially leaves without hairs or bristles.

**glade** An open, often grassy space in a forest.

**gland** An organ or tissue that produces and secretes a substance that is used elsewhere in the organism. 1 The substance is secreted onto a surface (⋆exocrine gland e.g. sweat glands onto the skin; digestive juice glands into the stomach), or into the bloodstream (⋆endocrine gland e.g. ⋆pituitary gland; ⋆Islets of Langerhans). Glands may secrete directly, or through ⋆ducts. 2 In plants, glands produce nectar in flowers, digestive juices in carnivorous plants, gums and resins from stems, etc.

**glandular fever** Also called **infectious mononucleosis**. Most people are infected by the ⋆Epstein–Barr virus, causing only mild cold-like symptoms. A few suffer more severe symptoms including fever, swollen glands and tiredness but the disease is not highly infectious. There is an increase in the numbers of ⋆white blood cells during the disease which lasts up to a month.

**glans** The conical structure at the tip of the penis or clitoris.

**glaucoma** A common cause of blindness in adults, produced by excessive pressure of fluid inside the eyeball.

**glaucous** The covering of leaves with a greyish-blue 'bloom' (a powdery-waxy secretion), sometimes giving the leaves a grey or silvery appearance.

**glial cells** Also called **glia**. Supporting cells in the vertebrate nervous tissue. Glial cells give mechanical and metabolic support to the ⋆neurones, and are important in guiding the growth and connections of neurones in babies, as well as the repair of damaged neurones in adults. The ⋆Schwann cells that form the ⋆myelin sheath are special types of glial cells.

**gliding possum** Any of the tree-dwelling marsupial possums, in families ⋆Petauridae and ⋆Burramyidae, with parachute-like membranes that allow them to glide when their legs are stretched out. See also the sugar glider, greater glider, etc. See Appendix 1.

**global warming** Warming of the surface and lower atmosphere of a planet due to the ⋆greenhouse effect. Warming such as this could alter climates throughout the world, and also cause sea levels to rise at least 30 cm.

**Globigerina** A common genus of the order of protozoans called ★Foraminifera, whose calcium carbonate-rich skeletons sink after death to form huge deposits on the floor of the oceans. *Globigerina* ooze is a mixture of mud and these skeletons. See Appendix 1.

**globular protein** ★Proteins that fold to form compact three-dimensional shapes, many of which are roughly spherical. e.g. Haemoglobin. This contrasts to the linear shape of the ★fibrous proteins.

**globulin** Any of a group of relatively simple animal and plant ★proteins which are insoluble in water, but soluble in salt solution, and are ★coagulated by heat. Among the globulins are the main proteins in plant seeds, and the ★immunoglobulin of ★antibodies. See also gamma globulin.

**glomerule** A small compact cluster, especially of an inflorescence of flowers.

**glomerulus** A compact intertwined, often knot-like mass. e.g. The knot-like cluster of capillaries, almost enclosed by the expanded end (★Bowman's capsule) of the renal tubes of the vertebrate ★kidney. Only small molecules such as water, salts and sugar (glucose) can pass into Bowman's capsule. The useful parts are then re-absorbed into the bloodstream further along the renal tubes. See Fig. K1.

**glottis** 1 In vertebrates, the opening of the wind pipe (tracheae) into the back of the throat (pharynx). 2 In mammals, the opening between the muscular skin fold called ★vocal cords found in the voice box (larynx).

**glow-worm** 1 A firefly, especially the wingless female or luminous larvae. Actually the firefly is a beetle of Family Lampyridae, and is confined to locations north of Sydney. 2 The larvae of certain fungus gnats (Family Mycetophilidae) found near Sydney and parts of Tasmania.

**glucagon** A hormone produced by the ★Islets of Langerhans in the pancreas in response to low concentrations of glucose in the blood. It acts in the reverse manner to ★insulin, causing glycogen stored in the liver to be converted into glucose. Chemically, glucagon is a small protein.

**glucose** Also called dextrose. A six-carbon simple sugar (monosaccharide), $C_6H_{12}O_6$, found widely in nature as part of disaccharides (e.g. sucrose or table sugar), polysaccharides (e.g. starch, cellulose, glycogen). See Fig. C2. It is the first chemically stable product formed during ★photosynthesis, and the final product of carbohydrate ★digestion in vertebrates. During ★respiration, glucose is broken down, releasing energy in the process. See also carbohydrates.

**glutamic acid (glu)** Also called **glutamate**. One of the 20 ★amino acids that occur naturally in proteins. Glutamic acid is one of the first products of nitrogen ★assimilation, being converted into many other amino acids. It may also act as a ★neurotransmitter. See Fig. A6.

**glutamine (gln)** One of the 20 ★amino acids that occur naturally in proteins. Along with ★asparagine, it is important in removing free ammonia from the ★cytoplasm. See Fig. A6.

**gluteal** In the region of the buttocks.

**gluteus** Any of the three major muscles of the buttocks. e.g. gluteus maximus.

**glycerol** A colourless, odourless, and slightly sweet ★alcohol that forms an important part of many ★lipids (fats). When fats are digested, they are converted into fatty acids and glycerol. Chemically, glycerol is a triol (an alcohol with three –OH groups), $CH_2OHCHOHCH_2OH$.

**glycine (gly)** One of the 20 ⋆amino acids that occur naturally in proteins. A number of other molecules are made from glycine, and glycine may be an ⋆inhibiting ⋆neurotransmitter at certain ⋆synapses. See Fig. A6.

**glycogen** Often called **animal starch**. A ⋆polysaccharide made up of long branching chains of ⋆glucose molecules. Glycogen is the form in which animals, and some algae and fungi, store carbohydrates. In vertebrates, glycogen is mainly stored in the liver and muscles. It acts as a short term, readily available energy supply. Fats, the other form of storing energy, are much more long term and not as readily converted to glucose for respiration. See also glucagon.

**glycolysis** The anaerobic part of ⋆respiration in animals (i.e. functions without the need for oxygen). Glucose is converted to pyruvic acid with the release of some energy. If after glycolysis there is plenty of oxygen reaching the cells, the pyruvic acid is further converted into carbon dioxide and water in the ⋆Krebs' or tricarboxylic acid (TCA) cycle. If there is insufficient oxygen reaching the cells, then in animals, the pyruvic acid is converted to lactic acid. The lactic acid soon builds up to cause cramp, or 'stitch'. The lactic acid is removed after normal supplies of oxygen are restored. In plants, the pyruvic acid is converted to alcohol if there is insufficient oxygen (see fermentation). Chemically, glycolysis involves the sequential action of eleven enzymes, and produces two molecules of pyruvic acid, and two molecules of ATP for each glucose molecule involved. See Fig. G4.

**glycoprotein** Any of a group of naturally occurring compounds in which carbohydrate side-chains are covalently bonded to a protein 'backbone'. Important glycoproteins include certain enzymes and hormones, and various antigens occurring on the surface of cell membranes that are important in cell recognition (see histocompatibility antigen). Glycoproteins also form an important part of the cell walls of ⋆bacteria.

**gnarl** A knotty protuberance from a tree trunk or branch. Trees are said to be gnarled if their trunks/branches are extremely twisted.

**gnat** Also called **midges**. Any of the small two-winged insects, order ⋆Diptera, that look like tiny mosquitoes, but are always smaller and usually cannot pierce the skin with their mouthparts. They differ from mosquitoes in that they hold their forelegs aloft when standing, rather than their hind legs.

**Gnetopsida** A class of rare vascular plants. See Appendix 1.

**goanna** Any of the large monitor ⋆lizards, family Varanidae, especially the lace monitor *Varanus varius* which is common near many built-up areas. See Appendix 1.

**goblet cells** A pear-shaped surface (epithelial) cell that produces ⋆mucus (especially ⋆glycoproteins). In the intestinal tract, the mucus helps protect the protein walls from being digested. In the respiratory tract the mucus helps trap small particles that would otherwise clog the ⋆alveoli. The goblet cells on the surface of fish produce the mucus that gives them their 'slimy' feel.

**goitre** The enlargement of the ⋆thyroid gland. There are several causes. e.g. Lack of sufficient iodine in the diet to make the hormones normally produced by the thyroid.

**golden algae** See Chrysophyta.

**Golgi body** Also called the **Golgi apparatus**. These flattened disc-shaped

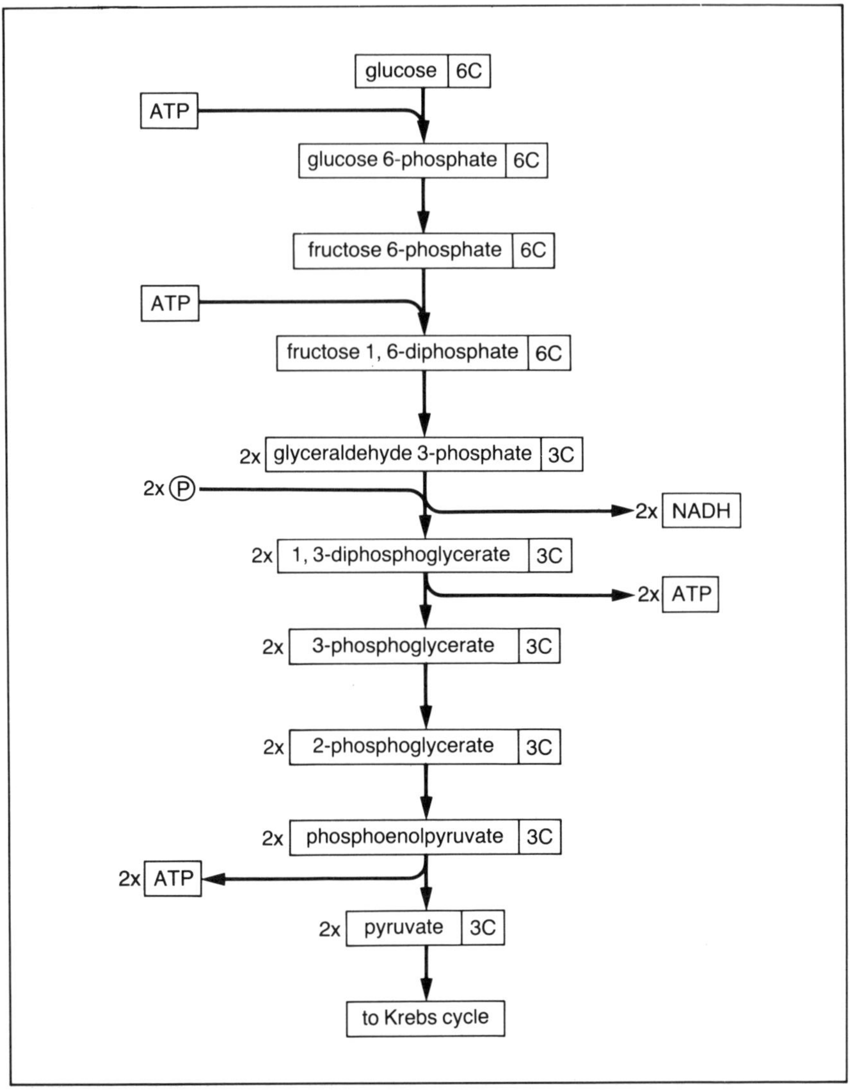


**Fig. G4 Glycolysis (respiration)**

and membrane-bound sacks, plus ⋆vesicles, function as a collecting and packaging centre for substances manufactured by the cell to be later released outside. Golgi bodies are common in ⋆goblet cells, ⋆neurones, and other cells producing secretions. See also exocytosis, endoplasmic reticulum, and Fig. C5.

**gonad** The reproductive organ which produces gametes such as sperm and ova. e.g. ⋆Ovary and ⋆testis. The gonads often produce hormones as well. e.g. The testes produce ⋆testosterone. The ovary produces ⋆progesterone and ⋆oestrogen.

**gonadotrophin** A group of several hormones produced by the front (anterior) lobe of the ⋆pituitary, or by the ⋆placenta, and which act on (stimulate) the gonads (ovaries and testes). e.g. ⋆FSH, ⋆LH. The output of these hormones is controlled by the ⋆hypothalamus.

**Gondwana** The ancient continent, believed to have existed during the Paleozoic and Mesozoic times, that eventually split up to form Australia, India, Antarctica, Africa, and South America. See plate tectonics and Fig. P7. Compare Pangaea and Laurasia.

**gonococcus** *Neisseria gonorrhoeae*, the bacterium that causes ⋆gonorrhoea. See below.

**gonorrhoea** One of the most common ⋆sexually transmissible diseases (STD), caused by infection of the kidney-shaped ⋆bacterium *Neisseria gonorrhoeae*. In women, long term infection can lead to blockage of the oviducts, and infection of a baby during birth, which may in turn lead to blindness. Compare syphilis.

**Goodeniaceae** The Goodenia family of dicot flowering plants. e.g. *Goodenia, Damperia*. See Appendix 1.

**Gorilla** A large usually ground-dwelling anthropoid ape of Africa, *Gorilla gorilla*. See Appendix 1.

**gout** An excess of uric acid in the blood that leads to a painful inflammation of the joints, especially those in the feet and hands.

**Graafian follicle** Spherical structures in the ⋆ovary inside of which the ⋆ovum matures. After the ovum is released into the ⋆oviduct, the remaining parts of the follicle become what is called the ⋆corpus leuteum, producing hormones such as ⋆progesterone that help suppress ⋆menstruation. The Graafian follicle is also believed to be a source of ⋆oestrogen in the ovary.

**gradient** A regularly increasing or decreasing change in environmental conditions. e.g. Temperature, ⋆salinity.

**graft** To join a piece of plant or animal tissue to a part of the body where it was not originally growing. The graft may be transferred from its normal position to a new position on the same body. e.g. Skin grafts take undamaged skin from say the thigh or buttocks, and attach it to a burnt or damaged visible area such as the face. The graft may also be between different organisms. e.g. With roses and fruit trees, shoots with desirable qualities may be grafted onto disease-resistant roots. With humans, skin tissue or even whole organs (e.g. heart, kidney) have been grafted from one person to another person (e.g. donated). However, grafts to other people often suffer ⋆rejection due to attacks by the ⋆immune system.

**Gram's stain** A stain that is able to distinguish two main groups of bacteria. Gram positive bacteria stain a deep purple colour. e.g. *Staphylococcus, Streptococcus*. Gram negative bacteria do not stain. e.g. *Gonococcus*, typhoid bacteria.

**Gramineae** See Poaceae.

**grana** Singular is **granum**. In ⋆chloroplasts, groups of flattened disc-like sacs (⋆vesicles) stacked like coins in a pile. Inside the grana, ⋆chlorophyll molecules are also stacked in layers. This is thought to increase the chance that a photon of light passing through a chloroplast will strike at least one of the many layers of chlorophyll molecules. See Fig. C6.

**granule** A minute grain or particle.

**graptolites** Fossilised skeletons of ⋆extinct marine colonial animals that are common in Cambrian, Silurian and Ordovician rocks. They are widely used in dating such rocks. The small polyp animals lived in little cups on branch-like stems. See Appendix 2.

**grass** Any of the monocot family called ⋆Poaceae which contains about 9000 different species. Apart from use as pasture and fodder, grasses are the main grain crops that supply mankind with food. e.g. Wheats (*Triticum*), maize (*Zea mays*), sugar cane (*Saccharum officinarum*). See Appendix 1.

**grass-tree** Also called **yakka**. Any of the woody stemmed trees in genera *Kingia* or *Xanthorrhoea*, monocot Family Xanthorrhoeaceae, with a tuft of grass-like leaves on top. Some specimens of slow-growing genus *Xanthorrhoea* are believed to be more than a thousand years old.

**grasshopper** Any of the terrestrial, herbivorous ⋆insects in the order ⋆Orthoptera which have large hind legs for hopping. One particular group, the plague locusts (e.g. *Chortoicetes terminifera*), have the ability to reproduce very rapidly under certain conditions, causing much damage to the vegetation. See Fig. I5.

**grassland** A natural area where the dominant vegetation is ⋆perennial grasses, with perhaps scattered trees mainly confined to water-courses. Usually there is insufficient rainfall for many trees to grow. See also prairie, pampas, steppe, savanna.

**graticule** A grid drawn onto a microscope slide to assist in the location/counting of specimens.

**gravid** Pregnant. Carrying eggs or young.

**great ape** One of the larger apes, placental mammal family Pongidae. e.g. Gorilla, orang-utan, chimpanzee, gibbon.

**green algae** Algae of the Phylum (Division) ⋆Chlorophyta. See Appendix 1.

**greenhouse effect** 1 The increase in temperature in a glass-walled building (a greenhouse) caused by radiant energy from the Sun being able to enter through the glass, while heat inside is unable to leave. 2 A similar effect that may cause the average temperature of the Earth's atmosphere to rise. Radiant energy from the Sun is able to pass through the atmosphere to the Earth. However, the heat energy radiated back by the Earth has a greater wavelength, and is absorbed by atmospheric carbon dioxide and water vapour, thus warming the atmosphere.

**greens** A political movement dedicated to the preservation of the environment.

***Grevillea*** Any member of the mainly Australian genus *Grevillea*, Family Proteaceae, including the ornamental Spider flowers, and *Grevillea robusta* (Silky Oak), which is used for timber.

**grey matter** Any part of the vertebrate ⋆central nervous system that contains the cell bodies, ⋆dendrites, and ⋆synapses. Because its nerve fibres are not covered in ⋆myelin, the tissue is grey in colour. In the brain, grey matter covers most of the surface (⋆cortex), as well certain concentrations (called ⋆nuclei) which occur deep inside. In the ⋆spinal cord, the grey matter fills the interior. Most of the brain's coordinating work is performed in the grey matter. Compare white matter.

**groin** The lowest part of the ⋆abdomen where it joins the legs.

**grooming** An animal cleaning fur or feathers by licking, scratching, rubbing, picking off dirt or parasites, etc. An animal may groom itself, or may groom others of the same species. Social grooming is common amongst the primates, helping to increase the cohesiveness of the group.

**ground cover** 1 The area of ground covered by the foliage of a plant; 2 Used to refer to low spreading plants that cover the ground.

**ground meristem** The central part of the topmost growth points of a plant (⋆apical meristem). Cell division in this meristem produces the ground tissues (see below).

**ground tissue** A 'packing' tissue made up of ⋆parenchyma with other tissues embedded within it. The cells are living, but have no special structures.

**ground water** Underground water filling the spaces between the grains in the rock and soil. The upper surface of the ground water is called the water table.

**growth** The increase in bulk (dry weight and/or size) of an organism as the cells begin to specialise and organs form. Growth usually involves cell division and cell expansion. In most plants growth occurs at the ⋆meristem where cell division occurs. In animals, most body cells are able to undergo cell division.

**growth habit** The characteristic shape and appearance of a mature plant. e.g. Canopy tree, small shrub, prostrate herb, tall grass, etc.

**growth hormone** Also called **somatotrophin** (STH). A protein hormone secreted by the front (anterior) part of the ⋆pituitary which stimulates skeletal and general body growth.

**growth ring** Also called an **annual ring**. One of the many concentric rings of wood seen in a cross section of a stem of a dicot or gymnosperm. These rings arise due to differences in growth rates where there are marked seasonal climatic changes. Wood formed early in the growth season has larger cells, and is thus lighter in colour than the smaller cells formed later in the season. The new wood is called ⋆secondary ⋆xylem and ⋆phloem. See also wood, dendrochronology, and Fig. W2.

**growth substance** Less correctly called **plant hormones**. Substances produced by a plant that help co-ordinate its growth and reproduction. See auxin, abscisic acid, florigen, gibberellin, kinin, ethylene.

**grub** The ⋆larvae of certain insects, especially some moths and beetles. e.g. Witchetty grub is the larvae of a number of different types of wood-boring moths and beetles.

**guanine** One of the four nitrogen-containing organic bases that makes up the 'letters' of the genetic code of DNA and RNA. Chemically guanine is a ⋆purine, and helps form the ⋆nucleotides that make up the ⋆nucleic acids. See Fig. B3.

**guano** A natural manure composed mainly from the droppings of seabirds, often found on oceanic islands.

**guard cell** One of the pair of kidney-shaped cells that surround a ⋆stoma, or pore in the surface of a leaf. A change in the amount of water in the guard cells causes them to change shape (⋆turgid cells open the stoma, ⋆flaccid cells close it) thus controlling the entry of carbon dioxide and loss of water vapour from the leaf. See Fig. S9.

**gullet** A passageway for food. e.g. Oesophagus or food pipe in animals.

**gum** 1 The soft fleshy tissues surrounding the base of the teeth. 2 A substance that swells in water to form a ⋆gel, or a sticky or slimy solution. Gums that ooze from trees often harden on contact with air. Also called resin, kino, etc.

**gumnuts** The hard fruits of a eucalypt that contain the seeds. The numbers, arrangement, shape, etc. of the gumnuts are important in identifying species of *Eucalyptus*. See Fig. G5.

**gumtree** Any tree of the genus *Eucalyptus*, especially those with a smooth bark. e.g. Sydney Blue Gum, *E.*

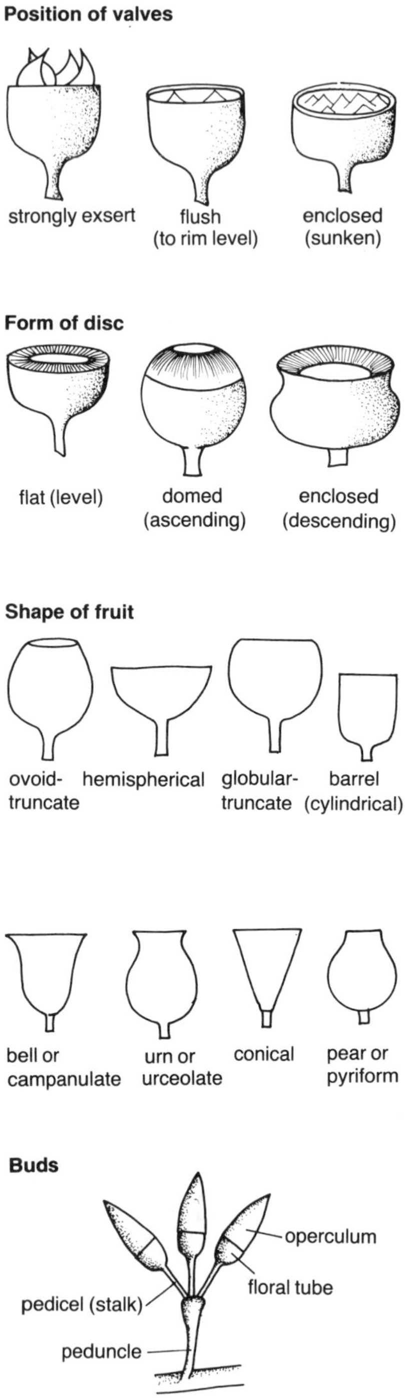


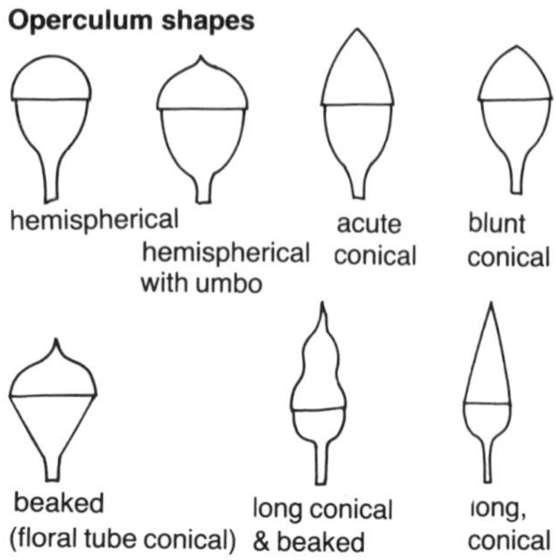


**Fig. G5 Eucalypt buds and fruit**

*saligna*. The Tasmanian Blue Gum, *E. globulus*, is the floral emblem of Tasmania.

**gut** The ★alimentary canal, especially the ★intestines.

**guttation** The loss of water in an uninjured plant in the form of droplets, especially at the tips or along the edges of the leaves.

**Gymnospermae** A grouping of non-flowering vascular plants that reproduce using cones which contain the developing seeds. The three main groups are the ★cycads, ★ginkgoes, and ★conifers (e.g. pines, cedars). Compare Angiospermae. See Appendix 1.

**gynoecium** The female parts of the flower. e.g. The ★carpels or pistils considered collectively. Compare androecium.

# Hh

**habit** 1 In plants, a characteristic form of growth. 2 In animals, a fixed pattern of behaviour, often instinctive.

**habitat** The place or environment in which a particular organism normally lives. e.g. Rock platform, creek, desert. Compare biome and ecosystem.

**habituation** A type of reflex learning where an animal is subjected to a continuous repetition of a stimulus without ⋆reinforcement, thus producing a diminished response to it. Without reinforcement, the behaviour is gradually lost.

**haem** A complex, roughly ring-shaped chemical structure (called ⋆porphyrin) that contains iron. An important part of proteins such as ⋆haemoglobin, ⋆myoglobin, and many ⋆cytochromes.

**haematoxylin** A blue dye that, usually in association with other chemicals, is used to ⋆stain nuclei and cellulose cell walls blue.

**haemocoel** The ⋆body cavity in arthropods and molluscs through which blood flows. Hearts usually pump the blood through a blood vessel to one end of the body where it is released into the body cavity to seep back amongst the organs to the other end. Compare circulation.

**haemoglobin** The red pigment of vertebrate red ⋆blood cells that is responsible for carrying oxygen and some carbon dioxide. See Fig. H1. It also occurs in other animals (e.g. earthworms), and in the root-nodules of nitrogen-fixing ⋆legumes.

Haemoglobin is normally bright red, but turns bluish when oxygen is lost. It consists of a protein (polypeptide) part called globin, which is connected to the iron-containing pigment called haem (see above). The haem is the oxygen carrying part, while some carbon dioxide can be carried at the ends of the protein (polypeptide) chains. Compare myo-globin. See respiratory pigments.

**haemophilia** A genetic disease in which the blood does not readily clot, resulting in excessive bleeding. Haemophilia is a ⋆sex-linked disease determined by a recessive gene on the X chromosome.

**haemorrhage** A usually considerable loss of blood due to bleeding.

**haemorrhoid** Also called piles. Dilated (swollen) ⋆veins around the rectum.

**hair** 1 In mammals, the thread-like growths from the ⋆skin. Each hair consists of two parts—the shaft that extends above the skin's surface, and the root which is embedded in the skin, within the hair follicle (see below). The lower end of the root is enlarged into a knob called the hair bulb. Above the surface of the skin, hair cells become filled with a tough protein called ⋆keratin that helps strengthen and water-proof the shaft. Hair colour depends on the amount of the pigment ⋆melanin that is present, and also on the presence of air bubbles. The number of air bubbles increases with age, and are responsible for the white colour of hair in older people. Hair helps prevent loss of heat (by trapping a layer of insulating air), and may also have a sensory role. For example, cats' whiskers help them judge the width of an opening. 2 Any thread-like outgrowths from the surface cells of a plant. Such hairs, also called ⋆trichomes, can be made of one or more cells. ⋆Root hairs absorb water and dissolved salts from the soil. Some hairs produce pain-causing chemicals e.g. Stinging plants. Others are sensory. e.g. Insectivorous plants such as the sundew. Some leaves are covered with hairs that trap a layer of air next to their surface. This air can become saturated with water, thus reducing ⋆transpiration.

**hair follicle** The layer of cells and connective tissue that forms a small sack surrounding the root of a ⋆hair.

**hakea** Any shrub or tree of the Australian genus *Hakea*, Family Proteacea, which produce their seeds in hard woody fruit.

**half life** The time taken for one-half of a sample of unstable atoms to decay by radioactive means. e.g. Carbon-14 has a half-life of 5730 years. Thus if we had 1 gram today, in 5730 years there would be 0.5 grams, and in a total of 11 460 years (2 × 5730) there would be 0.25 grams, and so on. Knowledge of half-lives can allow the age of certain specimens containing radioactive materials to be determined. See also radioactive dating, radiocarbon dating.

**hallucination** Distortion of normal sensory experience. Sensory experiences for which there is no appropriate external source. e.g. Hearing voices when no-one has spoken. Hallucinations can be due to mental problems, or due to drugs.

**hallucinogen** Drugs that produce distortions of the senses (hallucinations). e.g. LSD, mescaline.

**halophyte** Plants which live in places of high salinity (salt concentration) such as salt marshes,

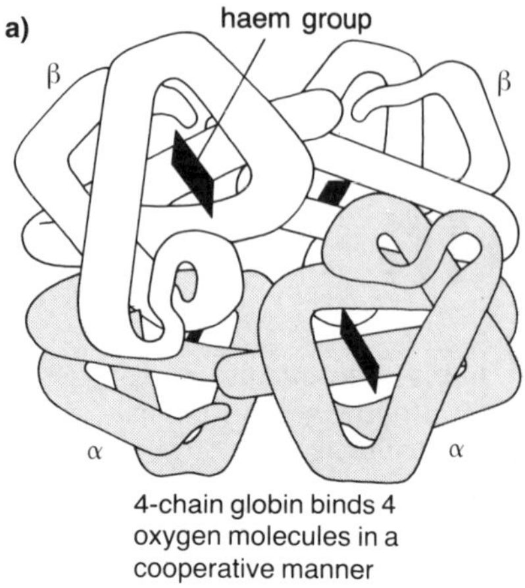


**Fig. H1 Haemoglobin (a) Three-dimensional shape (b) Haem group**

and the mud flats of estuaries. Compare xerophyte, hydrophyte, mesophyte.

**haltere** A sense organ of the ⋆Diptera (two-winged) flies which allows them to remain stable in flight. In place of the hind wings are two club-shaped structures containing many nerves, that oscillate rapidly during flight.

**hamstring** ⋆Tendons above the back of the knee.

**hand** In humans and the primates, the end part of the forearm that includes the wrist, palm and digits (fingers). See also pentadactyl limb.

**haploid** A single set of unpaired chromosomes in a nucleus. e.g. Sperm and ova in humans, as well as the gametes of most plants and animals. Normally, all other cells are diploid. However, normal cells of some protozoa, and the males of some insects (e.g. bees), are haploid. In plants that undergo ⋆alternation of generations, the ⋆gametophyte is haploid. The spores that produce the gametophyte are also haploid. Compare diploid. See also life cycle.

b)

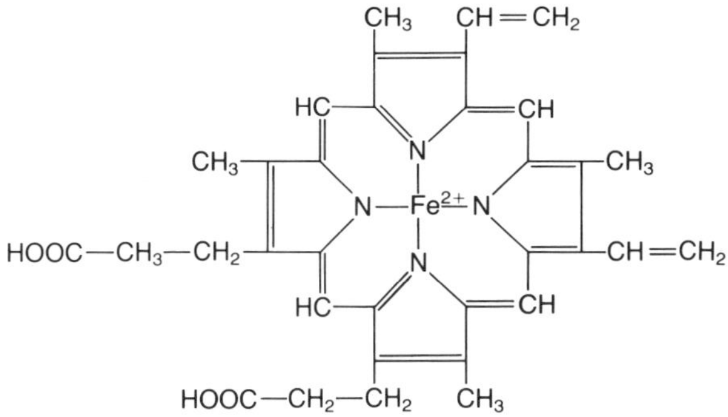


**haptotropism** See thigmotropism.

**hard drugs** Dangerously addictive drugs. e.g. Heroin. Compare with soft drugs such as aspirin.

**hardwood** 1 In botany, the wood produced by woody dicotyledons. Compare softwood, which is the wood of gymnosperms. 2 In Australia, woods produced mainly from eucalypts.

**Hardy–Weinberg Law** Under stable conditions, alternative alleles of a gene occur with constant frequency. The conditions needed for this to occur include: **a** a very large population, **b** random mating, **c** no natural selection because the population has been living under the present conditions for a very long time, **d** no mutations, and **e** no migration. These circumstances never exist perfectly.

**hare** A rabbit-like placental mammal, but with bigger ears. Introduced into Australia, *Lepus capensis* occurs in eastern areas.

**harelip** A human birth defect (⋆congenital deformity) where usually the upper lip has a vertical fissure or groove resembling a similar groove in a ⋆hare.

**hashish** A dark brown resin collected from the flowering tops of the ⋆cannabis (Indian hemp) plant that is usually smoked. Compare marijuana.

**Haversian canals** Very narrow (about 50µm in diameter) channels through the ⋆bones, carrying blood vessels and nerves, thus allowing communication between the ⋆marrow and the exterior of the bone. See also osteocytes.

**hay fever** An ⋆allergic reaction to pollen. Compare asthma.

**HCFC** A group of compounds where a hydrogen atom replaces a chlorine atom in a ⋆chlorofluorocarbon (CFC). They are less damaging to the ⋆ozone layer than CFCs.

**head** 1 In animals, the most anterior (forward) or uppermost part of the body, containing the brain, many sense organs, and openings to the digestive and respiratory systems. 2 In plants, a compact mass of the plant. e.g. The head of a cabbage is a compact mass of foliage leaves. The head of a cauliflower is a compact mass of flower buds. The head of a sunflower is also a compact mass of small flowers.

**health** Soundness and fitness of the body, including freedom from injury or disease.

**hearing** The detection of sound waves, and the interpretation by the brain (perception) of the nerve impulses so produced. See also ear.

**heart** A hollow, muscular organ or structure which, by regular squeezing (rhythmical contractions) pumps the blood through the body. See Fig. H2.

In vertebrates, used blood enters the thin-walled right ⋆atrium (on the right side as far as the owner of the heart is concerned) from the body, and then passes down into the right ⋆ventricle. When the muscular ventricle contracts, blood is pumped to the ⋆lungs via the pulmonary ⋆artery (see circulation). When the now oxygenated blood returns, it enters the left atrium, and then passes into the left ventricle. This chamber has even thicker walls, which when it contracts, is strong enough to pump the blood to the whole body through the arteries. The heart and arteries contain ⋆valves to prevent the backward flow of the blood. The rate of contraction is under indirect nervous control (see pace-maker). Worms such as the annelids (including earthworms) may have one main and several smaller supporting hearts. Insects have a long heart which contracts as a wave. This pushes the blood forwards to be released through pores into the body cavity (⋆haemo-coel). See also systole, diastole, cardiac muscle.

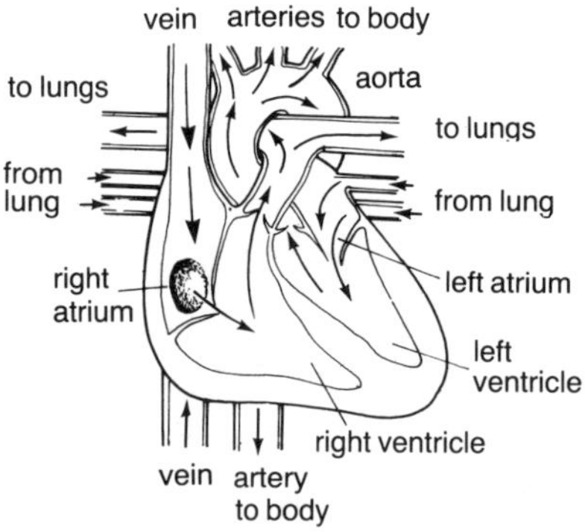


**Fig. H2 Heart**

**heart attack** Due to a coronary thrombosis (myocardial infarction). It is caused by a blood clot (⋆thrombus) becoming lodged in the ⋆coronary artery. This cuts off the blood supply to heart muscles resulting in an area of infarction, or death of tissues.

**heartburn** A mild to severe burning sensation in the upper abdomen or beneath the breastbone. It is usually due to the very acid contents of the stomach re-entering the food pipe (oesophagus), often following a heavy meal.

**heartwood** The central mass of non-living ⋆xylem tissue in the trunks of trees that provides mechanical support. The heartwood is usually harder and more resistant to decay than ⋆sapwood because it has been infiltrated by minerals and resins. The heartwood also preserves a record of its growth cycles in its ⋆growth rings. See also wood and Fig. W2.

**heath** In Australia, an area of land (heathland) covered by low, small-leaved and drought-resistant shrubs. Natural areas of heath are usually exposed to drying winds or growing on sandy and porous soils. In some areas, the lack of sizeable trees has more to do with constant burning than with poor soils. Typical plants of Australian heaths are dwarf *Banksia*, and species in the genera *Epacris* and *Leucopogon*. Compare scrub.

**heather** Any of the low evergreen shrubs of the Family Ericaceae, especially *Calluna vulgaris*, the common heather of England and Scotland, with small pinkish-purple flowers.

**heathland** See heath above.

**Heidelberg jaw** A fossil of an early human's lower jaw found near Heidelberg, Germany, in 1907.

**heliotropism** Also called ⋆**phototropism**.

**helix** A spiral such as would be formed if a cord was wrapped around a cylinder. See also double helix and Fig. D3.

**helper T-cells** See T-cells.

**Hemichordata** A subphylum of small, worm-like marine ⋆chordates that are called ⋆acorn-worms. See also Appendix 1.

**Hemiptera** Also called **bugs**. An order of insects that include aphids, vegetable bugs, bed bugs, scale insects, and leaf hoppers. As bugs feed by sucking the juices from plants, they are of enormous economic importance. They have two pairs of wings, the front ones thicker than the rear wings. See Appendix 1.

**hemorrhage** See haemorrhage.

**hemp** A bushy plant, *Cannabis sativa*, extensively used as a source of fibre for ropes, twine, and other products. Hemp plants are also a source of ⋆cannabis, ⋆hashish and ⋆marijuana.

**heparin** A substance which helps stop blood clotting by preventing formation of ⋆thrombin, and by acting against thrombin already formed. Heparin is an ⋆anticoagulant that exists in most human tissues, but especially the lungs and the liver. See also mast cell.

**hepatic** Concerning the liver. e.g. Hepatic portal veins carry blood from the intestines to the liver.

**Hepaticopsida** The ⋆liverworts, a class of plants in the Phylum (Division) Bryophyta, with many features similar to the ⋆mosses. See Appendix 1.

**hepatitis** Usually a virus-caused inflammation of the liver. Hepatitis A is spread by contamination of water and food by human wastes (faeces). Hepatitis B is a different virus spread mainly by the transfer of blood from one person to another, and has been more common amongst intravenous drug users and homosexuals. It is classed as a ⋆sexually transmissible disease (STD). Hepatitis C is mainly spread through blood and is a major threat to blood supplies. There are also non-viral forms, such as amoebic, alcoholic, toxic, and syphilitic.

**herb** Flowering plants with a green, non-woody stem.

**herbaceous** Having the nature of a herb.

**herbarium** A collection of dried plant specimens arranged in systematic order. Compare botanical gardens.

**herbfield** An alpine plant community of short ⋆herbaceous plants, usually growing in the form of a 'mat'. They survive under the snows of winter, and even in summer the ground beneath them is usually damp and cold.

**herbicide** Also called **weedkiller**. An agent that kills or inhibits the growth of herbs, especially weeds. e.g. 2,4-D and 2,4,5-T.

**herbivore** An animal that feeds mainly or entirely on plants or plant products. Compare carnivore, omnivore.

**hereditary** Capable of being transmitted genetically, through sperm and ova, from parents to offspring. Compare genetics.

**heredity** 1 The transmission of genetic characteristics from parents to offspring. 2 The capacity of an individual to develop traits or features that are present in parents or ancestors.

**hermaphrodite** Also called **bisexual**. 1 An animal which possesses both male and female sex organs (testes and ovaries). e.g. Earthworm. 2 A flower that contains both stamens and pistils. e.g. *Hibiscus*.

**hernia** Also called a **rupture**. The protrusion of an organ or part of an organ through a weak spot in tissues which normally contain it. e.g. In men, part of the intestines can protrude through the muscles of the abdomen wall into the sack holding the testicles.

**heroin** A ⋆narcotic ultimately derived from the opium poppy. ⋆Morphine, the chief ingredient in opium, is first extracted, and then chemically converted into heroin. A dangerous addictive drug.

**herpes** Any of several viruses which cause inflammation of skin or mucous membranes, often with blisters. 1 **Herpes simplex type 1** produces cold sores, especially on the lips. 2 **Herpes simplex type 2**, or **genital herpes**, is a ⋆sexually transmissible disease (STD) and much more serious, causing recurrent pain when the infected person is under stress.

**herpetology** The study of amphibians and reptiles.

**heterochromatin** The ⋆chromatin in a chromosome which stains more deeply than adjacent regions, and contains relatively few genes.

**heterogamy** Also called ⋆**alternation of generations**.

**heterosis** Also called ⋆**hybrid vigour**.

**heterotroph** An organism which is unable to make its own food, and thus depends on other organisms as a source of food. e.g. All animals, fungi, most bacteria, and insectivorous plants (partly). Compare ⋆autotrophs.

**heterozygous** 1 An organism in which the two ⋆alleles for a particular gene are different. e.g. In Mendel's peas, the gene for height would have the alleles Tt (tall). Compare homozygous. 2 A ⋆hybrid organism with different alleles for many genes.

**hexose** Describes a group of ⋆monosaccharides which have six-carbon atoms. e.g. Glucose, fructose, galactose. Most polysaccharides are made from combinations of many hexose sugars. See carbohydrates and Fig. C2.

**hiatus hernia** A ⋆hernia resulting from the stomach being pushed up through the diaphragm into the chest cavity.

**hibernation** Passing the winter in a ⋆dormant or lethargic state. This reduced physiological activity helps conserve energy when little food is available. Compare aestivation, dormancy.

**hiccups** A series of uncontrolled contractions of the ⋆diaphragm, and the associated sudden closure of the ⋆glottis.

**hierarchy** See peck order.

**hindbrain** Also called the **rhombencephalon**. The hindmost of the three divisions of the brain of an embryo vertebrate which develops into the ⋆medulla and ⋆cerebellum.

**hindgut** The end section of the ⋆colon in vertebrates, which is mainly involved with absorption of water.

**hip** 1 The ripe fruit of a rose. 2 The side part of the body near the hip joint.

**hip bone** The end part of the ⋆pelvic girdle.

**hip girdle** Also called the ⋆**pelvic girdle**.

**hirsute** Hairy and shaggy. Covered with hair, bristles, etc.

**Hirudinea** A class of segmented worms in the phylum Annelida that includes the ⋆leeches. See Appendix 1.

**hispid** Rough due to the presence of stiff hairs, bristles, etc.

**histamine** A chemical that is released from tissues when they are injured, and causes dilation (enlarging) of blood vessels which leads to ⋆inflammation. This ⋆amine is made from the amino acid histidine, and is often associated with various ⋆allergic reactions. See also antihistamine, mast cell.

**histidine (his)** One of the 20 common ⋆amino acids found in proteins. It is also used to make (a ⋆precursor) histamine (see above).

**histocompatibility antigen** ⋆Glycoproteins on the surface of cells that act as ⋆antigens, thus starting the ⋆immune response that leads to the rejection of ⋆grafts and ⋆transplants.

**histology** The study, usually with a microscope, of the structure of ⋆tissues and their organisation into ⋆organs.

**histones** A class of relatively simple proteins with unusually large proportions of the amino acids arginine and lysine in their structure. They are associated with DNA in the chromosomes of ⋆eukaryotic cells, and may be involved in turning off genes.

**HIV** Short for **human immunodeficiency ⋆virus**, the basic cause of ⋆AIDS. This virus invades certain ⋆T-cells, thus weakening the ⋆immune system. This allows other infections to become established, eventually killing the infected person. See also retrovirus.

**HL-A system** The group of ⋆histocompatibility antigens found in humans that promote the greatest immune response.

**hoary** Covered with very short, hard to see hairs, thus giving a whitish appearance.

**Hodgkin's disease** A cancer of the lymphatic system associated with the ⋆Epstein–Barr virus.

**holdfast** An organ of attachment or an anchoring structure, as found in various algae (especially seaweeds), protozoans, and parasitic worms.

**Holocene** Also called **recent**. The geological period since the last ice-age, starting about 10 000 years ago. See Appendix 2.

**Holothuroidea** A class of echinoderms that includes the ⋆sea-cucumbers. See Appendix 1.

**home range** The territory about which an animal normally moves in search of food, shelter and mates.

**homeostasis** The steady state, or state of constancy, of the internal environment of an organism. Any changes to internal temperature, chemical composition, etc. brings a response by the body to correct it. e.g. If the amount of glucose in the blood increases after a meal, extra ⋆insulin is secreted to have it removed to the liver and stored as glycogen. If internal temperature increases, the body may be stimulated to sweat, thus cooling the body by evaporation.

**homeothermic** Also spelt ⋆**homoiothermic**.

**hominids** Man (*Homo*), and human-like fossils such as *Australopithecus*.

**hominoid** Primates including humans and the ⋆anthropoid apes such as orang-utan, gorilla, chimpanzee and gibbon.

***Homo*** A genus of primates that includes *Homo sapiens*, or modern humans, and fossils such as Cro-magnon Man; *Homo neanderthalensis* and fossils called Neanderthal Man; *Homo erectus*, fossils called Peking and Java Man; and *Homo habilis*, a fossil intermediate between *Homo erectus* and *Australopithecus*.

***Homo erectus*** The name means 'upright Man'. A species of fossil humans, bigger than *Australopithecus*, but smaller than modern humans. Their brain volume ranged from 750 to 1200 $cm^3$, they more-or-less walked upright, and their remains have been associated with use of stone tools and the first use of fire (⋆Acheulian ⋆culture). See also Peking and Java Man.

***Homo habilis*** The name means 'handy' or 'tool Man'. A species of fossil humans, very similar to *Australopithecus*, but a little taller and with a slightly larger brain volume.

**Homo neanderthalensis** Neanderthal Man, now thought to be a branch group of early humans rather than a subspecies of *Homo sapiens*.

**Homo sapiens** The name means 'wise' or 'intelligent Man'. The species name for modern humans, as well as fossils such as ★Cro-magnon Man.

**homogeneous** Of the same kind or nature.

**homoiothermic** Also called **homeothermic** and **'warm-blooded'**. Having a constant body temperature independent of changes in the surroundings. e.g. Mammals and birds. Compare poikilothermic.

**homologous** 1 Similar in fundamental structure, position, and development but not necessarily in function. 2 Similarity of structure due to a common ancestry. Compare analogous.

**homologous chromosomes** A pair of chromosomes that have the same ★genes at corresponding points (see locus) along their length. The ★alleles, or forms of the gene, may differ on each of the chromosomes. ★Diploid cells have all their chromosomes in such pairs. During ★meiosis, homologous chromosomes pair at prophase, when ★crossing over may occur. See Fig. H3.

**homozygous** 1 An organism in which the two ★alleles for a particular gene are the same. e.g. In Mendel's peas, the gene for height could have the alleles TT (tall) or tt (short). Compare with ★heterozygous. 2 A pure-bred organism with the same alleles for many genes.

**honey flower** Also called ★**mountain devil**.

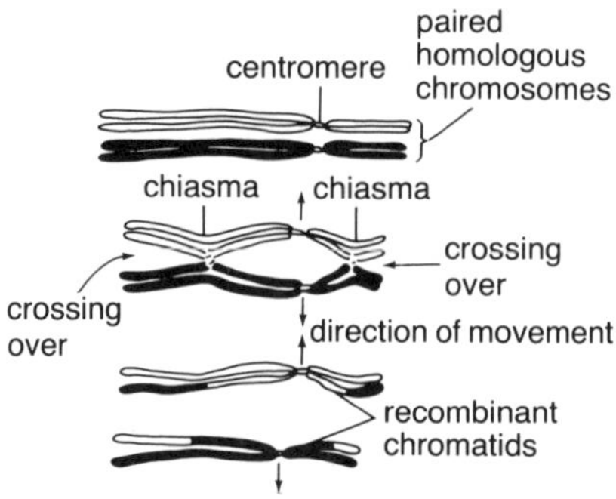


**Fig. H3 Homologous chromosomes**

**honeydew** 1 A sweet substance passed from the anus of aphids, leafhoppers, and other insects which is sought by insects such as ants. 2 A sweet liquid that exudes from the leaves of some plants in hot weather. 3 A type of melon, *Cucumis melo*, similar to 'rock melons' but with a greenish flesh.

**hookworm** A small, parasitic round worm, phylum ★Nematoda, which lives in the intestines of humans and other animals. See Appendix 1.

**Hoop pine** See pine.

**hormone** A chemical produced in minute amounts in one part of an organism and transported to other parts of the body where it exerts its effects. In animals, hormones are produced in ★endocrine glands and released into the bloodstream for transport to other parts of the body. e.g. See insulin, adrenalin, oestrogen. 'Plant hormones' are more properly called plant ★growth substances as they often exert their effects where they are produced, and not elsewhere in the plant. See also endocrine system, hypothalamus, pituitary, neurohormones.

**horn** 1 A stiff pointed projection from the head of a vertebrate. 2 The tough substance composed of ★keratin which covers, or forms the horns, hooves, claws or nails of animals.

**horse** A large, solid-hoofed and herbivorous placental mammal, *Equus caballus*, in order Perissodactyla. See Appendix 1.

**horticulture** The science of growing fruit, vegetables, flowers, etc.

**host** 1 An organism upon or in which another organism (often a ⋆parasite) lives and for which it provides protection, food, or both. e.g. A person can be host for a ⋆tapeworm. Relationships are not always parasitic. e.g. Sharks are hosts for remoras. 2 The recipient of a graft in tissue or organ transplants. Compare ⋆intermediate host and ⋆primary host.

***Hovea*** An Australian genus of dicots, Family Fabaceae, bearing clusters of small pea-shaped flowers.

**human** Any member of the genus ⋆*Homo*.

**human immunodeficiency virus** See HIV.

**humerus** The single long bone in the arm of a human that extends from the shoulder to the elbow. In other animals, the corresponding bone of the forelimb. See Fig. S3.

**humour** Any plant or animal fluid.

**humus** Dark material is the upper layer of the soil formed from decaying vegetable and animal matter. Humus provides nutrients for plant growth, and increases the water-holding capacity of the soil.

**Huntington's syndrome** A rare but fatal inherited disease where the victim suffers progressive loss of mental ability due to brain-cell death in the region of the ⋆basal ganglia, along with the reduction of some ⋆neurotransmitters and the build-up of another, ⋆dopamine. The symptoms are most common in middle age.

**Huon Pine** See pine.

**husbandry** The business of a farmer, usually associated with the raising of animals.

**hybrid** 1 An organism resulting from fertilisation by gametes which differ by one or more genes. A ⋆heterozygous organism. 2 More generally, the offspring of a cross between two different species, races, or varieties. A crossbred animal or plant. Crossing different species usually results in sterile offspring. e.g. Crossing a horse and a donkey to produce the ⋆sterile mule. When different races or varieties are crossed, the offspring may show certain advantages called hybrid vigour (see below).

**hybrid vigour** Also called **heterosis**. Qualities such as an increase in size, resistance to disease, numbers of offspring produced, etc., resulting from crossbreeding individuals of different races or varieties.

**hybridisation** The act or process of producing a hybrid (see above). Includes ⋆cross-breeding, and ⋆cross-pollination.

**hydatid worm** A minute ⋆tapeworm of four segments which infests dogs, cats and other carnivores. If its larvae develop in humans or wild or domestic animals, a fluid-filled hydatid cyst can form that contains developing worm larvae.

***Hydra*** A genus of small freshwater cnidarians (coelenterates) in the order ⋆Hydrozoa. A ring of sting-containing tentacles surrounds a mouth that leads to a digestive cavity. Wastes leave through the same opening. Has the appearance of a small ⋆sea-anemone. See Fig. P8 and Appendix 1.

**hydranth** A small colonial ⋆polyp.

**hydration** The act of incorporating or combining with water to form a hydrated material.

**hydrocarbon** A class of chemical

compounds containing only the elements carbon and hydrogen. e.g. Methane $CH_4$ is often a product or rotting vegetation in swamps. Ethene $C_2H_4$ is produced by fruits as they ripen. Oil (petroleum) is a complex mixture of hydrocarbons.

**hydrogen bond** A weak chemical bond between two molecules due to the attraction of a hydrogen atom with a partial positive charge to an oxygen or nitrogen atom with a partial negative charge. Although such bonds are much weaker than ⋆covalent or ⋆ionic bonds, they are of extreme importance to life. e.g. The two strands of the DNA molecule are held together by numerous hydrogen bonds.

**hydrogen (H)** An element that is a colourless, odourless gas. Hydrogen explodes when mixed with oxygen and ignited. Occurs in water ($H_2O$), and all ⋆organic molecules such as carbohydrates, proteins, lipids and vitamins.

**hydrolysis** 1 A chemical reaction in which water is a reacting agent (reactant). 2 A reaction in which a compound on reacting with water is split into two parts, one combining with an H from the water, the other with an OH. e.g. ⋆Digestion of food is often a hydrolysis reaction aided by (⋆catalysed) enzymes in the digestive juices. See also conversion of ⋆ATP to ADP.

**hydrophilic** 'Water-loving' substances. Such substances can dissolve in water and other ⋆polar solvents because hydrophilic molecules are themselves polar. e.g. Alcohol ($C_2H_5OH$) and soap molecules. Compare hydrophobic.

**hydrophobic** 'Water-hating' substances. They are poor at mixing with ⋆polar water molecules because hydrophobic molecules are non-polar. e.g. Hydrocarbons and fats (lipids). Compare hydrophilic.

**hydrophyte** A plant which lives in water or a moist environment. Compare xerophyte, halophyte, mesophyte.

**hydroponics** The growth of plants in ⋆aqueous solutions containing ⋆essential inorganic substances.

**hydrotropism** The growth of plant roots towards water or moisture. Generally, the growth response towards water is greater than the growth response to gravity (⋆geotropism). See also tropism.

**Hydrozoa** A class of cnidarians (coelenterates) which includes the solitary *Hydra*, the colonial *Obelia*, the Portuguese man-of-war, 'bluebottles', and other examples. See Appendix 1.

**hymen** A membrane that partly blocks the opening to the ⋆vagina.

**Hymenoptera** An Order of insects which have two pairs of membrane-like wings (attached to each other by small hooks) and undergo complete ⋆metamorphosis. The group includes ⋆wasps, ⋆ants, ⋆bees, ⋆ichneumon wasps, and sawflies. See Appendix 1.

**hyperplasia** An increase in size of an organ due to an increase in the number of cells. Although the organ is enlarged beyond normal, the tissues retain their normal arrangement.

**hypertension** Abnormally high ⋆blood pressure in the arteries.

**hypertonic** A solution which has a concentration such that it gains water by ⋆osmosis across a ⋆semi-permeable membrane from another less concentrated solution on the other side. e.g. A cell placed is freshwater swells due to such osmosis. Compare hypotonic, isotonic.

**hyphae** Singular is **hypha**. One of the branched thread-like filaments of ⋆fungi. The fruiting body of a fungus (e.g. the cap of a mushroom) is only a small part of the whole organism.

Hyphae spread through the surrounding soil etc. to obtain nutrients by breaking down the organic matter that is present. See also heterotrophism.

**hypodermic** Used to describe the placing of medical remedies under the skin. e.g. Hypodermic injections.

**hypogyny** Attachment of the sepals, petals and stamens below the ovary of a flower. The ovary is then called superior. See Fig. F2.

**hypothalamus** A region within the ⋆brain located above, and connected to, the ⋆pituitary. It is the major coordinating centre for many bodily functions. See Fig. B7.

The hypothalamus controls the ⋆autonomic nervous system, with separate centres for both the ⋆sympathetic and ⋆parasympathetic sections. The hypothalamus integrates the ⋆nervous and ⋆endocrine systems of the body by controlling secretions from the pituitary (e.g. a hormone from the hypothalamus controls release of ⋆ACTH from the pituitary). It also includes regions that initiate and control sleep, feeding, drinking, and certain aspects of behaviour associated with aggression and reproduction. The hypothalamus is richly supplied with blood vessels since it must also measure and regulate the amount of water in the blood, and the temperature of the body. Compare thalamus.

**hypothesis** A temporary working explanation based on accumulated facts, that explains a particular phenomena, but which must be further tested by experimentation. If not found valid, it should be modified or discarded. In biology, most often applied to phenomena that are unable to be observed directly as they are too small or take too long. Generally, a hypothesis becomes a ⋆theory as it successfully survives more and more tests. Compare law.

**hypotonic** A solution that has a concentration such that it loses water by ⋆osmosis across a ⋆semi-permeable membrane to another more concentrated solution on the other side. e.g. A cell placed in a concentrated salt solution shrivels up due to such osmosis. Compare hypertonic, isotonic.

**Hyracoidea** An order of placental mammals that includes the hyraxes and coneys. See Appendix 1.

# Ii

**IAA** Short for **indole acetic acid**. One of the main ★auxins, a group of plant ★growth substances. See Fig. A11.

**ice age** The periods of Earth history when it is believed that large sections of the far northern and far southern hemispheres were covered by huge glaciers. In Australia, during the last ice age (about 10 000 years ago), parts of south-eastern Australia near Mt Kosciusko and parts of Tasmania are believed to have been covered by glaciers.

**ichneumon** ★Wasps in the insect order ★Hymenoptera. They are important in biological control as they lay their eggs in the larvae of many insect pests. See Appendix 1.

**ichthyology** The scientific study of fish.

**identical twins** See twins.

**IgA** Short for the complex ★antibody molecule called ★**immunoglobulin A**. This class of antibody molecules are abundant in tears, ★colostrum, and other secretions. It may provide an external defence against viral infections.

**IgD** Short for the complex ★antibody molecule called ★**immunoglobulin D**. This class of antibody molecules is found in B-★lymphocyte cells and is involved in their ★differentiation.

**IgE** Short for the class of complex ★antibody molecule called ★**immunoglobulin E**. This class of antibody molecules is responsible for many ★allergic reactions. They are attached to the surfaces of many cells, and when stimulated by an invading ★antigen, they stimulate the cells to produce chemicals that make it easier for antibodies and white blood cells to reach the site of invasion.

**IgG** Short for the class of complex ★antibody molecule called ★**immunoglobulin G** This class of antibody molecules are most abundant in blood, being produced in the latter stages of an antibody response. It is active against bacteria, viruses and toxins.

**IgM** Short for the class of complex ★antibody molecule called ★**immunoglobulin M**. This class of antibody molecules is most abundant in the blood, being produced in the early stages of an antibody response to an infecting organism.

**igneous** Describing rocks formed by volcanoes. Volcanic igneous rocks are those that have solidified on the surface and have small crystals. e.g. Basalt. Plutonic igneous rocks have solidified underground and have larger crystals. e.g. Granite.

**ileum** The lower part of the vertebrate ★small intestine, between the ★jejunum and the start of the ★large intestine. It is mainly responsible for absorption of digested food.

**ilium** The dorsal (upper or back part) of the bone of the ★pelvic girdle. It is the point where the ★sacrum of the vertebrae (backbone) attaches to the pelvic girdle.

**imago** The adult sexually mature form of an insect. Compare nymph.

**imbricate** With edges overlapping, as with tiles on a roof.

**immigration** Coming into a new place of residence.

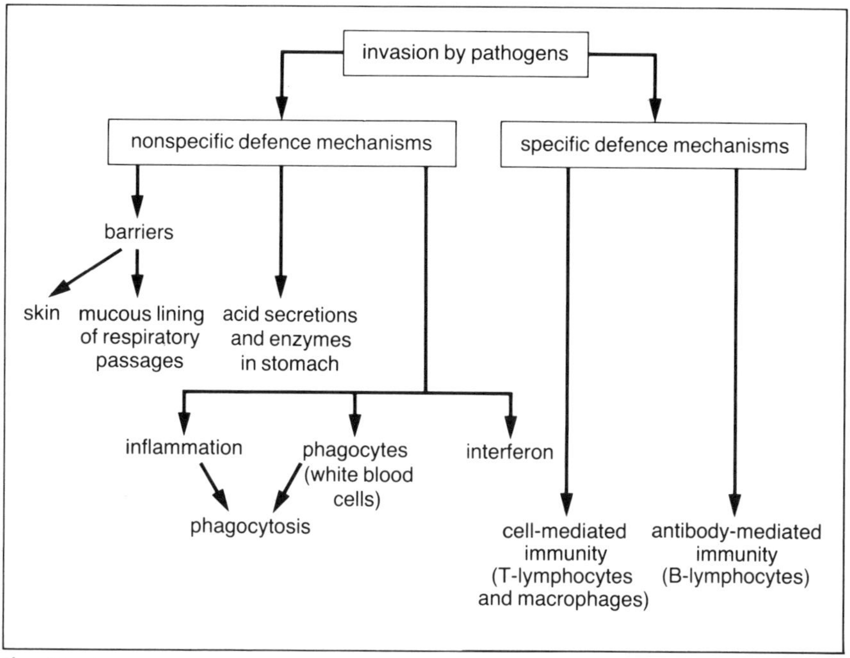


**Fig. I1 Immune response**

**immune response** Also called **immune reaction**. The production of certain *B-cells (white blood cells), *antibodies, and special chemicals such as *interferon, to combine with and/or remove an *antigen (invading toxin or organism). See Fig. I1. The response to an initial invasion of an antigen may be slow, leading to a period of sickness before the immune system takes charge. However, any subsequent invasion is more quickly dealt with. Hence we rarely suffer twice from diseases such as measles or chicken pox. See also immune system, immunity and clonal selection.

**immune system** The system of special cells and molecules produced by a body to protect itself from invading disease causing organisms or *toxins. See immune response, immunity, immunological memory, immunisation, immunological tolerance, immunosuppression.

**immunisation** The act of protecting someone from disease by the use of a *vaccine (to produce active immunity) or a *serum (to produce a passive immunity). See immunity below.

**immunity** The ability to resist infection, or to overcome the effects of infection.

There are two main types of immunity in animals: **1 Nonspecific immunity**. Defences that protect us from any pathogen that invades the body. The first line of defence is the physical barrier of skin and various *mucous membranes. If pathogens get beyond these barriers, the second line of defence includes the removal (by *engulfing) of invading *antigens by special white blood cells called *phagocytes, *inflammation and the action of chemicals such as the *interferons and *lysozymes. In each case the agent is not specific against that particular invasion, but removes any antigen of the appropriate class.

**2 Specific immunity**. The third line of defence a body has against a pathogen is the production of ⋆lymphocytes and ⋆antibodies that must recognise the substance or cell to be attacked. Thus they are specific to each ⋆antigen. There are two main divisions: **a Humoral response**. Protein molecules called ⋆antibodies produced by ⋆plasma cells (a type of ⋆B-cell) attack antigens such as bacteria, viruses and ⋆toxins (e.g. venom) that are in the fluids of the body outside the cells. **b Cellular response**. ⋆T-cells of the immune system attack viruses or bacteria that have become established within cells of the animal body as well as against invading fungi and microscopic animals. T-cells will also attack transplanted tissues and help remove cancer cells. See also immune response.

**immunofluorescence** A technique for identifying antibodies or antigens in tissues. If attempting to detect a particular antibody in a tissue, a fluorescent dye may be attached to an antigen that is known to react with that antibody. The technique can be used in reverse, with the fluorescent dye being added to an antibody.

**immunoglobulin** The class of proteins that act as ⋆antibodies. e.g. See IgA, IgD, IgE, IgG, IgM above. These protein molecules consist of two parts. The constant part makes up most of the molecules, and is the same for any particular class of antibody (e.g. IgA and IgE). The smaller variable region allows antibody molecules to be produced in millions of different varieties, one able to fit every different type of antigen that could invade the body. See clonal selection and Fig. A7.

**immunological memory** The more rapid ⋆immune response to a second invasion by a particular ⋆antigen. Once an antigen has been responded to, the immune system appears to remember it the second time. Long-lived memory ⋆B-cells and ⋆T-cells exist. See Fig. I2.

**immunological tolerance** The inability to produce antibodies or T-cells in response to a particular antigen. After birth, the body's immune system learns to identify itself so that there

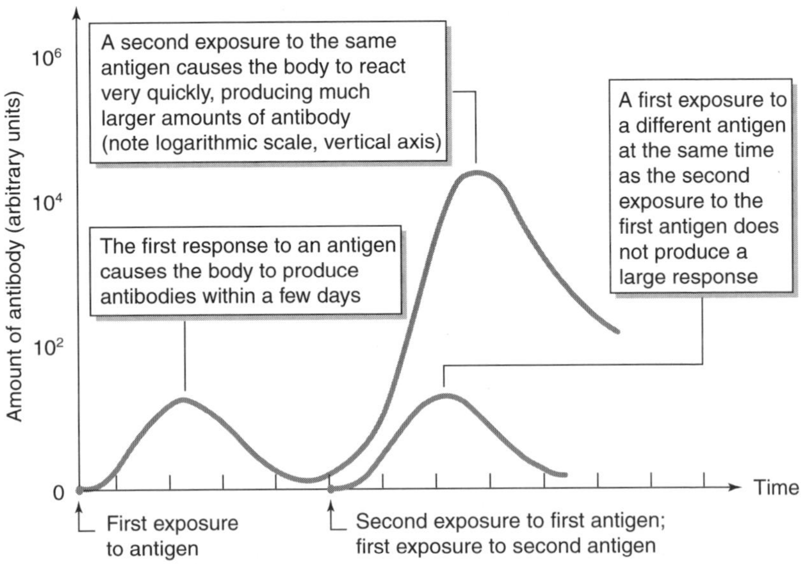


**Fig. I2 Immune system memory**

is no immune response to its own tissues. This protection can sometimes break down, resulting in ⋆auto-immune reactions. The initial 'learning' phase can also go wrong with the body thinking that certain invaders are part of itself, thus developing an immunological tolerance to an antigen.

**immunosuppression** The use of a drug or other agent (e.g. X-rays) to inhibit the ⋆immune response. This technique is often used during ⋆transplant operations when attempting to prevent the ⋆rejection (an immune response) of the new organ.

**implantation** The attachment of the developing embryo (⋆blastula stage) to the soft, folded and thickened walls (⋆endometrium) of the womb (⋆uterus) at the start of pregnancy.

**impotence** The inability of a male to produce or sustain an erection of the penis during sexual intercourse.

**impregnate** 1 To make pregnant or cause to conceive. To fertilise. 2 To saturate throughout. To fill all the spaces. To permeate.

**impressed** Sunken, as if by impression.

**imprinting** The development of a behaviour pattern in animals, especially birds, where birds reared from eggs treat any large moving objects that they first see as their parents. In nature it usually is their parents, but birds reared from eggs in captivity often imprint on their keepers. See also bonding.

**impulse** The signal or 'message' sent along a ⋆neurone (nerve fibre) in the form of an electrical-chemical change. Although electrical in nature, the nerve impulse is not the same as electricity. See Fig. I3.

As the impulse passes a given point, sodium ⋆ions pass through very small pores in the membrane into the neurone, whilst potassium ions pass to the outside. As it takes some time for the cell to pump the ions back to their original locations, no other impulse can pass for a few thousandths of a second (called the ⋆refractory period). Thus a series of impulses sent along a neurone will be spaced out, never closer than the refractory period. The energy for this movement is supplied by the cell along its length, not by the stimulus that started the impulse going. This ensures that the size of the impulse (amplitude of the ⋆action potential) does not change as it travels the length of the cell. See Fig. I3. Nerve impulses also follow the ⋆all-or-none rule. The stimulus must reach a certain intensity before a signal is sent. However, the larger the stimulus, the more impulses sent per second. Hence the difference between mild and intense pain would be the number of impulses reaching the brain per second, not on the size of the signals. The speed at which the impulse travels depends on the animal concerned, the diameter of the neurone, and on the amount of ⋆myelin sheath that is around the neurone. Impulses travelling along neurones with larger diameter and more myelin sheath, travel faster, moving at between 1 and 100 metres per second.

**inbreeding** The mating of closely related organisms. e.g. ⋆Self-pollination in plants. Brother–sister matings in animals. Since inbreeding increases the chance of harmful ⋆recessive genes appearing, matings between brothers and sisters, and even with close cousins, are outlawed in most human societies. With plants and animals, inbreeding may be necessary in the early stages of the development of a new variety.

**incisor** A sharp chisel-edged cutting ⋆tooth. e.g. The often two pairs of front teeth found in mammals, being especially prominent in rodents and rabbits.

**inclusion** A microscopically visible mass of non-living material found within the cytoplasm of a cell. e.g. Fat droplets, pigment granules.

**incompatibility** 1 In many flowering plants, the genetically controlled mechanism to prevent ⋆self-fertilisation. The incompatibility can be complete, where pollen from a flower will not grow on the ⋆stigma of the same or genetically identical flower. In other cases, pollen from a genetically identical flower grows much more slowly than that from other flowers, thus allowing self-pollination if cross-pollination fails. 2 In flowering plants, the failure of a graft between the ⋆stock and ⋆scion to unite. 3 In some fungi, a

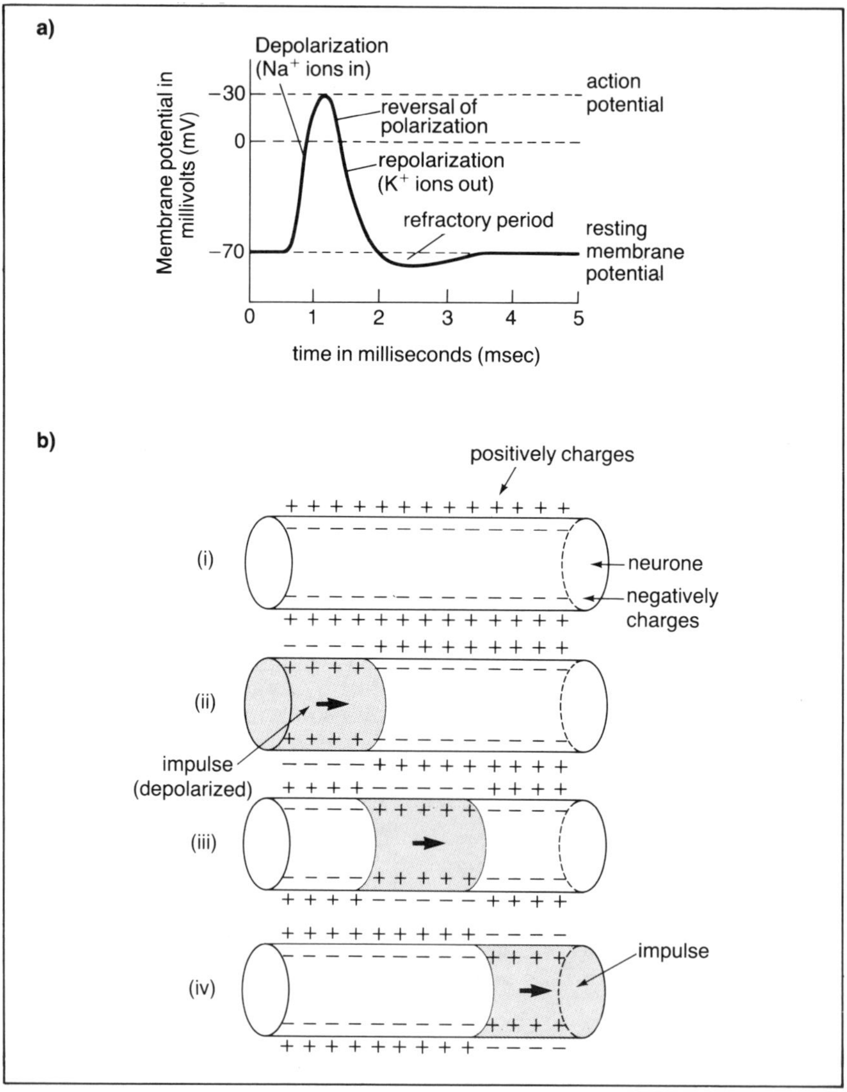


**Fig. I3 Impulse (a) Voltage changes (b) Movement of the impulse**

genetically determined mechanism to prevent sexual reproduction from occurring between individuals of the same strain. 4 In animals, the immune response that results in the rejection of a graft or the failure of a blood transfusion.

**incomplete dominance** Similar to **co-dominance**. In genetics, when two forms of a gene (★alleles) are present, but neither is dominant. The resulting offspring are intermediate in form. e.g. When homozygous, some flowers can be red (RR) or white (WW). When heterozygous for the flower colour gene, they are an intermediate pink (RW). In this type of dominance the characteristics are 'blended' and are not both present when viewed closely enough. Compare ★co-dominance.

**incontinent** Unable to control loss of urine or faeces, usually due to poor muscle control.

**incubation** 1 The act of sitting upon and hatching eggs by the heat of the body. e.g. Birds. 2 Artificially maintaining the conditions of warmth and moisture needed for the development of eggs, embryos, bacterial cultures, etc.

**incubation period** The period of an infectious disease between the infection and the appearance of the first symptoms.

**incurved** Bending or curved inwards or upwards.

**indehiscent** Not opening at maturity. Used to describe seeds.

**independent assortment** In the formation of ★gametes, each member of a pair of ★alleles (different forms of a gene) moves independently of each other, resulting in a random combination with other genes. e.g. If two pairs of alleles, Aa and Bb are present in an organism, then there are equal chances of forming any of the four possible types of gamete: AB, Ab, aB, ab. This can only take place when all the alleles A,a,B, and b are on different chromosomes (i.e. not linked). This rule is also known as Mendel's second law, and accounts for the 9:3:3:1 ratio in ★dihybrid crosses. However, ★crossing over and genes that are ★linked can upset this expected ratio.

**index finger** The forefinger, or first finger next to the thumb.

**Indian hemp** See hemp.

**indigenous** ★Native. Not imported or introduced. Growing naturally in a country or particular area.

**indigestion** Also called **dyspepsia**. Problems with the digestive system which cause heartburn, distension, excess gas (belching), stomach cramps, and fullness.

**indole acetic acid** Often shortened to ★IAA.

**inducer** A molecule that activates genes, perhaps by blocking the action of a ★repressor. See operon and translation.

**induction** 1 The process in an embryo where one tissue directs the ★differentiation of another. 2 The artificial starting of labour during pregnancy.

**indumentum** The hairy covering of plants.

**indusium** 1 Any covering of a ★sorus of ferns. 2 The pollen cup of certain dicots.

**industrial melanism** The appearance of dark (due to ★melanin) forms of an organism in industrial areas blackened by ★pollution. e.g. In Britain, the peppered moth, *Biston betularia*, is normally a light-coloured moth well camouflaged by day as it rests on shaded parts of trees in natural areas. Occasionally a mutation causes a dark form to be produced, but is soon eaten by birds as it is easily seen against the light background. However, in industrial areas, the light form is conspicuous to birds and soon eaten, while the dark form is camouflaged and survives to reproduce. This process of ⋆natural selection results in the light form of the peppered moth predominating in natural areas, whilst the dark form is most common in industrial areas.

**inermous** Lacking spines or prickles.

**infant** A child during the first few years of life.

**infection** The invasion of an organism by potentially disease-causing microbes such as bacteria, viruses, etc.

**infectious** Able to be spread from one to another and thus cause a new ⋆infection. Infectious diseases are caused by ⋆pathogens, e.g. cholera. Compare contagious.

**inferior ovary** The ⋆ovary is located below the other parts of the flower (e.g. petals, sepals). e.g. Fuschia, daffodil. See epigyny and Fig. F2.

**infertile** Unable to fertilise, or be fertilised. Incapable of reproduction. Barren, sterile.

**infestation** The state or condition of being overrun by uncontrolled numbers of disturbing or annoying organisms. e.g. Fleas, lice, mice.

**inflammation** Also called the **inflammatory response**. In animals, the early response at the site of an injury which makes the area unfavourable for infection to occur. This contributes to the defence of the organism. The blood flow to the area is increased by the widening (dilation) of the ⋆capillaries, but loss of blood through the wound is restricted by formation of blood clots (see ⋆fibrin). The extra flow of blood to the area brings blood proteins such as ⋆anti-bodies and ⋆interferon to the site where they seep through the capillary walls into the tissues. Leucocytes (or white blood cells, including ⋆macrophages and ⋆lymphocytes) migrate from the capillaries to the site of the injury. All this activity produces symptoms of pain, tenderness, redness, swelling and heat. See also abscess, arthritis and histamine.

**inflammatory response** See inflammation.

**inflorescence** A flower-bearing shoot (stalk) of a plant. The arrangement of flowers on such a stalk. See Fig. I4. Inflorescences are classified by the method of branching: **1 Indefinite or racemose**. The main axis increases in length by growing at the tip, giving rise to lateral (side) branches bearing flowers. e.g. ⋆raceme, ⋆panicle, ⋆corymb, ⋆spike, ⋆spadix, ⋆catkin, ⋆umbril, ⋆capitulum. See Fig. I4. **2 Definite or cymose**. The main axis ends in a flower and further development takes place by lateral branches, each of which behaves in the same way. e.g. monochasium, dichasium. See Fig. I4.

**influenza** Also called the **flu**. An acute highly infectious viral disease tending to occur in ⋆epidemics. Influenza A virus tends to cause the large epidemics while influenza B virus causes outbreaks between the epidemics. The viruses keep changing their type, thus evading the defences of the ⋆immune system. Hence the type of virus present one season may be quite different the next season. See antigenic drift.

**information coding** See genetic code.

**infusion** 1 The soaking of a substance in water or other solvent in order to extract its soluble components. 2 The introduction of medicinal liquid into a vein, artery or tissue.

**ingestion** The taking of food or water into the body.

**inhalation** The act of drawing air into the lungs, as in breathing.

**inheritance** All the characteristics

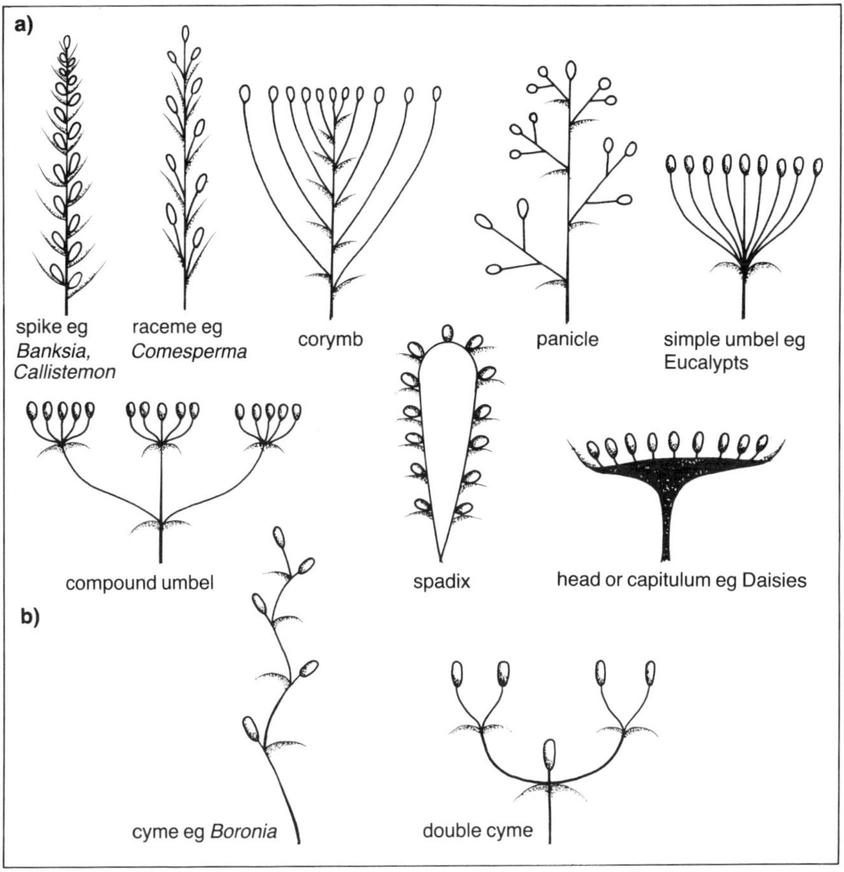


**Fig. I4 Inflorescence (a) Racemose inflorescence (b) Cymose inflorescence**

of an individual which were dependent on the genes received from their parents. These characteristics can be shown in family trees or ⋆pedigrees.

**inhibition** 1 **Nervous**. The prevention or reduction of ⋆effector (muscle, gland) activity by the action of certain nervous impulses. e.g. When a muscle such as the ⋆bicep in an arm is stimulated to contract by its ⋆motor nerve fibres, the balancing ⋆tricep muscle is made to relax by inhibiting the motor nerve cells connected to it. Inhibition also occurs in the ⋆autonomic nervous system. e.g. ⋆Parasympathetic nerves inhibit the heartbeat whilst ⋆sympathetic nerves stimulate the heartbeat. Nervous inhibition can occur in one of two ways: **a Presynaptic inhibition**. This inhibition occurs before an ⋆impulse reaches a synapse from an excitatory neurone. Another, but inhibiting neurone reduces the amount of excitatory ⋆neurotransmitter (e.g. acetylcholine) that can be released, thus decreasing the size of the possible response of the next neurone. **b Postsynaptic inhibition**. This inhibition occurs after an excitatory impulse has passed across a synapse. Normally the excitatory neuro-transmitter has to produce changes to overcome a negative potential of about 70 millivolts. See action potential. This may require that many excitatory neurones fire at about the same time before enough neurotransmitter is produced. Inhibition can occur here when the inhibiting neurone(s) make the negative potential needed to be overcome, even larger, say 80 millivolts. **2 Enzymes**. In ⋆feedback inhibition of cell reactions controlled by ⋆enzymes, if one of the products reaches too high a concentration, it may inhibit an earlier stage in the reaction.

**innate behaviour** Any ⋆inherited behaviour shown by most of the members of a species living under the same conditions. See also instinct.

**inner ear** The vertebrate organ of hearing and balance. See Fig. E1. **1** The part responsible for hearing is a snailshell-shaped structure called the ⋆cochlea. See also ear. **2** The part responsible for balance is called the ⋆vestibular system, and is divided into three parts. **a** The ⋆sacculus and ⋆utricle are both chambers containing sensors that help us keep our balance (equilibrium) when vertical. **b** The utricle is also important in helping us to respond to linear acceleration. **c** The three ⋆semicircular canals sit on top of the other parts listed. They are arranged at right angles to each other, one in each of the three dimensions of space. They help the body respond to rotational (angular) acceleration.

**inoculation** **1** The introduction of a microbe (e.g. bacteria, virus) into a suitable ⋆culture medium for growth. **2** Introduction of a weakened or inactive agent of disease into the body to produce a mild form of the disease, thus giving immunity. e.g. The use of cowpox vaccine to give immunity to smallpox.

**inorganic nutrient** Essential nutrients that are not carbon compounds. e.g. Water ($H_2O$), nitrates ($NO_3^-$), sodium ions ($Na^+$). A few simple substances that contain carbon are also classed as inorganic. e.g. Carbon dioxide ($CO_2$), carbonates ($CO_3^{2-}$).

**inrolled** The margins turning up and rolled inwards.

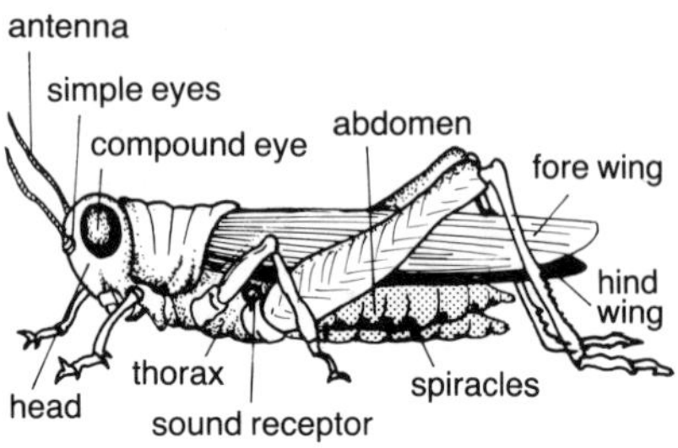


**Fig. I5 Insect**

**insect** Any member of the arthropod class Insecta. It is the largest class of living things, there being more species of insects than all others groups put together. Insect bodies are divided into three parts (head, thorax, abdomen), with three pairs of jointed legs being attached to the thorax. See Fig. I5. Many insects also have one or two pairs of wings attached to the thorax. Insect bodies are covered by an ⋆exoskeleton, and the young hatch from eggs to go through various stages of ⋆metamorphosis. There are two subclasses. **a** The Apterygota lack wings at all stages of growth, and include the silverfish and springtails. **b** The subclass Pterygota are the winged insects, and can be divided into the Exopterygota, where the young ⋆nymphs always resemble the adults as there is no pupal stage (e.g. locusts, bugs, dragonflies), and into the Endopterygota, where the young go through complete metamorphosis from larvae to pupa to adult (e.g. flies, butterflies, moths, bees, wasps, ants, beetles). See Appendix 1.

**insecticide** A chemical used to kill insects. e.g. DDT.

**insectivore** 1 Any animal adapted to feeding on insects. 2 Any member of the order of mammals called Insectivora, including the moles, shrews, and hedgehogs. See Appendix 1.

**insectivorous plants** Also called **carnivorous plants**. Plants which obtain part of their nutrients (especially nitrogen) from small animals, mostly insects, trapped by various mechanisms. Once trapped, the insects are digested by juices secreted by the plant, and the nutrients absorbed. e.g. Pitcher plant, sundew.

**inseminate** To introduce semen into a female to cause fertilisation. To impregnate.

**insemination** The act of inseminating (see above), especially applied to ⋆artificial insemination of animals.

**insertion** The place or method of attachment. e.g. In plants, leaves and parts of a flower. In animals, the attachment of muscles to bones.

**insight** A type of learning where there is a sudden grasping of a solution to a problem without the use of trial-and-error methods.

***in situ*** Latin for 'in place'. In its normal or natural position.

**inspiration** The act of drawing air into the lungs. ⋆Inhalation.

**instar** A stage in the development of an insect ⋆larva or ⋆nymph between two ⋆moults. Each species has a fixed number of normal instars.

**instinct** A natural inherited behaviour pattern, or series of patterns, where there is a tendency for the animal to produce predictable behaviour when subjected to a given stimulus. However, not all animal behaviour is totally genetically determined (instinctive). e.g. A bird will often sing if reared away from all other birds. However, the richness and complexity of the song can be increased by rearing it with other birds.

**insulation** Reducing the movement of heat. e.g. Layers of fat, or a covering of hair or fur.

**insulin** A protein ⋆hormone secreted into the bloodstream by the β-cells of the ⋆Islets of Langerhans of the ⋆pancreas in response to high levels of glucose in the blood. Insulin lowers the amount of glucose in the blood by stimulating its use by body cells. It also helps prevent the breakdown of ⋆glycogen (in the liver and muscles) to form more glucose. Compare glucagon.

**integration** The process by which the nervous system combines all the information from the senses and other ⋆receptors, and then produces the appropriate response from the muscles and other ⋆effectors. Integration permits coordination of behaviour, and is believed to occur at the synapses. See also facilitation, summation, inhibition.

**integument** 1 One of two layers surrounding the mature ovule of a flowering plant. In gymnosperms, there is only one such layer. These layers eventually mature to form the seed coat, or ⋆testa. 2 The outer surface covering of an animal. e.g. Cuticle of an insect, skin of a mammal, scales of a fish.

**intelligence** The capacity to profit by experience, through reasoning and the analysis and association of ideas. Often measured (very roughly) by ⋆IQ.

**intention movement** A particular behaviour of an animal that can be used to predict its next activity. e.g. The stages in nestbuilding by birds.

**intercalary** 1 Inserted between others. 2 Intercalary meristem is ⋆tissue that gives rise to new tissue by cell division and is located between tissues that are already mature (★differentiated). It is well away from the ★apical meristem where new tissue normally arises. e.g. The area of cell division at the base of a leaf, or near the ends of the ★internodes of grasses. In each case it allows for an increase in length.

**intercellular spaces** 1 In plants, air-filled spaces between the walls of neighbouring cells. e.g. The system that allows the entry of air into the ★cortex and ★pith. 2 In animals, the intercellular spaces are filled with fluid (★lymph).

**interferons** A group of small proteins produced by an animal cell to combat an infection by a ★virus. Interferons are not specific in their action like antibodies. Some interact with any virus. Interferons inhibit replication of the virus, especially in places where antibodies cannot reach. See also interleukins.

**interleukins** A group of proteins produced by various ★immune system cells which act on other cells in the immune system. Abbreviated as IL-1, IL-2, etc. e.g. IL-2 is produced by ★T-cells during an immune response and helps stimulate the production of more T-cells of the type needed to remove the invading ★antigen.

**intermediate filaments** Long fibres, part of a cell's ★cytoskeleton, that are 7 to 11 nm in diameter, and thus intermediate between ★microfilaments and ★microtubules. There are at least five types, all helping to provide a supporting frame-work within a cell. e.g. In muscle cells they hold the contractile parts in place. In epithelial cells they hold the nucleus in place.

**intermediate host** A ★host used by a ★parasite only during an immature, or ★larval, stage of the parasite's life cycle. e.g. The pig tapeworm *Taenia solium* must live part of its lifecycle in pigs, hence the danger to humans of eating undercooked pork.

**internal environment** The intercellular fluid that bathes the cells of the body. The mechanism of ★homeostasis keeps the composition of this fluid constant. e.g. pH, concentration of glucose and various ions. Abnormal changes in this composition can be serious, especially for brain cells.

**interneurone** Also called **intermediate** or **intercalary neurones**. A ★neurone located between ★sensory and ★motor neurones in the central nervous system. e.g. The neurones that connect sensory and motor neurones, in the spinal cord, as part of a ★reflex arc. See Fig. R2.

**internode** 1 In plant stems, the part between two successive nodes. e.g. With grasses the internodes are the parts between successive 'cracks' or divisions. 2 In animals, the internodes are the myelin-covered regions of a neurone, between the ★nodes of Ranvier.

**interoceptor** A sensory receptor located within the body. e.g. Receptors lining the stomach to detect the presence of food, and receptors in the muscles to detect the amount of stretch.

**interphase** The state of a cell when not undergoing cell division (e.g. ★mitosis, ★meiosis). Although a cell in interphase is often said to be 'resting', all it is resting from is cell division and is in fact very active.

**interpolate** To insert something between existing data. e.g. When graphing, only certain points are plotted from measurement. We interpolate new values when the plotted points are joined with lines. Compare extrapolate.

**interspecific** Between species. e.g.

Interspecific competition occurs between different species. See also ⋆intraspecific and ⋆Gause's principle.

**interstitial** Occupying small spaces between something else. e.g. Between soil particles, or between large cells.

**interstitial cells** The cells in the male testes located between the ⋆tubules that carry the sperm. These interstitial cells produce hormones such as ⋆testosterone.

**intestinal flora** Bacteria that normally inhabit the intestine, and are helpful to the body. Some of these bacteria produce vitamin K which the body cannot otherwise make, while others make gases (averaging 500 mL per day). About one-third of the ⋆faeces expelled from the body is bacteria (usually dead). The bacteria *Escherichia coli*, widely used experimentally, normally lives in the human intestine. See also normal flora.

**intestinal juices** Digestive juices secreted by the walls of the intestine. Intestinal juices contain enzymes that, with the help of ⋆pancreatic juices and ⋆bile, complete the final stages in digestion of protein, fat and carbohydrates. Intestinal enzymes include maltase, sucrase and lactase that help digest carbohydrates. Enzymes called peptidases help digest protein. About 2 to 3 litres of this slightly alkaline liquid is secreted each day.

**intestine** That part of the ⋆alimentary canal connecting the ⋆stomach to the ⋆anus (⋆crop to ⋆cloaca in some animals).

In humans it consists of the ⋆small intestine (⋆duodenum, ⋆jejunum, ⋆ileum) and the ⋆large intestine (⋆caecum, ⋆colon, ⋆rectum). The intestine is coiled to fit the abdominal cavity, as it is much longer than the body (about 8 metres in humans). The internal surface is usually folded, thus increasing the surface area for digestion and absorption. See Figs D2 and V3. The walls of the small intestine are covered with projections called ⋆villi which further increase the surface area. Near the stomach, the walls contain glands which produce digestive juices. Other digestive juices enter through ducts from the ⋆liver and ⋆pancreas. Smooth muscle in the walls churns up the contents, and ⋆peristalsis forces the food along the intestine, eventually to the anus. See also digestion.

**intracellular** Occurring within a cell or cells e.g. ⋆Catalase is an intracellular enzyme found inside liver cells.

**intramarginal** Situated within the margin, and near the edge, especially of leaves.

**intraspecific** Occurring among members of the same species. e.g. Intraspecific competition. Compare interspecific.

**intron** Also called **intervening sequences**. In ⋆eukaryotic cells, that part of a newly ⋆transcribed RNA molecule (from DNA in the nucleus) that contains a non-coding sequence of bases, which are then cut out by enzymes to convert the RNA transcript into messenger RNA. Compare exon. See also transcription and restriction enzymes.

**invagination** The folding inwards of part of a sheet of cells so as to form a pocket-like cavity. A common process during the ⋆development of animal embryos.

**invasion** In ecology, the expansion of a species' range into new territory.

**inversion** 1 A type of ⋆chromosome mutation where the order of genes is reversed. e.g. If the order was originally ABCDEFGH, inversion may produce the order ABFEDCGH. 2 See also temperature inversion.

**invertebrate** 1 Lacking a backbone or ⋆spinal column. 2 Any animal of the group called Invertebrata, those animals which are not members of the ★Vertebrata. e.g. *Amoeba*, sponges, jellyfish, worms, snails, flies, sea-stars.

***in vitro*** Latin for 'in glass'. e.g. An experiment done in the 'test tube', or some other artificial set-up. See *in vitro* fertilisation.

***in vitro* fertilisation** Called **IVF** for short. The fertilisation of an egg (★ovum) in a dish (*in vitro* means 'in glass') by sperm. 'Test tube' baby may be the popular term for IVF, but is incorrect. The sperm and eggs are mixed in a Petri dish, and after fertilisation, the embryo is allowed to grow for about a week. It is then placed into the mother's womb (★uterus) to complete development.

***in vivo*** Latin for 'in life'. e.g. An experiment performed in a living organism.

**involucre** A whorl, or several whorls, of ★bracts surrounding an ★inflorescence, flower or cone.

**involuntary** Not under the control of the will. e.g. The contraction of some muscles is beyond our control, including those in the heart and iris. See autonomic nervous system and reflex arc.

**involute** Rolled inwards or upwards.

**iodine (I)** A non-metallic element needed to make the thyroid hormones. Lack of iodine in the diet leads to the disease ★goitre. Iodine is also used as an ★antiseptic and in the chemical test for starch (where iodine is yellow-brown if no starch is detected but blue-black if starch is present).

**ion** An atom, or group of atoms, that are electrically charged as a result of the loss or gain of electrons. e.g. Chloride $Cl^-$, sulfate $SO_4^{2-}$, magnesium $Mg^{2+}$, ammonium $NH_4^+$.

**ionic bond** A ★chemical bond formed due to the attraction between ions of opposite charge. i.e. positive attracts negative. Often the positive ion is a metal and the negative ion is a non-metal. However, inside cells the positive and negative ions can be complex arrangements of carbon, hydrogen, oxygen and other atoms, e.g. ions derived from ★amino acids. Compare covalent bond.

**Iridaceae** The iris family of monocot flowering plants. It includes *Patersonia, Gladiolus, Crocus,* and onion 'grass'. See Appendix 1.

**iridium** A metallic element, with the highest density of all the elements. It is an extremely rare metal on Earth, but more common in comets and asteroids. Hence, when high concentrations of iridium were found in rocks at the ★Cretaceous-★Tertiary boundary, support was given to the theory that ★dinosaurs became extinct due to collisions of asteroids or comets with the Earth.

**iris** 1 The structure that controls the amount of light reaching the ★retina of the eye in vertebrates (e.g. humans, cats) and ★cephalopods (e.g. squid, octopus). In humans the iris is the coloured part of the eye, blue eyes containing a little melanin, brown eyes a lot. The hole in the centre of the iris is called the ★pupil, and is made larger in dull light by muscles which form a radial pattern in the iris. Muscles around the inside edge of the iris, next to the pupil, close the pupil in bright light. See ciliary muscles, eye, and Fig. E9. 2 A cultivated flowering plant of the genus *Iris*.

**iron (Fe)** A metallic element that plays an important role in the ★haemoglobin of the blood, and in various ★enzymes.

**ironbark** Any members of the genus *Eucalyptus* that have a thick, dark, highly furrowed and very hard bark, such as *E. paniculata*, or Grey Ironbark of coastal New South Wales.

**irradiate** To expose to radiation. e.g. Light, heat (infra-red), ultraviolet, or *radioactive particles.

**irregular** Used to describe flowers where one or more of the petals is different in size or shape from the others.

**irritability** The ability of living matter to react or respond to a *stimulus.

**irruption** In ecology: 1 An unpredictable, irregular, large-scale *migration into an area. e.g. Insects, birds. 2 An abrupt increase in population numbers.

**Islets of Langerhans** Groups of cells scattered throughout the *pancreas. Alpha cells secrete the hormone *glucagon into the blood stream, while beta cells secrete the hormone *insulin.

**isobilateral** Having the same structure on both sides, especially used to describe leaves. Bifacial.

**isogamy** The condition where the gametes are similar, not being recognisably male or female. Found in some algae, fungi and protozoans. Compare sperm and ova.

**isolating mechanism** A mechanism which prevents breeding between different populations so that the genes of each do not mix. Isolating mechanisms are important in the development of new varieties, stopping genes favourable in one population from mixing with neighbouring populations, thus establishing a difference. There are a number of such mechanisms: **a Geographical isolation**. Populations cannot mix due to separation by mountain ranges, valleys, roads, oceans, etc. **b Behavioural isolation**. Populations of similar animals breed at slightly different times, have slightly different courtship displays, etc. **c Physiological isolation**. Populations are separated because sperm may not fuse with ova, genitals vary in shape, offspring are hardy, but sterile, etc. See also evolution.

**isoleucine (ile)** One of 20 or so naturally occurring amino acids found in proteins. Isoleucine can be converted to *acetyl CoA. See Fig. A6.

**isomer** A molecule with the same molecular formula as another (i.e. the same number of atoms of each type), but with different structural formula (i.e. the atoms are arranged differently). e.g. Glucose and fructose both have the formula $C_6H_{12}O_6$, but the atoms are arranged differently. See Fig. C2.

**isometric** The contraction, without shortening, of a muscle. Compare isotonic below.

**Isopoda** The order of *crustaceans which includes pill bugs, wood lice (garden slaters), water-slaters, etc. These small flattened arthropods are roughly oval in shape, and have flexible segmented bodies. See Appendix 1.

***Isopogon*** A genus of dicot shrubs in the family Proteaceae. The flowers are often called drumsticks. See Appendix 1.

**Isoptera** An order of soft bodied social insects, called *termites or 'white ants'. See Appendix 1.

**isotonic** 1 Two solutions having the same concentration. Compare hypertonic and hypotonic. 2 A muscle that shortens as it exerts a steady force. Compare isometric above.

**isotope** Atoms of the same chemical element (i.e. have the same numbers of electrons and protons), but which have different numbers of neutrons. e.g. Carbon-12 (normal) compared to carbon-14 (radioactive).

**IVF** See *in vitro* fertilisation.

**ivory** A variety of *dentine that makes up the tusks of the elephant, walrus, etc.

# Jj

**jabiru** Australia's only stork, a large-billed ★bird often called black-necked stork, in family Ciconiidae.

**jacana** A bird in family Jacanidae with very long toes to enable it to walk on water weeds, thus giving it popular names such as lotus bird, lilytrotter, or Christbird.

**jarrah** A large tree, *Eucalyptus marginata*, of Western Australia noted for its dark durable timber.

**jaundice** A yellow discolouration of the skin and tissues produced by ★bile pigments in the blood. The excess bile pigments in the blood can be due to many factors including liver disease and blocked bile ducts. Newborn babies often suffer a temporary jaundice as they have an excess of red blood cells which are destroyed in the first few days of life.

**Java Man** Skeletal remains of ★*Homo erectus*, found on the island of Java.

**jaw** 1 In vertebrates, a bony or cartilaginous structure, the ★mandible or ★maxilla, in which the teeth are embedded. 2 More generally, the bones and associated structures which surround the mouth and enclose the mouth cavity.

**jejunum** The middle section of the ★small intestine extending from the ★duodenum to the ★ileum. Digestion and absorption of food take place here. See Fig. D2.

**jellyfish** Any of the free-swimming ★cnidarians (coelenterates), Class Scyphozoa, shaped like a saucer or umbrella, with tentacles hanging from underneath, and having a jelly-like consistency. e.g. *Chironex fleckeri*, the highly venomous box jellyfish or 'sea-wasp'. Most jellyfish, however, are harmless.

**jerboa** *Antechinomys laniger*, now known as the ★**Kultarr**. It resembles a rodent from North America with the same name.

**joey** A young kangaroo or wallaby, especially while still in the pouch.

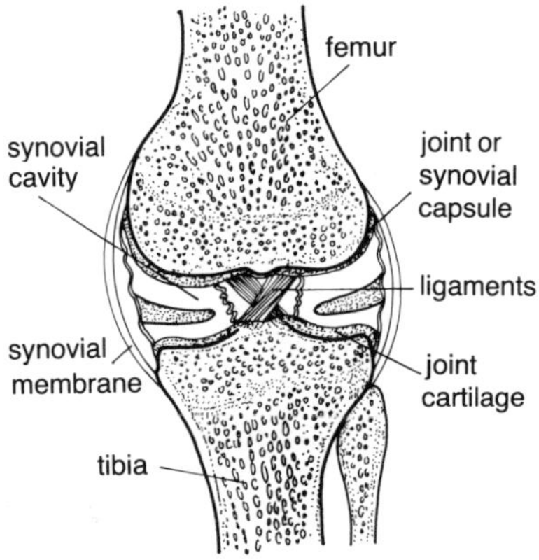


**Fig. J1 Joints**

**joints** 1 A region between two parts or segments of the body of an organism, or of its limbs or appendages. 2 In vertebrates, the point of contact between two ★bones. See Fig. J1. The amount of movement possible at a joint varies greatly. There is no movement possible between the bones of the skull. Some movement is possible between the vertebrae (bones of the spine), while a lot of movement is possible between limb bones. Where such movement can occur, the surfaces of the joint are covered with a smooth ★cartilage, and lubricated by fluids in the joint cavity. The joint is held together by various muscles and ⋆tendons. Joints where movement is free include ball-and-socket joints such as those at the shoulder, and between the legs and pelvic girdle, and hinge joints such as at the knee and elbow.

**Joule** The ⋆SI unit of energy. A Joule is the energy needed to move a force of 1 Newton for a distance of 1 metre in the direction of the force. (1 ⋆calorie equals 4.2 Joules.)

**jugular vein** In vertebrates, the main ⋆vein that returns blood from the head, and particularly from the brain.

**jungle** A general term used to refer to very dense, almost impenetrable, vegetation. Most common in tropical regions. Compare forest.

**junk DNA** A common term for the large sections of DNA that do not appear to have any function. These lengths of DNA have a simple non-coding repetitive sequence of bases.

**Jurassic** The middle period of the Mesozoic era. See Appendix 2.

**juvenile** A young immature organism.

**juvenile hormone** One of a group of hormones produced in insects which promotes the growth of larval (caterpillar, grub) structures, and inhibits ⋆metamorphosis. Normally in balance with the ⋆moulting hormone (ecdysone), which promotes metamorphosis into the adult stage. As the larva matures, the production of juvenile hormone decreases, the excess moulting hormone causing metamorphosis.

# Kk

**kangaroo** The largest of the ⋆macropod family of herbivorous Australian marsupials, with powerful hind legs used for movement by jumping. It forms part of Australia's coat-of- arms. See Appendix 1.

**kangaroo paw** Any member of the genus *Anigozanthos*, dicot flowers of Western Australia that resemble the paw of a kangaroo, especially the red and green *A. manglesii*, which is the floral emblem of Western Australia.

**Karri** A valuable hardwood tree of Western Australia, *Eucalyptus diversicolour*.

**karyotype** The number and general appearance (size, shape, banding) of a set of ⋆chromosomes within the nucleus of a typical body cell of an organism.

**kauri** A tall ⋆conifer, *Agathis australis*, from New Zealand, yielding valuable timber and resin. See pine.

**keel** Also called a **carina**. The keel of the breastbone (sternum) of a bird which provides a large area for the attachment of powerful flight muscles.

**keeled** A leaf that has a prominent midrib on the underside, reminiscent of the keel of a boat.

**kelp** Any of the large flat brown seaweeds of the Phylum (Division) Phaeophyta, or brown algae. The 'bull kelp' from the Antarctic coasts are among the most massive of seaweeds. See Appendix 1.

**keratin** A fibrous protein with high sulfur content present in animal structures such as horn, nail, claws, beaks, hair, wool and feathers. Keratin is also present in the cornified surface cells of the skin, thus providing waterproofing. This process, called cornification or keratinization, occurs as the keratin gradually replaces all the cytoplasm present in the surface cells.

**kernel** 1 The entire grain or seed of a cereal such as wheat or corn. 2 The inner, and usually more edible, part of a seed with a hard outer coating. e.g. The kernel of a nut.

**ketone** A carbon compound with the functional group –CO–, located within the molecule, not on the ends. e.g. Acetone is $CH_3COCH_3$. Also called alkanones.

**key** A method used to help identify plants and animals. Often in the form of a ⋆dichotomous key in which identifying features are arranged in a tabular form such that one is always presented with two alternative choices. e.g. Part of a key to identify chordate classes:

1 Feathers present . . . . BIRD
  No feathers . . . . . . . . Go to 2
2 Hair or fur present . . MAMMAL
  No hair or fur . . . . . . Go to 3
3 Skin

**kidneys** The paired organs in vertebrates which filter wastes from the blood, especially those wastes containing nitrogen, such as ⋆urea, thus forming ⋆urine. See Fig. K1.

The basic unit of filtration is called the ⋆nephron. The nephron starts with the ⋆Malpighian body, a mass of capillaries called the ⋆glomerulus surrounded by the ⋆Bowman's capsule. Blood flows through capillaries into the Bowman's capsule where most of the dissolved chemicals and smaller proteins pass through tiny pores to enter the *tubule of the nephron. The remaining liquid along with the larger proteins continue along the capillary. Once in the tubule, the useful parts are actively reabsorbed into the capillary, leaving only wastes and water. The tubule then forms the *loop of Henlé, where most water is reabsorbed into the surrounding capillaries. Animals which are adapted to survival in deserts have relatively longer loops of Henlé. The kidney in mammals contains an outer *cortex where the Malpighian bodies are concentrated, and an inner *medulla containing the tubules. These tubules drain into the *pelvis from which the ureter leads to the *bladder. See Fig. E8. Because the kidney helps control the concentration of various chemicals in the bloodstream, it is classed as an organ of *osmoregulation.

**kidney stone** Crystalline lumps that form most often in the pelvis of the kidney, and cause extreme pain as they pass through the ureter.

**killer T-cells** See T-cells.

**kin selection** The *selection that results from animals being prepared to help those related to themselves more than those not so closely related. Since related organisms share more *genes than those which are unrelated, it is thought by some scientists to be a way in which the genes ensure their own preservation. See also altruişm, sociobiology.

**kinaesthetic** Detecting movement. e.g. The sense organs in vertebrate muscles, tendons, and joints which help detect the position and movement of those parts.

**kinase** A substance which converts an inactive form of an enzyme into its active form. e.g. Enterokinase converts inactive *trypsinogen to the digestive enzyme *trypsin which helps split

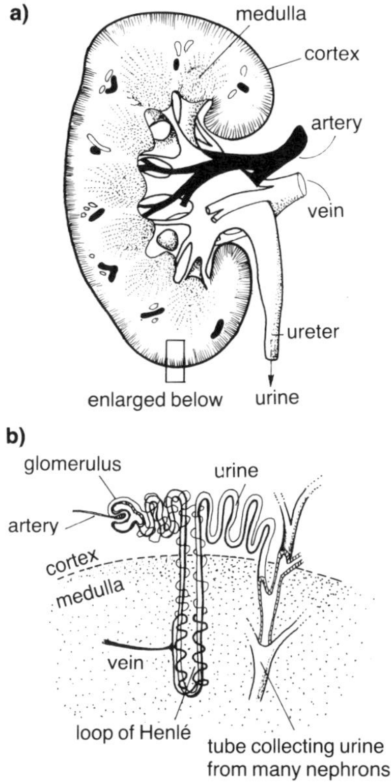


**Fig. K1 Kidney (a) Cross-section (b) Nephron**

protein molecules in the alkaline conditions of the small intestine.

**kinesis** The rate of movement produced by a stimulus. e.g. Speed, rate of turning, etc. It does not involve direction. Compare taxis and tropism.

**Kingdom** The five major categories into which living things are divided. e.g. Plantae, Animalia, Fungi, Monera and Protista. See Appendix 1.

**kingfisher** Any member of the bird family Alcedinidae, many members of which dive into water to catch their prey. e.g. Azure kingfisher. However, others gather their food on the floor of forests. e.g. *Kookaburra.

**kinin** 1 In plants, one of a group of plant ★growth substances that promotes cell division, cell enlargement, and helps initiate the development of buds and roots. e.g. Naturally occurring kinins are called ★cytokinins. A commonly used synthetic kinin is kinetin. 2 In animals, one of a group of small proteins (polypeptides) that occur in the blood and cause the widening (dilation) of blood vessels, perhaps as part of the inflammatory response.

**kino gum** The ★gum produced by certain trees, including *Eucalyptus camaldulensis*, or ★River Red Gum, and used in medicine and tanning.

**kinship** The possession of a common ancestor in the recent past. See kin selection.

**kiwi** Any of the several flightless birds, genus *Apteryx*, which occur naturally in New Zealand.

**knuckle walking** A mode of walking of four-limbed (★quadruped) animals where they balance on the projecting joints of the fingers (knuckles) rather than their front legs or feet. e.g. Gorilla and chimpanzee.

**koala** The usually slow, tail-less, tree-living marsupial *Phascolarctos cinereus*, renowned for its restricted diet of a few species of *Eucalyptus* leaves. The koala is the animal emblem of Queensland. See Appendix 1.

**Koch's postulates** A series of steps that must be followed to show that a particular micro-organism is responsible for a particular disease. These steps were recorded by Robert Koch (1843–1910).

**kookaburra** Either of two species of large Australian ★kingfisher, especially *Dacelo novaeguineae* or Laughing Kookaburra, renowned for its 'laughing' call.

**Kowari** A small carnivorous marsupial of central Australia, *Dasyuroides byrnei*, in the family Dasyuridae. See Appendix 1.

**Krebs' cycle** Also called the citric acid cycle, a part of ★respiration. A complex series of enzyme-controlled reactions in which pyruvate, a product of carbohydrate metabolism in the ★glycolysis reaction (Fig. G4), is broken down to form carbon dioxide ($CO_2$), while NAD is converted to NADH, and FAD is converted to $FADH_2$. During the ★electron-transport chain that follows, most of the energy carried by the NADH and $FADH_2$ is stored in the bonds of ATP (Fig. E3). The enzymes needed for the Krebs' cycle are located in the gel-like matrix within the cavity of the ★mitochondria. Pyruvate is converted to 2-carbon ★acetyl CoA, which then combines with 4-carbon oxaloacetic acid to form 6-carbon citrate. A number of steps then remove 2 carbons one at a time (as $CO_2$) to leave 4-carbon succinyl CoA, which is then converted through a number of steps back to 4-carbon oxaloacetic acid. At various points along the cycle, NAD is converted to NADH, and FAD to $FADH_2$. For each acetyl CoA at the start of the cycle, 12 ATP molecules are produced. See Fig. K2.

**krill** Shrimp-like ★crustaceans of the Order Euphausiacea, genus *Euphausia*, existing in enormous numbers in Antarctica, and making up the main food for ★baleen whales. See Appendix 1.

**Kultarr** A small carnivorous marsupial in the family ★Dasyuridae, *Antechinomys laniger*, sometimes called the Jerboa-marsupial. See Appendix 1.

***Kunzea*** A genus of dicot shrubs with ★ericoid leaves in family Myrtaceae. Common in coastal localities. See Appendix 1.

**Kurrajong** A tree, dicot Family Sterculiaceae, that is often grown in gardens, *Brachychiton populneus*. Related to the bottle-tree and flame-tree.

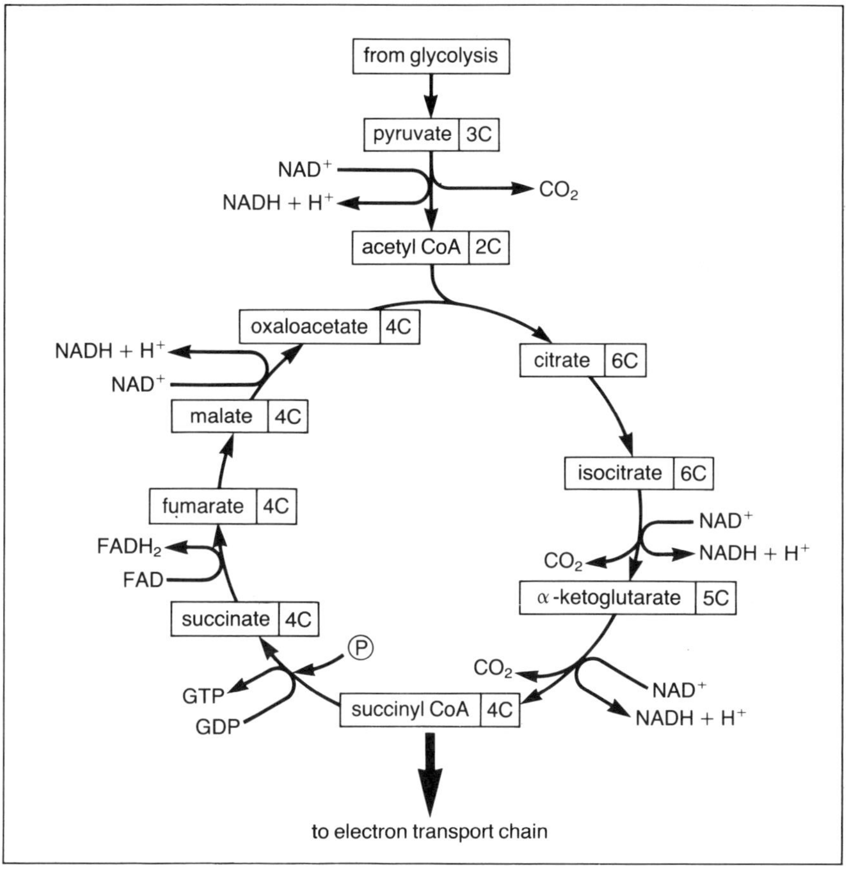


**Fig. K2 Krebs' cycle (respiration)**

**kuru** A disease thought to be caused by a ⋆prion. Once transmitted during cannibalistic rituals, it is gradually disappearing as cannibalism dies out. Compare scrapie and mad cow disease.

**kwashiorkor** A protein-deficiency disease occurring often in children raised in tropical countries on cereal diets after breast feeding has stopped. Certain amino acids needed to make protein are missing from the diet.

# Ll

**labellum** The lip or lower petal of an orchid flower.

**Labiatae** See Lamiaceae.

**labium** 1 A lip or liplike structure. 2 Folds of skin lying on each side of the opening to the *vagina. 3 The lower 'lip' of an insect or crustacean. See mouthparts and Fig. M4.

**labour** Childbirth. There are three stages: dilation (stretching or enlarging) of the cervix; expulsion of the baby; expulsion of the placenta. See also reproduction.

**labrum** A flat plate of *exoskeleton, hinged to the head of an insect above its mouth, thus forming an upper 'lip'. See mouthparts and Fig. M4.

**labyrinth** The canals in the bone surrounding the *inner ear of vertebrates.

**laceration** A wound caused by the tearing of *tissues.

**lacewing** See antlion.

**lachrymal duct** A tube that drains tears from the inner corner of the eye to the back of the nose.

**lachrymal gland** The tear glands found in the eyelids of terrestrial vertebrates. e.g. In humans and other mammals, it lies beneath the upper eyelid, and produces a watery and slightly *antiseptic liquid (tears) which moistens and cleans the *cornea of the eye.

**lactase** An enzyme found in the digestive juices of the intestines that breaks down (by *hydrolysis) the double sugar (disaccharide) lactose (milk sugar) into the simple sugars (monosaccharides) glucose and galactose. See also digestion.

**lactation** The secretion of milk from the *mammary gland.

The system of milk ducts in the breast enlarges during pregnancy under the influence of *oestrogen. Milk-secreting glands increase in number under the influence of *progesterone. During labour, a pituitary hormone called oxytocin is produced which causes the release of colostrum and milk. First to be produced is *colostrum, which is not true milk but a secretion containing protective antibodies from the mother, as well as fats and sugars. True milk production begins when *prolactin, a pituitary gland hormone, acts upon the breasts already prepared by oestrogen and progesterone.

**lacteals** The *lymph vessels which drain the *villi of the small intestine of vertebrates. The fatty acids and glycerol produced by the digestion of fats are absorbed into the cells lining the small intestine where they recombine to form fats. These fats then form droplets called *chylomicrons which are released into the lacteals. See also digestion and Fig. V3.

**lactic acid** An organic acid, $CH_3CH(OH)COOH$, formed by the action of bacteria on lactose (milk sugar), and responsible for the souring of milk. It is also produced in mammals during the breakdown of glucose when oxygen is in short supply (anaerobic conditions). e.g. After running a hard race muscular pain is caused by a build-up of lactic acid. See also glycolysis.

**lactic acid bacteria** A large group of spherical and rod-shaped ★bacteria that produce ★lactic acid during respiration. e.g. Genera such as *Streptococcus*, *Diplococcus*, and *Lactobacillus*. Lactic acid bacteria are of economic importance in the production of cheese, yoghurt, etc., from milk. The female vagina, between puberty and menopause, also contains lactic acid bacteria; the lactic acid they produce presumably helping to create an environment unsuitable for more dangerous bacteria.

**lactogenic hormone** Also called ★**prolactin**.

**lactose** A double sugar (disaccharide), made of the simple sugars (monosaccharides) glucose and galactose, and found in milk. See carbohydrates and Fig. C2.

**lacuna** 1 In animals, a small space or cavity, especially one in ★cartilage or ★bone, which contains a cell. 2 An air space or cavity in the stem of certain plants.

**lacustrine** Living or occurring in a lake.

**Lagomorpha** An order of placental mammals that includes the ★rabbits and ★hares. They differ most obviously from rodents in having four upper incisors instead of two. See Appendix 1.

**Lamarckism** Also called the 'theory of ★acquired characteristics'. A once popular theory of ★evolution proposed by Jean Baptist Lamarck (1744–1829) in the early 1800s. The theory proposes that characteristics acquired by an organism during its life, by use and disuse, or from the environment by say an accident, can be inherited by future generations. e.g. Ancestors of the giraffe stretched their necks to reach the leaves in trees, and this longer neck was inherited by their offspring. However, the theory could not explain all features, e.g. why tigers developed stripes. Nor could a mechanism be found for such environmental factors to influence the genes which transmit the genetic information. Compare Darwinism and Pangenesis.

**lamella** 1 A thin, platelike structure or layer. 2 In chloroplasts, the layered arrangement of ★chlorophyll designed to increase the chances of trapping photons of light as they pass through. The lamella are stacked to form ★grana. 3 The thin wall (middle lamella) which cements together the cellulose cell walls of adjacent plant cells. 4 The gills under the cap of a mushroom or toadstool. 5 One of the concentric layers of bone surrounding the ★Haversian canals which carry the blood vessels and nerves.

**Lamiaceae** Also called the **Labiatae**. The mint family of dicots, which contains many aromatic herbs including mint, sage, thyme and lavender. See Appendix 1.

**lamina** The flattened blade of a leaf.

**lamp shells** A member of the phylum ★Brachiopoda. See Appendix 1.

**lampbrush chromosome** Large chromosomes found in amphibian eggs, with ★DNA side loops that produce a brushlike appearance when seen through a microscope. They represent sites of active synthesis of RNA.

**lamprey** A jawless, eel-like vertebrate of class ★Agnatha. See Appendix 1.

**lancelets** Marine fish-like members of the subphylum Cephalochordata. These free-swimming transparent animals have a ★notochord and ★gill slits. See Appendix 1.

**lanceolate** Leaves shaped like the head of a lance. See Fig. L1.

**lancet** A small, pointed, double-edged knife used to prick the skin to produce a drop of blood.

**land bridges** Narrow connections of land that are believed to have once existed between present-day continents, across which certain plants and animals are thought to have extended their distribution or range. Such connec-tions may have formed when sea levels were lower during the ice ages.

**large intestine** The end part of the intestines where water from undigested food is reabsorbed into the bloodstream to leave behind the much drier ⋆faeces. Bacteria are also present, converting some material to gases and producing some vitamins. The large intestine has a diameter of about 7 cm, and a length of 1 to 1.5 m. The walls are able to stretch as it fills with slow-moving undigested food. At the beginning of the large intestine is the ⋆caecum, a small bag-like structure in humans. In herbivores it is much larger and filled with bacteria that help in the breakdown of cellulose. Attached to the caecum is the 2 to 3 cm long ⋆appendix. The main part of the large intestine is called the ⋆colon. At the end of the colon is the ⋆rectum where faeces are stored before being ejected through the ⋆anus, a process called ⋆egestion. See digestive system and Fig. D2.

**larva** The young and immature form of an organism which is unlike the adult. The larva hatches from the egg, and after a period of feeding etc., undergoes ⋆metamorphosis to form the adult. e.g. The tadpole is the larval stage of a frog. A maggot is the larval stage of a fly. A witchetty grub is the larval stage of a certain beetle. A caterpillar is the larval stage of a butterfly. Many marine molluscs, worms, and echinoderms have small transparent ⋆planktonic larval stages. See also life cycles.

**laryngitis** Inflammation of the larynx (voice box), during which it is difficult to speak above a whisper, if at all.

**larynx** Also called the voice box. The cartilaginous structure containing the vocal folds (vocal cords), located between the base of the tongue and the beginning of the windpipe (trachea). See also epiglottis and Fig. D2.

**latent heat** The amount of heat per gram of material given out or absorbed during a change of state. e.g. When water evaporates from perspiration, it changes from liquid to gas, absorbing heat from our bodies and thus cooling us down. Water is very useful in our bodies as it does not readily boil, nor does it readily freeze. Compare specific heat.

**lateral** Located to the side. Directed away from the side.

**lateral line system** A system of sense organs present in aquatic vertebrates (jawless fish, fish, amphibians), arranged in lines down each side of the body, and in a complicated pattern of lines on the head. These sense organs detect changes in pressure, especially those due to movement and low frequency sounds, thus assisting fish to swim in schools.

**laterite** Highly ⋆leached soils found in higher rainfall areas. Most of the minerals, and much of the silica has been removed, leaving many oxides of aluminium and iron (which gives the soil a red colour). Rather poor soils.

**latex** A white milky juice produced by a number of plants. e.g. Dandelion. The juice coagulates (forms a rubbery substance) on contact with air. Natural rubber is formed from a latex produced by the rubber tree.

***Latimeria*** The genus of the 'living fossil' called coelacanth, a fish very similar to numerous fossils found in the Cretaceous and thought to be extinct until specimens were caught off southern Africa in 1938.

**Laurasia** The ancient land mass

comprising the present-day continents of North America, Europe and Asia. Like Gondwana, Laurasia was once a single land mass that has been split apart by *continental drift (*plate tectonics) since the Cretaceous. See Fig. P7.

**law** A principle or rule that summarises a large number of observations. e.g. Mendel's laws of genetics. Compare theory and hypothesis.

**L-DOPA** Short for **L dihydroxyphenylalanine** (thankfully!). This amino acid is important in the production (synthesis) of the hormones *dopamine, *noradrenaline, and *adrenaline.

**leaching** The removal of the soluble parts of a mixture when a solvent passes through it. Even only slightly soluble materials (e.g. silica) are removed given enough time and water.

**leaf** The main organ for photosynthesis in green plants, usually a lateral outgrowth from a stem. Leaves come in many shapes and sizes, often an important point in identification. See Fig. L1.

A leaf consists of the leaf base by which it is attached to the stem, the leaf stalk (or *petiole), and the flattened leaf blade (or *lamina) where most of the leaf functions occur. Although most plants have many leaves, *deciduous plants lose their leaves each year. A leaf usually consists of conducting tissues (veins with *xylem and *phloem), and photosynthetic tissue (the *mesophyll) where the *palisade cells are arranged under the top surface of the leaf to catch as much light as possible, while the *spongy mesophyll on the underside contains many air spaces to allow for the movement of gases. See Fig. S9. The surface cells (*epidermis) are often covered with a layer of wax called the *cuticle to help reduce the loss of moisture from the surface. The epidermis, especially on the underside of the leaf, contains many pores called *stomata that control the entry and exit of gases from the leaf. Not all leaves follow this pattern. In many plants they may be cylindrical, or even absent altogether. In *Eucalyptus*, the leaf normally hangs vertically so that palisade cells are found under both sides of the leaf, with the spongy layer between. Some plants have leaf-like structures which, however, have a different internal structure. e.g. Seaweeds, moss.

**leaf hopper** Small two-winged insects in order *Hemiptera, many looking like tiny *cicadas. See Appendix 1.

**leaf scar** A scar or marking showing where a leaf-stalk was attached to a stem.

**learning** Acquiring new patterns of behaviour in an animal, or modification of existing patterns of behaviour, as a result of experience. Types of learning include *conditioning and *imprinting.

**leatherwood** An Australian tree, *Eucryphia lucida*, renowned for the distinctive, highly flavoured honey produced from its flowers.

**lecithin** A fatty substance (*phospholipid) present in all animal and plant cells, being an especially important part of the cell membrane.

**lectin** A protein found in plants that acts like an antibody when injected into animals, causing the clumping of red blood cells. In plants it may help *symbiotic bacteria recognise host roots. e.g. Nitrogen-fixing bacteria.

**leech** Any of the annelid worms of the class *Hirudinea. Leeches are *hermaphrodites with suckers for attachment located at both ends. Some are blood-sucking parasites of vertebrates, but most are carnivorous predators. See also anticoagulant and Appendix 1.

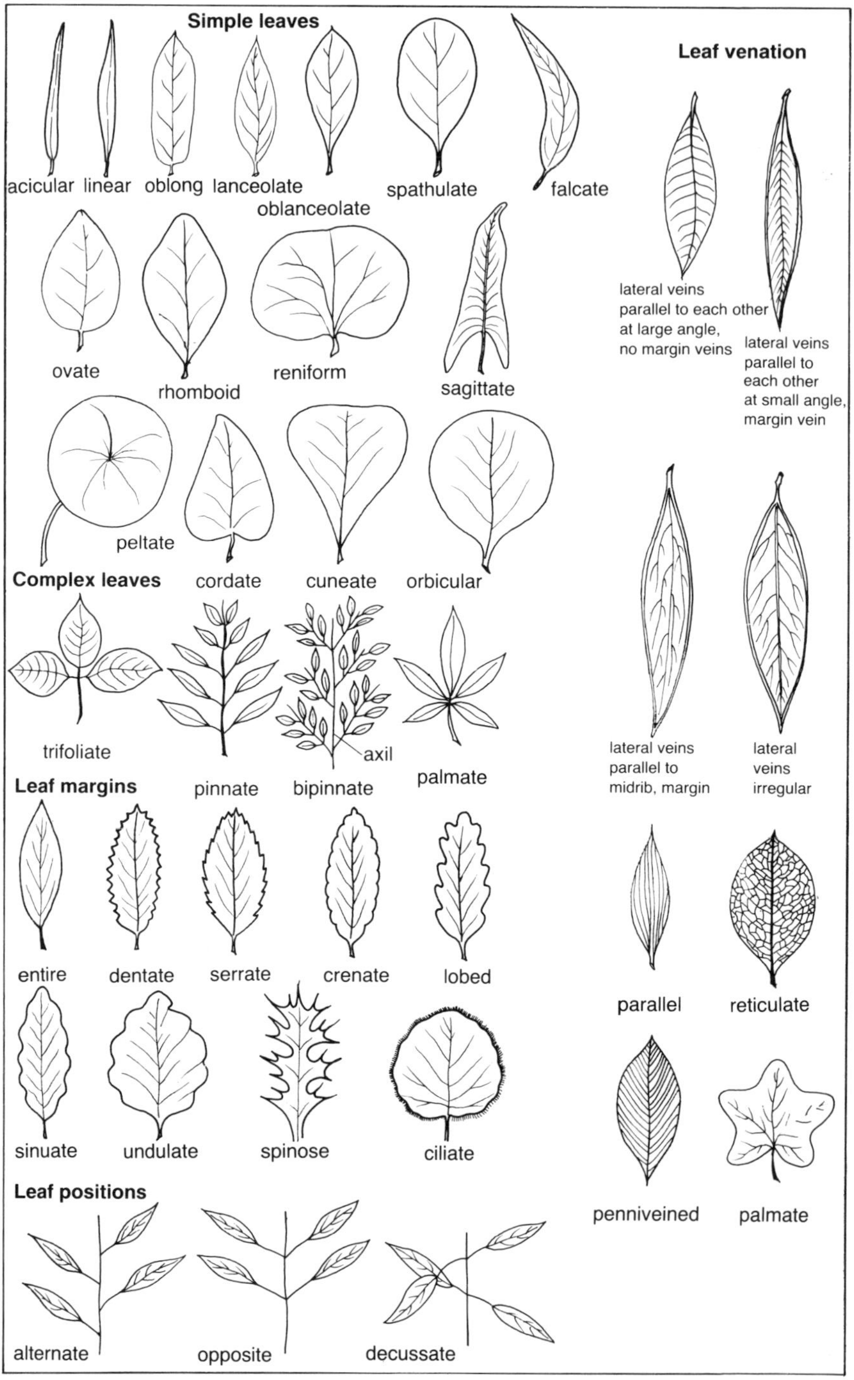


**Fig. L1 Leaf shapes**

**legless lizard** Any member of the ★lizard family Pygopodidae, a group of nocturnal snake- and worm-like lizards confined to Australia and New Guinea. Can be told apart from snakes in having ears, and tiny but apparently useless legs towards the rear of their body. See Appendix 1.

**legume** 1 Any member of the dicot plant family Leguminoseae, including wattle (*Acacia*), peas, beans, clover, peanuts, alfalfa, etc. Many such plants are used in agriculture to improve the nitrogen content of the soil, as their roots contain nodules in which live ★nitrogen-fixing bacteria. See Appendix 1. 2 A pod. A dry, ★dehiscent fruit (one that splits open at maturity) which develops from a flower with a single ★carpel, and which when ripe, splits open along both sides to form two halves, as seen with bean and pea pods.

**Leishman's stain** A blue coloured ★stain which colours red blood cells pink and white blood cells a deep blue.

**lens** In the eye of vertebrates, a transparent structure behind the pupil which focuses the image onto the light-sensitive ★retina. See Fig. E9. In land-living organisms, the ★cornea and lens function together to focus the image. The lens changes its shape to achieve this, being fatter when focusing on close objects, and thin when focusing on distant objects. In aquatic organisms, such as fish and amphibians, the lens is moved in-and-out like the lens of a telescope or binoculars to achieve focus. The lens of vertebrates is made of long thin cells containing a ★collagen-like protein called crystallin.

**lenticel** A small opening (a ventilating pore 0.5 to 1 mm in diameter) containing loosely packed cells with spaces between, found in the roots or bark of the stem of woody plants, and which enables gases to be exchanged with the atmosphere. See also wood.

**Lepidoptera** The order of insects that includes the ★butterflies and ★moths. See Appendix 1.

**leprosy** A chronic disease due to *Mycobacterium leprae*, a bacterium similar to that which causes tuberculosis. People with nerve leprosy develop discoloured skin areas that no-longer have the sense of touch because the nerves are dead. Ultimately fingers, toes, etc. may shrivel. People with skin leprosy develop thick swellings which grossly distort their features. There is much debate as to how contagious leprosy really is.

***Leptospermum*** A genus of dicot plants, family Myrtaceae, often called ★tea-tree. See Appendix 1.

**leptotene** A stage during the early prophase of the first division of ★meiosis where the chromosomes are a tangle of long, slender filaments.

**lesion** The alteration of tissues or their function due to injury or disease. e.g. Pimples, fractures, scratches, wounds, an abscess, or a wart could be classed as a lesion.

**lethal** Something that can cause death. Deadly.

**leucine (leu)** One of the 20 common amino acids found in proteins. It is the only amino acid that cannot be converted into glucose by the body. See Fig. A6.

**leucocyte** Also called ★**white blood cells**.

**leucoplast** In plants, a colourless body within a ★cell which stores starch, oil or protein. Usually found in cells not exposed to light, such as roots and internal stem tissues.

**leukaemia** A ★malignant, often fatal, disease of the blood-forming organs such as the bone marrow. There is usually a large overproduction of white blood cells with resulting

★anaemia, and enlarged spleen and lymph glands. There are many different forms of leukaemia.

**LH** Short for **luteinizing hormone**, produced by the anterior (front) lobe of the ★pituitary. In females, it helps stimulate (with FSH) secretion of ★oestrogen by the ★Graafian follicle, and promotes ★ovulation. After the follicle has released its egg into the ★oviduct, LH helps it to become a yellow body called the ★corpus luteum, which in turn produces ★progesterone. Progesterone is a hormone that helps prepare the lining of the womb (★uterus) to receive the egg should it be fertilised. In males, LH helps stimulate the secretion of ★testosterone from the ★interstitial cells of the ★testes, which are found between the sperm-producing cells.

**liana** A climbing plant (vine) with a woody stem found in tropical and subtropical forests. Lianas use other plants for support as they grow towards the sunlight in the canopy of the rainforest, but they are not parasitic.

**lice** Plural of **louse**. Any of the wingless, sucking, parasitic insects such as: head, body, crab lice (true lice, order Anoplura), and similar organisms such as biting bird lice, book lice, wood lice, etc. See Appendix 1.

**lichen** An organism formed by the ★symbiotic association of an ★alga and a ★fungus, and thus looks different from either of them. The alga produces food by ★photosynthesis, while the fungus absorbs water and minerals from the usually hard surface upon which it grows. e.g. Rock, bark or trees, soil. Some lichens form crusts on rocks (called crustose); others have flat, crinkly, leaf-like structures (called foliose); others are branched and upright (called fruticose).

**life** Those properties that distinguish living organisms from non-living matter. e.g. ★Metabolism, growth, ★irritability, inheritance, movement and reproduction.

**life cycle** The stages of development through which an organism passes, such as from egg to adult, to producing eggs again. In the best-known cycles, an organism gives rise to organisms similar to itself. e.g. Vertebrates, flowering plants. Other cycles produce organisms quite different from the adult, which then undergo ★metamorphosis to produce the mature form. e.g. Frogs, beetles, butterflies. In both the previous cases, all the stages in the life cycle have the same number of chromosomes in each cell (except sex cells). Some life cycles show ★alternation of generations, where the different stages can have different appearances as well as different numbers of chromosomes in their cells. e.g. Algae, moss, ferns, protozoa. Thus in plants such as moss and ferns, the ★sporophyte produces ★haploid spores which germinate and grow into the haploid ★gametophyte plants, which in turn produce sperm and ova. After ★fertilisation, a new ★diploid sporophyte is again produced.

**ligament** A band of dense fibrous connective tissue (bundles of ★collagen fibres) that helps hold two bones together at a ★joint. The ligament helps stop movements in certain directions, thus preventing ★dislocations. See Fig. J1.

**ligase** Any of a group of ★enzymes which helps a chemical bond to form. In the process, energy is supplied by converting ★ATP to ADP. e.g. Glutamine synthetase ★catalyses the formation of the amino acid glutamine from another amino acid called glutamic acid. In this reaction an $-OH$ in glutamic acid is replaced by an $-NH_2$. See also DNA ligase.

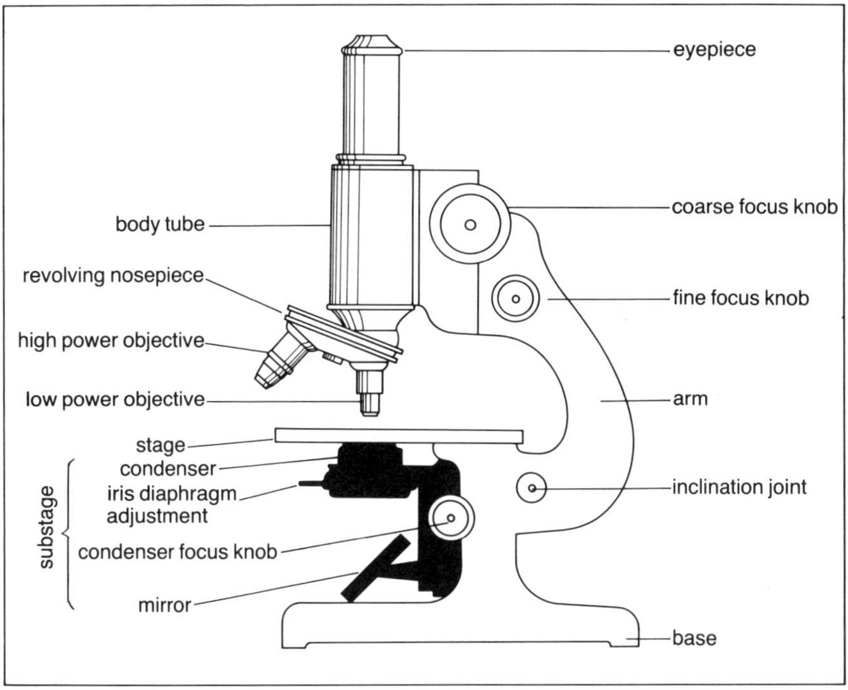


**Fig. L2 Light microscope**

**light** 1 Radiant energy in the form of electromagnetic waves with wavelength from about $4 \times 10^{-7}$m (violet) to about $7 \times 10^{-7}$m (red). **2** Light can also be considered as made of photons ('particles') travelling at the speed of light, about $3 \times 10^{8}$ metres per second. In aquatic environments, the available light decreases with the depth of the water and blue and violet ends of the spectrum penetrate deeper than the red end.

**light microscope** Also called an optical microscope (Fig. E2). **1 Monocular/binocular microscope** (Fig. L2). Transparent ⋆lenses are used to produce enlarged images of ⋆specimens, which must be very thin for light to pass through them. At best, a light microscope can distinguish points that are 3 μm apart (where 1 μm= $10^{-6}$m). The human eye can distinguish points that are about 80 μm apart (called the ⋆resolving power). A simple microscope has just one lens, and can be better called a hand lens or magnifying glass. Compound microscopes have two lenses in use at one time, one at each end of the tube. The one closest to the eye is called the ocular or eyepiece, while the one closest to the specimen is called the objective lens. Most such microscopes have several objective lenses on a rotating turret. Focus is achieved by moving the tube closer to, or further from, the specimen, or moving the stage that carries the specimen, towards or away from the objective lens. There is often two separate focusing knobs—one for rapid or coarse focusing; the other for fine or accurate focusing. Under the stage is the iris diaphragm which adjusts the amount of light reaching the specimen from the light source (which may be separate or built-in). There may also be a condenser lens under the stage to focus the light onto the specimen. Binocular microscopes are similar, but light from the objective lens is split, and travels to two eyepieces, one for each eye. **2 Stereo microscopes** have two tubes, but differ from binocular microscopes by having an objective lens and an eyepiece for each tube. They achieve only moderate magnification, but allow for a three-dimensional view which is useful when performing dissections, or examining small but not microscopic specimens. e.g. Insects, flower parts. Light is shone onto the surface of the specimen.

**light reaction** See photosynthesis and photosystems 1 and 2.

**lignin** A complex three-dimensional carbohydrate polymer making up about 25% of wood of trees. Lignin is also found in the thickened cell walls of *sclerenchyma, and mature *xylem vessels and *tracheids. It increases the strength of such tissues, making them more resistant to compression and tension. The polymer has a variable composition of sugars, phenolic compounds, alcohols and aromatic amino acids.

**lignotuber** The woody fire-resistant base (partly or wholly underground) of some plant stems. e.g. Certain species of *Eucalyptus*, especially the *mallee.

**Liliaceae** The lily family of monocot flowering plants. e.g. Onion, garlic, tulips, lilies, asparagus, *christmas bells. See Appendix 1.

**limiting factor** An environmental factor which limits the growth, distribution or population density of an organism. e.g. In deserts, soils may be high quality, but too little water provides a limiting factor influencing the growth, distribution and numbers of an organism.

**limnology** The study of the life in 'standing' inland waters (lakes, ponds), and all the factors that influence such life.

**limpet** A marine *gastropod mollusc (univalve) with a low cone-shaped shell. Often found on rock platforms. See Appendix 1.

**linkage** The tendency of certain genetic traits to remain together and be inherited together as a group. They fail to show *independent assortment. It is known as *sex linkage when these traits are carried on one of the sex chromosomes. See linked genes, below.

**linked genes** *Genes which are usually inherited together as a group due to them being together on the same *chromosome. See also linkage, sex linked and crossing over (Fig. H3).

**lipase** An enzyme that causes the break-up (by *hydrolysis) of fats into fatty acids and glycerol. In vertebrates it is secreted by the pancreas and walls of the intestines. See also digestion.

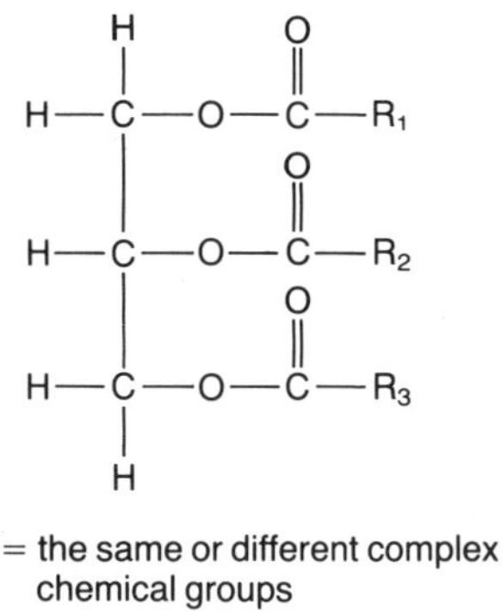


**Fig. L3 Lipids**

**lipid** Any of a large group of organic compounds which occur widely in living organisms, and which are insoluble in water but soluble in organic solvents such as alcohol and ether. When broken down (by *hydrolysis), they mostly yield *fatty acids and *glycerol. Lipids vary from simple lipids such as fats, oils and waxes, to complex lipids such as phospholipids, glycolipids, terpenes, and steroids. Compare carbohydrates and proteins. See Fig. L3.

**lipoprotein** A simple protein to which a ⋆lipid molecule has been attached. They occur in the blood and lymph, where they help transport the lipids absorbed from the small intestine to the liver, and from the liver to the site where fat is deposited (e.g. adipose tissue).

**liposome** A cell-like sphere containing liquids that is enclosed by an artificial ⋆membrane made of two layers of ⋆lipid. Used to study membranes and in medicine to deliver a drug into the body.

**lithosol** 'Skeleton ⋆soils' found over broken bedrock in mountainous and hilly regions. Heavy rain usually removes the soil soon after it forms.

**lithosphere** The rocks and other solid material that make up the Earth's crust.

**litter** 1 A number of young animals produced at one birth. 2 The scattered dead and decomposing leaves, and life within them, on the floor of a forest.

**littoral** 1 Relating to the seashore, especially the region between low and high tide, exposed alternately to air and water. See Fig. O1. 2 In lakes, the region between the shoreline and the outer limit for growth of rooted plants.

**liver** The largest organ of the vertebrate body and which has been described as the 'chemical factory' of the body. Below is only a partial list of its functions.

The liver functions as a gland by producing ⋆bile, which is stored in the ⋆gall bladder before being discharged into the small intestine when fats are present. However, the main functions of the liver are the ⋆metabolism of fats, proteins and carbohydrates, and the storage, filtering and detoxification of blood: **a** The liver helps maintain blood sugar levels, and is one of the main sites (along with muscles) where glucose is converted and stored as ⋆glycogen. Excess glucose is converted to fats. When glucose levels in the blood fall, glycogen in the liver is converted back to glucose. **b** Liver has an important role in protein synthesis. Excess amino acids are also broken down (deaminated) in the liver to yield energy, the nitrogenous wastes forming ⋆urea. **c** The liver is important for the synthesis, storage and utilisation of fats. **d** The liver inactivates many chemicals, including alcohol and drugs, and helps regulate hormones by removing any excess. It helps to detoxify many poisons. **e** The liver is the storage site for many fat-soluble vitamins such as vitamins A, D and E, as well as many minerals. **f** The liver produces prothrombin, fibrinogen and other blood plasma proteins needed for ⋆clotting. **g** Red blood cells are broken down in the liver, and the haemoglobin pigments present are used to produce bile.

**liver-fluke** A common ⋆parasitic fluke (flatworm in phylum ⋆Platyhelminthes) which lives in the bile ducts of sheep, cattle, etc., causing 'liver-rot' and great economic loss. An ⋆intermediate host, a snail living in wet grass, is needed for completion of the life-cycle.

**liverwort** A group of green plants similar to, but simpler than, the ⋆mosses, and in the plant Phylum (Division) ⋆Bryophyta. See Appendix 1.

**living fossil** Any organism which is alive today, but is very similar to fossils of organisms which were once thought to be extinct. e.g. ⋆*Latimeria* (a coelacanth) was only known as a Cretaceous fossil until specimens were caught off the coast of southern Africa in 1938. ⋆Ginkgo trees are a well-known plant example.

**lizard** Any ★reptile of the order Squamata, suborder Sauria, which includes the ★geckos, ★skinks, ★monitors, ★dragons and ★legless lizards in Australia. Most have four legs, a tail, and their body is covered by smooth dry scales. Their tongue is not forked, they have ears, and their eyelids are movable. See Appendix 1.

**loam** A loose soil type containing clay, sand and considerable organic material.

**lobe** A rounded part or projection of an organ e.g. the ear lobe.

**lobster** In Australia, a large marine crustacean in order ★Dècapoda and family Palinuridae which lacks the large claws present in north Atlantic lobsters which are in family Homaridae. Compare crayfish. See Appendix 1.

**loculus** Plural is **loculi**. A compartment, cavity or chamber.

**locus** Plural is **loci**. The position of a ★gene along a ★chromosome. There may be several different ★alleles that could be found at a particular locus. e.g. With blood groups: A, B and O.

**locust** 1 Any of the ★grasshoppers with short antennae, especially the plague locust such as *Chortoicetes terminifera*. 2 The cicada, family Cicadidae, is often incorrectly called a locust.

**long sighted** A condition in which light from near objects is focused behind rather than upon the light-sensitive ★retina. Close objects are blurred while distant objects are distinct. Thus glasses or contact lenses are often needed for close-up work such as reading. Compare short sighted.

**long-day plants** Plants which flower only when subjected to a certain minimum (often 12 hours or more) of daily illumination (e.g. sunlight), including most summer ★annuals in temperate regions. Recent research has shown that they should be called 'night-short' plants, as they must have no long periods of darkness for them to flower. Compare day-neutral and short-day plants. See also photoperiodism.

**longitudinal** Relating to length or the long axis of an organism. e.g. A longitudinal section (LS) is a cut along the length of an organism.

**loop of Henlé** The U-shaped part of the tubule in the ★kidney where water is reabsorbed into the blood capillaries. Organisms which live in deserts often have loops that are much longer, and thus absorb more water than similar organisms living in moister environments. See Fig. K1.

**lorikeet** Any member of the nectar-feeding parrot family Loriidae, especially the rainbow and scaly-breasted lorikeets that frequent gardens and tourist parks.

**louse** Plural is ★**lice**.

**lowan** The ★mallee fowl.

**LSD** Short for **lysergic acid diethylamide**, a crystalline solid. LSD is a powerful and addictive (psychological ★dependence) drug that causes ★hallucinations. Long term effects can include depression and other psychiatric complications.

**LTH** Short for **luteotropic hormone**. Also called ★prolactin or lactogenic hormone.

**lucerne** Any of the fodder ★legumes of family Papilionaceae. e.g. Townsville lucerne.

**luciferin** A substance present in the light-producing organs of ★bioluminescent organisms which, when oxidised, yields light energy. e.g. ★Fireflies. See also luminescence below.

**lumbar** Relating to the waist or loins. e.g. Lumbar vertebrae are those vertebrae near the waist, below those supporting ribs.

**lumen** 1 A cavity. A space within a tube or sac. 2 In plants, the cavity within the cell walls once the ⋆protoplasm has disappeared. e.g. Tracheids.

**luminescence** A giving out of light at relative low temperature, and thus not due to the effects of heating. e.g. ⋆Phosphorescence, ⋆bioluminescence, ⋆fluorescence. See also luciferin.

**lung** One of the pair of large spongy respiratory organs in vertebrates where oxygen from the air enters the bloodstream and carbon dioxide leaves (Fig. L4).

In humans, air enters the lungs to fill the space created when muscles cause the chest and ⋆diaphragm to expand. After entering the nose, the air passes through tubes that become gradually narrower (called, in order, ⋆trachea, ⋆bronchi, ⋆bronchioles), to enter tiny 0.1 mm diameter sacs at the ends called ⋆alveoli. The alveoli are lined by a moist membrane, the thinnest and most delicate membrane in the body. Oxygen passes from the air, across this membrane, to the blood in the capillaries. At the same time, carbon dioxide passes from the blood to the air in the alveoli. Then the muscles of the chest and diaphragm contract, forcing out much of the air in the lungs.

**lung book** The respiratory organ of air-breathing ⋆arachnids (e.g. spiders), where the thin, leaf-like membranes look like the pages of a book.

**lungfish** Any of several long tropical freshwater fish in the order Dipnoi which can breathe air through a lung-like structure (see air bladder), as well as absorb oxygen from water through their gills. This air-breathing ability helps them survive in stagnant waters during droughts. e.g. Queensland lungfish.

**luteal** Relating to, or produced by, the ⋆corpus luteum.

**luteinizing hormone** Called **LH** for short. See above.

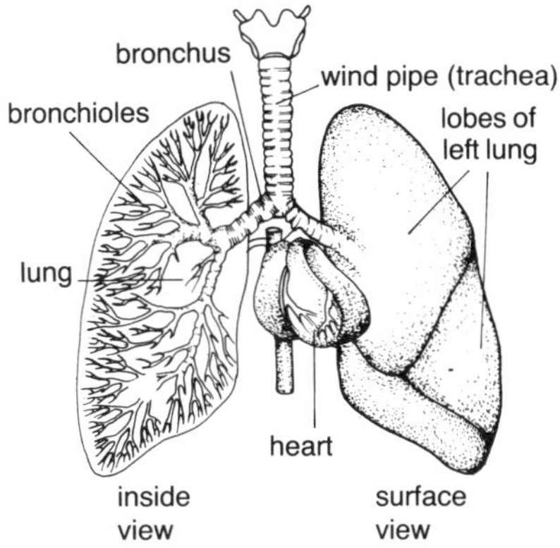


**Fig. L4 Lungs**

**luteotropic hormone** Called **LTH** for short. Also called ⋆**prolactin** or **lactogenic hormone**.

**Lycopsida** The class of vascular plants which includes the ⋆clubmosses and quillworts. See Appendix 1.

**lymph** The colourless fluid in the ⋆lymphatic system. This fluid drains from between the body cells, having originally seeped out of blood vessels. Lymph has the same concentration of salts as blood ⋆plasma, but lower concentrations of blood proteins. The blood cells present are mainly ⋆lymphocytes.

**lymph heart** Any of the two-chambered pumping structures found in the lymphatic vessels of amphibians, and some fish and reptiles. Not present in mammals.

**lymph node** Also called a **lymph gland**. An oval structure of ⋆lymphatic tissue located along the main ⋆lymph vessels. Usually several lymph vessels enter a lymph node, but only one exits. As ⋆lymph passes through, white blood cells (⋆macrophages) help clear it of any microbes that may be present. As a result, a lymph node may become enlarged and enflamed. e.g. Swollen 'glands'. Microbes and other antigens that enter lymph nodes also meet the lymphocytes that are present: B-cells are stimulated to produce antibodies; T-cells respond by changing into 'killer', 'helper' and 'suppressor' T-cells.

**lymph vessel** Any of the tubes that carry ⋆lymph from the tissues of the body back to the veins. The smallest vessels lack the smooth muscle fibres and valves present in the larger lymph vessels.

**lymphatic system** A one-way system of tubes (lymphatic vessels) that drain the fluids which have seeped from the blood vessels, back into the blood stream. The smallest lymphatic vessels are thin-walled tubes, but the larger collecting vessels have smooth muscle fibres in their walls, and valves to prevent reverse flow. Contraction of surrounding muscles provides most of the pumping force. The lymphatic system drains into a large vein at the base of the neck.

**lymphatic tissue** Also called **lymphoid tissue**. The location of ⋆lymphocyte maturation. Sometimes specific organs can be identified. e.g. Tonsils, adenoids, spleen, thymus, and lymph nodes. In other areas, the lymphatic tissue is spread amongst other tissues. e.g. In the respiratory and digestive tracts.

**lymphocyte** A type of white blood cell or ⋆leucocyte. These cells are round in shape (6–8 µm across in people), with a large nucleus and little cytoplasm. They make up about 25% of the white blood cells. They are produced in the red bone marrow, but finish maturing in ⋆lymphatic tissue. Most are not ⋆phagocytes, but are able to move around. When an ⋆antigen is detected, they rapidly increase in numbers. There are two main types. ⋆T-cells (T-lymphocytes) are involved in helping recruit other cells such as ⋆macrophages to attack invaders and B-cells to produce antibodies. T-cells also carry out cell-mediated immunity by developing into 'killer', 'helper' and 'suppressor' T-cells. ⋆B-cells (B-lymphocytes) carry out antibody-mediated immunity by developing into antibody-producing cells (⋆plasma cells) or memory cells. See also B-cell, T-cell and immune system.

**lymphoma** Any ⋆malignant tumour made up of lymph tissue. e.g. Hodgkin's disease.

**lyrebird** Either of two, large, fowl-like ground-dwelling birds (genus *Menura*), of south-eastern Australia. They are famous for their mimicry of sounds during their call, and the spectacular display of the males during which they spread out their long, lyre-shaped tails.

**lysine (lys)** One of the twenty ⋆amino acids occurring naturally in proteins. Lysine is converted by bacteria into polyamine cadaverine, one of the main components of the bad smell around rotting animal flesh.

**lysis** The destruction of cells by agents such as antibodies, snake venom, or even ⋆hypotonic (very low concentration) salt solution.

**lysosome** Membrane-bound bodies (organelles) that contain chemicals for the breakdown of materials, usually within the cell. e.g. Worn out organelles, material taken in by ⋆endocytosis, and self-destruction of the cell when it is worn out. ⋆Phagocytes in particular have many lysosomes for the breakdown of materials that they ⋆engulf. See also autolysis.

**lysozyme** An enzyme that catalyses the destruction of the cell walls of many bacteria (often during ⋆phagocytosis), causing them to split open. Lysozymes are found in tears, saliva, nasal secretions, and egg white.

# Mm

**macerate** 1 To soften by soaking in water, with or without heating, so that a substance can be separated into its parts. 2 To treat a tissue so that it separates into its cells.

**macroevolution** ⋆Evolution that suggests major changes in organisms to account for the origin of the higher classification groups such as classes and phyla. Compare micro-evolution.

**macromolecule** A large molecule, with ⋆molecular weight of several thousand or more, composed of many smaller units (⋆monomers) linked together to form a ⋆polymer. e.g. Proteins, fats, polysaccharides (starch, cellulose), nucleic acids.

**macronutrient** An element essential for the life of an organism, and required in relatively large amounts. There are twelve macronutrients: carbon, hydrogen, oxygen, nitrogen, sodium, chlorine, sulfur, phosphorus, potassium, magnesium, calcium, and iron. Compare micronutrients.

**macrophage** Also called **monocytes**. A type of ⋆phagocyte (and hence ⋆leucocyte) found widely in the vertebrate body. Their main function is to engulf and remove worn-out cells and other debris, and invading bacteria and toxins that have been immobilised by antibodies. Macrophages occur in connective tissues, and quickly move to sites of injury. ⋆Pus is partly the accumulation of macrophages as they deal with invading bacteria, etc. Macrophages also occur in the blood vessels of the lymphatic system, spleen, bone marrow, liver, etc.

**Macropodidae** The family of large-footed, hopping, herbivorous marsupials that includes the ⋆kangaroos, wallabies, rat-kangaroos, tree-kangaroos, etc. See Appendix 1.

**macroscopic** Large enough to be seen by the unaided eye. Compare microscopic.

***Macrozamia*** Also called **Burrawang**. A genus of Australian cycads, a group of plants with palm-like leaves and large seeds on the female cones. The seeds are very poisonous, although the Australian Aborigines devised ways to carefully wash the poisons from the crushed fruits so that they could be eaten. See also Appendix 1.

**macrurous** Long-tailed. Compare brachyurous.

**mad cow disease** See bovine spongiform encephalopathy.

**Madagascan realm** A ⋆biogeographical realm off the east coast of Africa with its distinctive fauna of ⋆Prosimians. See Fig. B5.

**Magdalenian** A cultural period in the stone age (upper Palaeolithic) in which ⋆Cro-magnon Man flourished with extensive industry and artistic achievements. See culture.

**maggot** The legless ⋆larva of a ⋆fly (e.g. Bushfly, housefly), especially when living in decaying matter.

**magnesium (Mg)** A soft silvery and chemically very reactive metal that exists as the $Mg^{2+}$ ion in living things. Magnesium is an essential part of the ⋆chlorophyll molecule and a number of enzyme systems.

**magnetic resonance imaging** See MRI

**magnification** 1 Increase in the

apparent size of an object, as when seen through a ★lens, or series of lenses as used in a ★light microscope. **a** Magnification of a single lens

$$M = \frac{\text{image size}}{\text{size of original object}}$$

**b** Magnification of a series of lenses = magnification of the first lens × magnification of the second lens ($M = M_1 \times M_2$). **2** See also biological magnification.

**magpie** Any of the species of largish black-and-white Australian birds in the genus *Gymnorhina*, and is normally found on grassland in more open areas. The magpie is shown on the Australian coat-of-arms representing South Australia.

**mahogany** **1** Any of certain tropical American trees such as *Swietenia mahogani* and *S. macrophyla* with its hard reddish wood used in making furniture. **2** The common name for a number of plants in Australia, including some *Eucalyptus* species. e.g. *E. pellita* or Red Mahogany found along the east coast.

**major histocompatibility complex** Called MHC for short. A group of ★genes that code for specific proteins that are found on the surface of cells. Because of the large number of these genes, and their many ★alleles, different animals of the same species are likely to have quite different proteins on their cell membranes. This is what causes rejection of organ transplants between humans. MHC proteins are also involved in communication between B-cells and T-cells during the ★immune response.

**malaise** Listless, tired, irritable and generally distressed feelings, often before an infection that produces a fever.

**malaria** An infection that produces recurring chills and then fevers, caused by a protozoan parasite (★Plasmodium) transmitted to humans by the bite of certain types of mosquito (e.g. *Anopheles*). The parasite invades and destroys the red blood cells. See also Sporozoa.

**male** **1** In animals, the sex which produces ★sperm cells. **2** In plants, that organism, or part of an organism, which produces ★gametes that move to achieve fertilisation.

**malignant** Life-threatening and invasive. Usually refers to the growth and spread of ★cancers. Compare benign.

**mallee** **1** A growth form where trees (especially *Eucalyptus*) form a number of woody stems arising from a large underground root (often ★lignotuber). There is no trunk (or ★bole) to the usually fairly short tree. Some trees always grow in the mallee form (e.g. White mallee *E. dumosa* of western Victoria), while others only take this form under difficult environmental conditions. e.g. Scribbly gum *E. haemastoma* near Sydney, especially after a bushfire. Compare tree. **2** Also used to classify plant communities dominated by plants of the mallee form, especially in Victoria.

**mallee fowl** A large mound-building fowl-like bird *Leipoa ocellata*, in family Megapodiidae, living in dry ★mallee areas. See also megapode.

**malleus** Also called the **hammer**. The first of the three tiny bones in the middle ear called the ★ossicles. i.e. The one nearest the eardrum. See Fig. E1.

**malnutrition** Lack of proper nutrition due to deficiencies in the diet. e.g. Lack of enough carbohydrate for energy, protein for growth and repair, vitamins, or minerals.

**Malpighian body** The filtering part of the ★nephron in the ★kidney. It is composed of the tangled mass of capillaries called the ★glomerulus surrounded by ★Bowman's capsule. See Fig. K1.

**Malpighian layer** The layer of cells (epidermis) under the surface of the skin (dermis) that contains the pigment ⋆melanin, and where active cell division (mitosis) occurs to replace the cells of the dermis that are being constantly worn away.

**Malpighian tubules** The thread-like tubes in the abdomen of insects, arachnids and myriapods, which open into the hindgut. The tubules extract wastes from the blood as it passes along the body cavity, and pass them to the hindgut to exit with the faeces.

**maltase** The enzyme which helps the breakdown (by ⋆hydrolysis) of the double sugar (disaccharide) ⋆maltose to form two molecules of glucose.

**maltose** A double sugar (disaccharide) obtained from the breakdown (by hydrolysis) of starch. It is composed of two glucose molecules chemically bonded together.

**Malvaceae** The mallow Family of dicot flowering plants, which includes the well known hibiscus, hollyhock and cotton. The floral emblem of the Northern Territory is the hibiscus-like *Gossypium sturtianum*, also a member of this Family. See Appendix 1.

**Mammalia** A class of 'warm-blooded' (⋆homoiothermic) vertebrates, usually with an insulating covering of hair or fur, and having sweat and ⋆sebaceous (a secretion resembling fat) glands in the skin. Mammals have mammary glands that secrete milk for the young, and have a four-chambered heart that keeps oxygenated and deoxygenated blood completely separate. The three main groups of mammals are the egg-laying ⋆monotremes, the ⋆marsupials whose young are born only partly developed, and the ⋆placentals which make up the largest group. See Appendix 1.

**mammary gland** The milk-producing gland characteristic of all mammals. There may be one or more pairs, depending on the type of mammal. See lactation for details of how the milk is produced.

**mammoth** An extinct elephant-like mammal with huge tusks and long, thick hair. The remains of these placentals have been found frozen in Siberia and Alaska.

**man** 1 Any member of the genus ⋆*Homo*. A human being. 2 The adult male of genus *Homo*.

**mandible** 1 In vertebrates, the lower jawbone. Compare maxilla. 2 In invertebrates, one of the pair of mouthparts used for tearing, or crushing food. e.g. Grasshopper. See Fig. M4.

**mangrove** 1 A tree or shrub that grows in both brackish and marine waters, especially in muddy estuaries, and sometimes has erect aerial roots (called ⋆pneumatophores). e.g. Trees in the genera *Avicennia* and *Rhizophora*. These trees form an important habitat as nurseries for young fish, and as a barrier in protecting the shoreline from wave-caused erosion. 2 The name of the vegetation type dominated by mangroves.

**mantids** See mantis below.

**mantis** Also called **mantids**. A family of usually long, thin, carnivorous insects well known for their triangular heads, large forelegs, and the way they devour their prey. e.g. Praying mantis. See Mantodea below, and Appendix 1.

**mantle** 1 In molluscs, the outermost layer of the soft body, and the flaps of tissue extending from it. The mantle secretes the shell (when present). The mantle cavity, between the mantle and the body, contains the gills. A similar structure is found in ⋆brachiopods. 2 In cephalopods the mantle forms the muscular body wall and is important in movement and respiration.

**Mantodea** The order of insects which includes the praying *mantis etc. See Appendix 1.

**manus** The end part of the forelimb of vertebrates. In humans it comprises the hand; in most other vertebrates it forms the forefoot. See Fig. P2.

**marginal** Located near the edge.

**marihuana** See *marijuana below.

**marijuana** Also spelt **marihuana**. A *narcotic formed from the dried leaves and flowers of the Indian hemp plant, **Cannabis sativa*.

**marine** Related to, or living in, the seas and oceans.

**marron** A large freshwater *crayfish, a *crustacean, found in Western Australia, *Cherax tenuimanus*.

**marrow** The soft jelly-like tissues filling the spaces within *bones. Red marrow produces red *blood cells, some white blood cells (*leucocytes) and some *platelets, and fills all the bone cavities of the young. In adults this is gradually replaced by fatty yellow marrow, with red marrow being confined to the spongy ends of the long bones, and in the vertebrae, ribs, sternum and skull.

**marsh** A waterlogged area. A swamp.

**marsupial** Any member of mammalian order Marsupialia, which includes the koala, wombat, kangaroos and wallabies, and possums of Australia, and the opossum of the Americas. Females bear young that are only partly developed. The young transfer to a pouch (*marsupium) which encloses the nipples, and where development is completed. See Appendix 1.

**marsupial cat** See **quoll**.

**marsupial mole** A small, blind, burrowing marsupial living in arid sandy regions of Australia, *Notoryctes typhlops*. See Appendix 1.

**marsupial mouse** Any of the many types of small Australian carnivorous marsupials which resemble mice. e.g. *Mulgara, *Dibbler, *Antechinus, *Dunnart. See Appendix 1.

**Marsupialia** See marsupial above.

**marsupium** A pouch on the abdomen of most *marsupials, and the *echidna. It covers the mammary glands, and in it the young complete their development.

**mast cell** Any of the many special cells found in connective tissue which store *histamines (released during injury) and *heparin (an anticoagulant). These substances are thought to be released if the immediate area becomes injured, inflamed, or antibodies are released.

**mastectomy** Surgical removal of the breast, usually due to cancer.

**masticate** To grind, chew, and crush food in preparation for swallowing.

**Mastigophora** Also called **Flagellata**. A class of *protozoans with one or more *flagella that are used for movement. See Appendix 1.

**mastitis** Inflammation of the breast, from bacterial infection or other causes.

**mastodon** Any of the fossil elephant-like mammals (but different from *mammoths) of the Miocene.

**mastoid** A lump on the human skull just behind the ear. It contains air spaces that connect to the middle ear.

**mating** 1 The act of sexual intercourse. 2 To place animals together so that breeding can occur.

**matrix** The substance in which cells are embedded. e.g. In cartilage, or connective tissue.

**matter** The material from which the physical objects of the universe are made. e.g. Atoms, molecules.

**maturation** The process of maturing, or developing into adulthood. Completing natural growth and development to adulthood.

**maxilla** 1 The upper jawbone of vertebrates. Compare mandible. 2 One of the paired ⋆mouthparts in arthropods, located behind the mandibles. See Fig. M4.

**maxillipeds** Crustacean ⋆mouthparts, 1 to 3 pairs of which exist behind the ⋆maxilla.

**mayfly** Short-lived insects in order Ephemeroptera, whose ⋆larvae are more often seen (in freshwater streams) than is the adult with its large membranous wings. See Appendix 1.

**maze** A complicated interconnected series of pathways and blind alleys, connecting the start and finish, through which animals are run. The number of attempts that they take to learn the exact route is used as an indication of their intelligence.

**mealy** Covered with coarse flour-like powder.

**mean** The average of a number of numerical values.

**measles** An infectious disease caused by a virus. See also Rubella.

**mechanoreceptor** A ⋆receptor or sense organ that responds to mechanical stimuli, either directly (e.g. touch, pressure), or indirectly (e.g. sound, equilibrium, position).

**median** 1 Located in the middle of the organism being studied. 2 In statistics, in a series of numerical values, the point above which the number of figures is the same as the number below it.

**medication** The use of medicines to help treat disease.

**medicine** Any substance used in the diagnosis, treatment, or prevention of a disease.

**medium** Plural is **media**. The food or materials prepared for the growth and culturing of bacteria and other microbes. e.g. An ⋆agar-based ⋆culture medium prepared in a Petri dish.

**medulla** The central part of an organ, as compared to the outer ⋆cortex. e.g. The medulla of the kidney and adrenal glands.

**medulla oblongata** The lowest or rear-most part of the vertebrate ⋆brain which connects directly to the ⋆spinal cord. It contains centres which help control external respiration, blood pressure, etc., via the ⋆autonomic nervous system. See Fig. B7.

**medullary ray** Also called a pith ray or primary ray. Any of the sheets of ⋆parenchyma cells that extend radially from the central ⋆pith (medulla) of a stem out between the ⋆vascular bundles. They are used for storage and the radial transfer of food. See also ray.

**medusa** The free-swimming, bell- or umbrella-shaped stage in the lifecycle of many ⋆cnidarians (coelenterates). Most jellyfish are the medusa stage. Compare with polyp.

**mega-** Prefix (often shortened to M) meaning one million or $10^6$.

**megaphyll** Also called a **macrophyll**. A large leaf with a complex vein pattern and internal structure, and often divided into a number of leaflets. e.g. The leaves of ferns, flowering plants. Compare ⋆microphyll.

**megapode** Any of the large fowl-like birds of Australia and nearby areas which construct a mound for display (e.g. ⋆lyrebird) or to incubate their eggs (e.g. ⋆mallee fowl, brush turkey).

**megaspore** Also called a **macrospore**. In some plants, the larger of the two types of ⋆spore produced. Often grows into the female ⋆gametophyte. Compare microspore.

**meiosis** The special type of cell division which produces the gametes (sperm and ova). See Fig. M1.

Four daughter nuclei are produced during meiosis, each containing half the number of chromosomes normally found in a body cell. Cell division may follow. Thus the normally ⋆diploid (2n) cells become ⋆haploid (n), allowing the normal chromosome number to be restored during ⋆fertilisation. During meiosis, genetic diversity is increased by ⋆crossing over (Fig. H3), and by ⋆independent assortment of the chromosomes. Although meiosis is one continuous process, it is convenient to divide it into a number of stages.

**Division one** The first cell division results in two cells each with only one of each type of chromosome. Thus the paired (⋆homologous) chromosomes are separated. The stage *before* division can be seen taking place is called ⋆interphase. During interphase the DNA and other cell structures are doubled. Although the chromosome has doubled, the two ⋆chromatids are still held together at the ⋆centromere (Fig. I13). The first division that follows is divided into four stages: **1 Prophase I**. The ⋆chromatin condenses to form visible chromosome threads, with the ⋆centromere visible. It then divides to form chromatids held together at the centromere. The nuclear membrane begins to disappear. The similar (homologous) chromosomes then line up next to each other, forming ⋆bivalents. During this time the corresponding chromatids may come into contact, forming temporary points of attachment called ⋆chiasmata. This alignment is very precise, and crossing over may now take place during which there is an exchange of chromosome material. This stage can be divided into five phases called ⋆leptotene, ⋆zygotene, ⋆pachytene, ⋆diplotene, and ⋆diakinesis. **2 Metaphase I**. The nuclear membrane disappears, the homologous pairs line up along the equatorial plane of the cell, and the centromeres become attached to the ⋆spindle. In animal cells the ⋆centrioles become visible. **3 Anaphase I**. Each member of the homologous pair of chromosomes is pulled by the spindle to opposite ends of the dividing cell (the ⋆poles). The sister chromatids remain joined at the centromere. (Compare this with anaphase I of ⋆mitosis.) Just which of the pair of homologous chromosomes end up in any particular cell is a matter of chance. It depends on how it arrived and was lined up along the equatorial plate during metaphase I. This chance process is called ⋆independent assortment. **4 Telophase I**. The movement of the chromosomes is completed. In some organisms two cells form, and there is an interphase, before the second division commences. In other organisms prophase II begins after only a very short telophase.

**Division two** This division resembles mitosis except there is no doubling of the DNA. In division 2 the chromatids separate. Again, it can be divided into four phases. **1 Prophase II**. The chromosomes recondense, and the nuclear membrane disappears, if these have reformed in division 1. New spindle fibres begin to form. **2 Metaphase II**. The chromosomes in each nucleus line up along the equatorial plane. **3 Anaphase II**. The sister chromatids, held together by the centromere, separate, and each resulting single chromosome moves towards opposite ends of the cell (poles). **4 Telophase II**. The spindles disappear and the nuclear membrane forms around each set of chromosomes. The cell may then complete its division, with a dividing cell membrane (and cell wall in plants) forming separate cells. The result is the production of from one to four gametes (in humans, four sperm but only one ovum—see polar bodies).

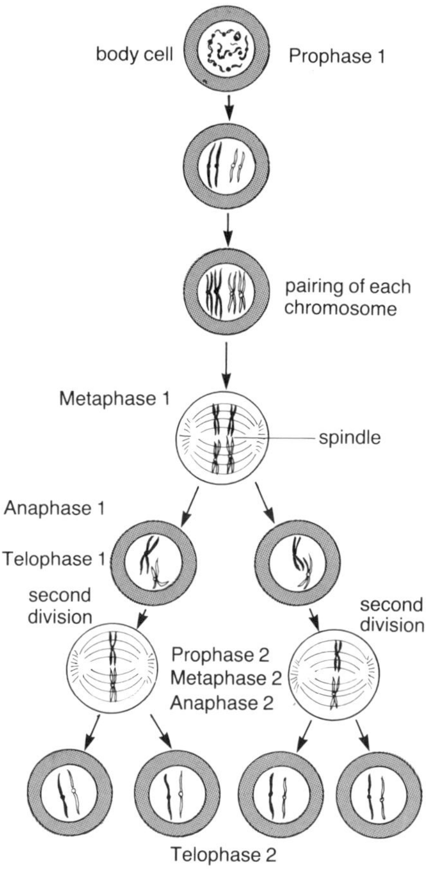


**Fig. M1 Meiosis**

**Meissner's corpuscle** See corpuscle and touch.

***Melaleuca*** A genus of dicot shrubs and trees, family Myrtaceae, commonly called paperbarks because of the nature of their bark. See Appendix 1.

**melanin** A dark brown pigment present in the hair, skin, and iris of many animals. It occurs in granules within cells, different concentrations giving black, brown and yellow colouration. Also see ACTH and albinism.

**melanism** The existence of a population of organisms which are dark in colour due to large amounts of melanin in their ⋆surface layers. e.g. Black panthers are a form of leopard. Dark forms of common moths (see ⋆industrial melanism).

**melanoma** A malignant form of ⋆cancer, usually of the skin, that arises from pigment-producing cells.

**melanophore** A type of pigment-producing cell (⋆chromatophore) which contains ⋆melanin.

**melatonin** A substance extracted from the ⋆pineal body of the ⋆brain, and which helps control the concentration of melanin in ⋆melanophores.

**membrane** 1 A thin layer of ⋆tissue. 2 A thin layer or sheet which covers a part, lines a tube or cavity, or connects organs or structures. 3 The very thin sac-like structure that surrounds ⋆cells, preventing their liquid contents from escaping. See plasma membrane. 4 The even thinner covering within a cell that encloses ⋆organelles, the ⋆nucleus, and forms the ⋆endoplasmic reticulum.

**membranous** Thin and sheet-like. See membrane.

**memory** The brain's capacity to retain and recall impressions, words, and other environmental experiences detected by the senses.

**menarche** The beginning of ⋆menstruation and egg maturation by the ⋆ovaries at ⋆puberty. The number of days between menstruation is often irregular to begin with. Menstruation usually starts between the ages 11 to 15, although it can be earlier or later.

**Mendel's laws** The two laws proposed by Gregor Mendel (1822–1884) to explain the inheritance of genetic 'factors' (now called ⋆genes) from one generation to the next.

**1 The law of segregation** 'Factors'

that control the features of organisms occur within cells in pairs. When gametes (sperm, ova) are produced, these 'factors' separate, only one going to each gamete. We now know that the factors are ⋆genes that occur on the ⋆chromosomes of each cell. Since there are two of each chromosome, there must be two forms of each gene. During ⋆meiosis, only one of each chromosome pair (and hence only one form of each gene) ends up in the sperm and ova produced.

**2 Law of independent assortment** When many 'factors' or genes are considered, then they separate (or segregate) independently of each other. This is true when genes are on different chromosomes, but is not true when genes are ⋆linked together by being on the same chromosome.

**meninges** Membranes that cover the ⋆brain and ⋆spinal cord. **Meningitis** is an inflammation of these membranes.

**menopause** The time when the menstrual cycle ceases, and ovaries cease to produce egg cells, thus marking the end of the ability to reproduce. Menopause usually occurs in women between the ages of 45 and 50, but can begin before or after this time.

**menses** The approximately monthly discharge of blood and various tissues from the vagina. See menstrual cycle.

**menstrual cycle** A type of ⋆oestrous cycle found in old-world monkeys, apes and humans where there is a rapid breakdown of the wall of the uterus leading to loss of blood and tissues, called ⋆menstruation. In humans this occurs at approximately 28 day intervals. This cycle of build-up and thickening followed by the breakdown of the uterus walls ensures that an embryo has suitable linings to embed in, and thus allows a ⋆placenta to develop. The menstrual cycle is under the control of ⋆hormones. The animals remain sexually receptive throughout the cycle.

**menstruation** Also called **periods**. The discharge at approximately monthly (28 days for women) intervals of menstrual fluid made up of blood and tissues from the walls of the uterus. The wall linings disintegrate if the ovum produced by the ovary is not fertilised, and then begin to rebuild in preparation for the next ovum.

**meristem** Specific regions in a plant where active cell division (mitosis) takes place. The new cells formed then undergo ⋆differentiation to produce mature tissues. e.g. Epidermis, cortex, vascular tissue. In dicots such as woody stemmed plants, the main meristem tissues are found in the tips of the stems and roots (⋆apical meristem), between the xylem and phloem of vascular bundles (vascular ⋆cambium), and under the bark ( bark cambium), and in young leaves. In monocots such as grasses, there are meristems at the base of each ⋆internode of the stem.

**mesentery** 1 In vertebrates, a membrane (consisting of two layers of ⋆peritoneum) which keeps in place the stomach, spleen, intestines, etc. by attaching them to the internal body wall. It contains the nerves, and the blood and lymph vessels which supply these organs. 2 In invertebrates, a comparable structure that connects the gut to the body wall.

**mesocarp** The middle layer of the ⋆pericarp of a flowering plant's fruit. In fruits such as the peach, this is the fleshy part that we eat. It is not present in all fruits. See exocarp and endocarp.

**mesoderm** The middle layer of the three layers (⋆ectoderm, mesoderm, ⋆endoderm) that form the ball-shaped ⋆gastrula stage of the developing ⋆embryo. The mesoderm develops into cartilage, bone, muscle, blood, kidneys, and gonads.

**mesogloea** The jelly-like substance lying between the inner walls (near the central cavity) and outer walls (near the surface) of a sponge or cnidarian (coelenterate).

**Mesolithic** The period between the ★Palaeolithic and ★Neolithic periods of the Stone Age.

**mesomorphic** Having an intermediate size of body build. See ectomorphic and endomorphic.

**mesophyll** The tissues (★chlorenchyma) made up of many chlorophyll-containing cells located between the upper and lower surface layers (epidermis) of a leaf. There are often two layers of mesophyll, the upper ★palisade layer and the lower ★spongy layer. Some leaves (e.g. *Eucalyptus*) have the spongy layer located between palisade layers on both surfaces. See Fig. L1.

**mesophyte** A plant which inhabits areas where there is not too much or too little water. Most flowering plants from temperate regions are mesophytes. Compare halophytes, hydrophytes and xerophytes.

**mesothelium** A sheet of cells lining the internal cavities of the body (★coelom) where the lungs, intestines, etc. are located. e.g. ★Pleura, ★peritoneum, ★pericardium.

**Mesozoic** Also called the **Age of Reptiles**. This ★era occurred between the Paleozoic and the Cainozoic. See Appendix 2.

**messenger RNA** Called ★**mRNA** for short.

**messmate** The name given to a number of Australian trees in the genus *Eucalyptus*, including *E. obliqua*, or Messmate Stringybark of the tablelands of New South Wales.

**metabolic rate** The rate of metabolism as determined by the amount of food consumed, heat produced, or oxygen used per unit of time. e.g. ★Basal metabolism.

**metabolism** The sum or total of all the chemical reactions occurring within a cell or organism. This involves the breaking down of complex chemicals to produce simpler ones (★catabolism) thus releasing energy and nutrients (e.g. some amino acids) that the body cannot make itself. It also involves the building up of complex molecules from simpler ones (★synthesis or ★anabolism).

**metabolite** A substance that takes part in any of the chemical reactions that occur as part of metabolism. Metabolites can be produced during metabolism, they can be taken in from the environment, or can be one of the many short-lived intermediate chemicals produced during metabolism.

**metacarpals** The bones of the hand or forefoot located between the wristbones (carpals) and the finger bones (phalanges). Very long in four-footed running animals. See pentadactyl limb and Fig. P2.

**metameric segmentation** Also called **metamerism**. The arrangement of the organs of an animal in a series of similar, repeating units along the length of its body. e.g. Annelid worms (such as the earthworm) have a number of segments, inside each of which is a similar arrangement of muscles, blood vessels, ★ganglia, ★nephridia, and sometimes *gonads. Also seen in some arthropods.

**metamorphic** Describing rocks formed by subjecting other rocks to heat and pressure. e.g. Schist, slate, gneiss, marble, quartzite.

**metamorphosis** The change in shape or form of an animal as it progresses from larva to adult. e.g. Insects can change from egg to ⋆nymph to adult, or egg to ⋆larva to ⋆pupa to adult. Amphibians change from egg to ⋆tadpole to adult. Annelids and crustaceans also undergo such changes. Metamorphosis is under hormonal control. See moulting hormone and life cycles.

**metaphase** The second stage of ⋆meiosis and ⋆mitosis. In each case, the chromosomes align along the equator of the dividing cell, and a ⋆spindle forms. See Figs M1 and M3.

**metaplasia** Alteration of a normal cell or tissue from one form to another, often to an abnormal type.

**metastasis** The spread of a disease from one part of the body to another, by transfer of cells or organisms, carried by the blood or lymphatic systems. e.g. Often applied to ⋆cancers which spread throughout the body.

**metatarsals** Bones in the hind foot of four-footed (tetrapod) vertebrates, located between the ankle bones (tarsals) and the toe bones (phalanges). Very long in running animals. See pentadactyl limb and Fig. P2.

**metatheria** Another name for the ⋆marsupials. Compare eutheria and prototheria.

**methionine (met)** One of the 20 naturally occurring ⋆amino acids found in proteins. Also converted into important chemicals in male ⋆semen. See Fig. A6.

**methyl blue** A ⋆stain that can be used to colour a nucleus blue.

**methyl violet** A ⋆stain that can be used to colour a nucleus violet.

**methylation of DNA** The addition of methyl ($CH_3$−) groups to the DNA molecule. Inactive sections that are not being ⋆transcribed are heavily methylated.

**methylene blue** A ⋆stain that can be used to colour a nucleus blue, and certain bacteria red.

**MHC** See **major histocompatibility complex**.

**mice** Singular mouse. Small placental mammals of the rodent family Muridae, especially the genus *Mus*, including *M. musculus*, the house mouse.

**micro-** One millionth or $10^{-6}$. Abbreviated to the Greek letter μ (mu). e.g. The length of a bacterium is between 0.1 μm and 5 μm. Compare milli.

**micro-organism** Also called **microbe**. A small microscopic living thing. e.g. All the protozoans, bacteria, viruses, and the microscopic algae and fungi.

**microbe** See micro-organism.

**microbiology** The study of micro-organisms such as bacteria, protozoans and viruses.

**microbody** Very small (0.3 to 1.5 μm), roughly spherical bodies (organelles) found in the cytoplasm of various plant and animal cells. The enzymes that they contain are enclosed in a membrane. e.g. ⋆Peroxisomes.

**microclimate** The climate immediately around an organism in its ⋆microhabitat. e.g. The microclimate on a flat leaf is different from that within a curled-up leaf.

**microevolution** Very small changes resulting from evolution. e.g. Development of dark populations of the Peppered Moth (⋆industrial melanism); development of resistance to antibiotics by bacteria; development of resistance to insecticides by insect pests. Compare macroevolution.

**microfibril** A ribbon-like strand, 10 to 45 nm wide, made of cellulose molecules lying almost parallel to each other. Microfibrils are the basic structural unit of a ⋆cell wall. The spaces between the microfibrils are filled with substances like ⋆pectin, and with water, thus helping to hold the microfibrils together.

**microfilament** A fine filament of protein, about 6 nm across and usually made of ⋆actin, and present in the cytoplasm of most cells. Microfilament functions include ⋆cytoplasmic streaming, ⋆amoeboid movement, and changes in cell shape. They provide supporting cores for the ⋆microvilli lining the intestine walls; and when a cell splits in two, microfilaments form to pinch the two sides of the membrane together. Compare microtubules, intermediate filaments. See also cytoskeleton.

**micrograph** A photograph of an image produced by a microscope. e.g. An electron micrograph is produced by an ⋆electron microscope. A photo micrograph by a ⋆light microscope.

**microhabitat** The local and special environment in a small habitat or place where an organism can live. e.g. Under a piece of tree bark; under your armpit.

**microinjection** The use of very fine pipettes to insert material into cells. e.g. During ⋆genetic engineering, genes can be inserted into cells. See also vectors.

**micrometre** Also called a **micron**. One millionth or $10^{-6}$ of a metre. Abbreviated to μm.

**micron** A micrometre. See above.

**micronutrient** A chemical that must be obtained from the environment by living things, but which is required only in very small amounts. e.g. ⋆Trace elements, ⋆vitamins. Compare macronutrients.

**microphyll** The usually small leaf of plants such as club moss, horsetails, etc. There is only one central vein (vascular bundle) running up the center of the leaf, with the leaf having little if any internal structure (mesoderm). Compare megaphyll.

**micropyle** A small pore in the end of each of the ⋆ovules inside a plant ovary. The pollen tube usually enters the ovule through this pore so that the male nuclei can reach the ovum inside for ⋆fertilisation. This pore may persist in the seed that develops from the ovule, allowing water and air to enter during ⋆germination. See Fig. S2.

**microscope** An instrument used to produce a ⋆magnified image so that what is being observed looks bigger than it really is. See ⋆light microscopes and ⋆electron microscopes for details of operation. Light microscopes have a maximum useful magnification of around 1500×, and a ⋆resolution of about 250 nm. An electron microscope can magnify about 250 000×, and has a resolution of around 0.5 nm. See also phase-contrast, interference, polarising, ultraviolet, and oil immersion microscopy. Figs E2 and L2.

**microscopic** Something so small that a microscope is needed to see it clearly. Compare macroscopic.

**microsome** Small particle obtained from the cytoplasm of a cell after it has been broken up and the parts separated by ⋆centrifuging. Microsomes consist of small pieces of membrane (endoplasmic reticulum), often with a ⋆ribosome attached.

**microspore** In some plants, the smaller of the two types of ⋆spore that are produced. Often grows into the male gametophyte. Compare megaspore.

**microtome** An instrument for cutting extremely thin ⋆sections (3 to 5 μm thick) of tissues for mounting on microscope slides.

**microtubule** Straight hollow tubes of protein found in the cytoplasm of most cells. They have an outside diameter of about 22 nm, and along with *microfilaments and *intermediate filaments, help organise the cytoplasm and give shape to the cell. The *spindle that helps move the chromosomes during cell division is made from microtubules. Microtubules are also used in the construction of *centrioles, *cilia and *flagella. See also cytoskeleton.

**microvilli** Microscopic finger-like projections of cytoplasm, about 2 to 4 μm in length and 0.1 μm wide, which extend above the exposed part of surface (*epithelial) cells, thus increasing the surface area available for absorbing materials. Microvilli are especially common along the cells lining the intestines and respiratory tract. Areas of dense, regularly arranged microvilli are called *brush borders.

**micturition** Urination or the act of passing urine.

**midbrain** 1 In humans, a small area between the forebrain and the hindbrain. It relays nerve impulses between the forebrain and hindbrain, and between the forebrain and the eyes. 2 The middle region of the embryonic brain, and the areas that develop from it. Relatively much larger in animals other than humans.

**middle ear** The air-filled cavity between the *eardrum and the *inner ear (*cochlea). Sound is carried across the middle ear by the *ossicles. The middle ear is connected to the rear of the mouth (*pharynx) by the *Eustachian tube, thus allowing air to enter and leave, equalising the air pressure on either side of the eardrum. See ear and Fig. E1.

**middle lamella** The thin wall that cements together the cellulose *cell walls of adjacent plant cells.

**midge** Any of a variety of small two-winged insects, in order *Diptera. See Appendix 1.

**midrib** The central or main vein of a leaf.

**migration** The periodic movement of populations of animals between one region and another, usually associated with seasonal changes.

Migration occurs in many groups: Humpback whales migrate north along the eastern and western coasts of Australia to sub-tropical waters each autumn, and back to the Antarctic each spring. The Long-finned eel (*Anguilla reinhardtii*) in eastern Australian rivers migrates to the Pacific, probably near New Caledonia, where it matures before returning to reproduce. The Bogong Moth (*Agrotis infusa*) migrates from the western plains of eastern Australia into the cooler high country each summer. Birds such as the Scarlet Robin (*Petroica multicolour*) migrate at the beginning of winter from cold high country to the warmer valleys below. The Mutton Bird (Short-tailed Shearwater, *Puffinus tenuirostris*) migrates from Bass Straight Islands off Victoria to the northern Pacific off Alaska, and back again, each year. There are many well known examples from overseas.

**mildew** A white powdery growth on plants, wooden walls, cloth such as wool, etc., caused by the growth of a fungus. Downy (or false) mildews penetrate deeply into their hosts, while powdery (or true) mildews live on the surface of their hosts in a similar way to *moulds.

**milk** The whitish fluid secreted by the *mammary gland of a mammal for use as nourishment by the young. Human milk is about 87% water, 7% milk sugar (lactose), 4% fats, 1% protein, and a small amount of salts. The composition of cow's milk is similar to human milk, but has more protein (3.5%) and less fats (3.5%) and lactose (5%). However, most of the extra protein is ⋆casein which forms a tough, compact curd in the baby's stomach compared to the fine ⋆flocculent curd produced from human milk.

**milk teeth** Baby teeth. The twenty temporary teeth that are lost when the permanent teeth grow (erupt).

**millilitre** Written **mL** for short. One thousandth of a litre. Approximately one cubic centimetre.

**millipede** Any of the many segmented arthropods of the class Diplopoda, where most segments of the body have two pairs of legs. They feed on decaying vegetation. Compare centipede. See Appendix 1.

**Millon's test** A chemical test used to show the presence of protein. Millon's reagent (a mixture of chemicals including a mercury compound) reacts with the amino acid ⋆tyrosine when they are heated gently together in a water-bath. Millon's reagent turns proteins a reddish colour. Compare Biuret's test.

**milt** The ⋆sperm-containing fluid from a male fish.

**mimicry** The similarity of an organism to one of another species, where the mimic is thought to gain an advantage. e.g. Australian examples include stingless hover-flies that resemble honey bees. Stick-insects ⋆camouflage themselves by mimicking the twigs that they walk upon. Some spiders and mantids resemble the ants that they prey upon. Some colourful butterflies resemble other very nasty-tasting types, so that predators such as birds end up leaving both types alone. See also protective colouration, warning colouration.

**mimosa** A legume plant of the Family Mimosaceae, which includes the wattles (*Acacia*), and the touch-sensitive plant *Mimosa pudica*.

**Mimosaceae** See mimosa above.

**mineral** 1 Any of the class of inorganic substances (e.g. quartz, feldspar) that make-up rocks and have a definite chemical composition and crystal shape. 2 The inorganic chemicals needed by plants from the soil (e.g. nitrates, iron), and by animals in their diet (e.g. iron, calcium, magnesium).

**Miocene** An epoch of the Tertiary period preceding the Pliocene and following the Oligocene. See Appendix 2.

**miscarriage** The expulsion of the ⋆foetus from the womb (uterus) before the foetus can survive.

**missing link** In evolution, a hypothetical organism that is intermediate between two known types. e.g. Peking Man was once considered a missing link between apes and humans. It is now thought that apes and humans evolved from a common ancestor. Compare transitional forms.

**mistletoe** Any of the plants of the family Loranthaceae which grow ⋆parasitically on other plants. In Europe, *Viscum album* with its red berries is used as Christmas decoration. In Australia, there are many different types of mistletoe, and they are common on *Eucalyptus*.

**Mitchell grass** A drought-resistant native grass, genus *Astrebla*, common on the western plains of New South Wales and Queensland where it forms arid tussock grasslands.

**mite** A small arachnid (has eight legs). Some mites are ⋆parasitic on plants and animals, while others live in the soil and in other habitats. See Appendix 1.

**mitochondria** Singular is **mitochondrion**. Minute rod, thread or sphere-shaped bodies (1 to 3 μm across) in cells (except bacteria and blue-green bacteria) that are the site of aerobic ⋆respiration. Mitochondria have been called the 'powerhouse' of the cell as they move about the cell to supply energy (in the form of ATP) for the many chemical reactions that take place. See Fig. M2.

Mitochondria can only just be seen with a light microscope, the number per cell varying from 1 to 10 000, being more common in cells which use large amounts of energy (e.g. liver, muscle). Mitochondria have two surrounding membranes, the outer one being smooth, the inner one having folds called ⋆cristae to increase the surface area. Along the inside of this inner membrane the enzymes for ⋆electron transport are arranged in order. The cavity within the mitochondrion is filled with a gel-like matrix and contains the enzymes for the ⋆Krebs' cycle. It also contains circular strands of DNA, some ribosomes, and other materials needed to make proteins.

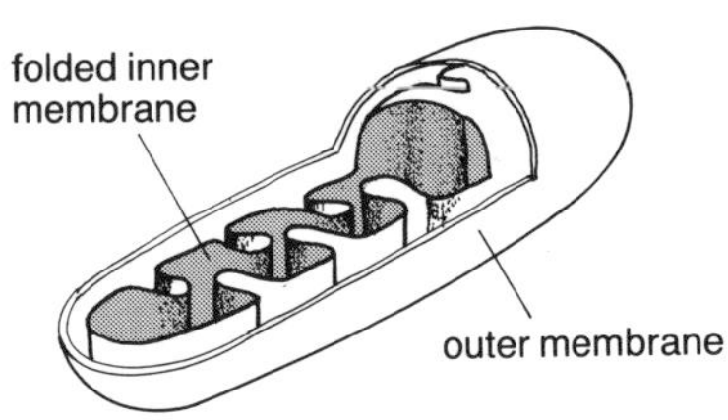


**Fig. M2 Mitochondria**

**mitosis** The normal process by which a cell nucleus divides into two. The daughter nuclei each contain the same number of chromosomes, and thus the same genes, as the parent nucleus. This nuclear division is usually followed by cell division. See Fig. M3.

Although really a continuous process, mitosis can be divided into four observable stages. The apparently resting stage, called interphase, is really very active as the DNA has already been duplicated before the first observable stage begins. **1 Prophase**. The chromosomes appear as pairs of long thin threads that condense to form pairs of ⋆chromatids attached together at the ⋆centromere (Fig. H3). Towards the end of prophase the ⋆nucleolus and nuclear membrane gradually disappear. **2 Metaphase**. The ⋆spindle forms, and ⋆centrioles in animals and some simpler plants. The chromosomes move to line up along the ⋆equatorial plate attached to their spindle fibres at the centromeres. **3 Anaphase**. The centromeres divide in two and the chromatids separate, being dragged to the ⋆poles of the dividing cell. **4 Telophase**. The chromosomes gradually lengthen and disappear as the nuclear membrane reforms and the nucleolus reappears. After telophase the cytoplasm is usually divided in two as new cell membranes (and cell walls in plants) form.

**mitral valve** Also called a **⋆bicuspid valve**.

**mixture** Occurs when two substances are added together and mixed, there being no chemical reaction between them. The components can be readily separated by physical means. e.g. Chromatography, filtration.

**model** A particular type of scientific ⋆theory where a process is 'pictured' by a physical analogy, or by mathematical equations, thus allowing predictions to be made and tested. e.g. Since the ⋆plasma membrane is only just visible even with an electron microscope, several models were proposed and tested in an attempt to explain its properties.

**molar** 1 One of the ⋆teeth used on grinding food. 2 A solution that contains one ⋆mole of the dissolved substance per litre of resulting solution.

**mole** The ⋆molecular mass of a substance expressed in grams. It is obtained by adding together the

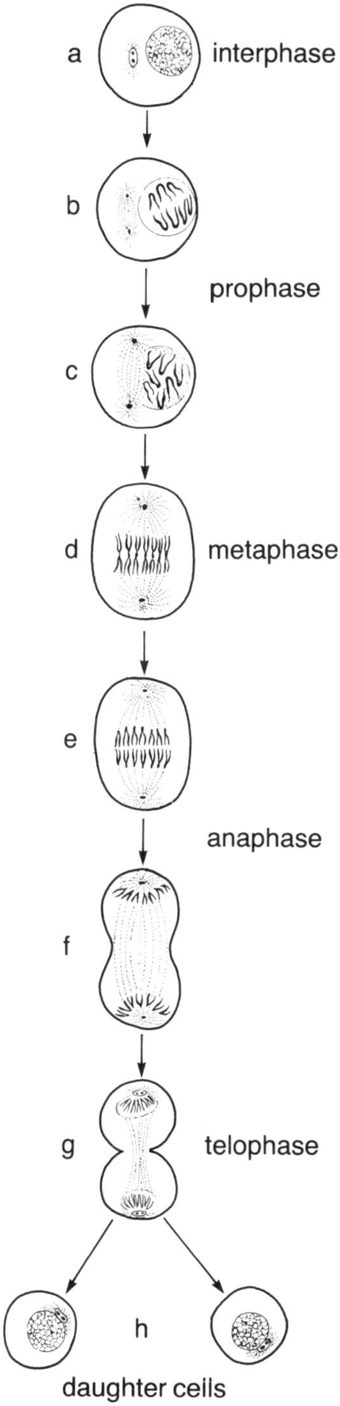


**Fig. M3 Mitosis**

⋆atomic masses of the elements that make up a ⋆molecule of the substance. (There are other definitions less useful to biology.)

**molecular mass** The mass of a molecule relative to, or compared to, the element carbon which is given the value 12.

**molecule** The smallest part of a compound, made up of a fixed number of atoms ⋆covalently bonded together.

**Mollusca** A phylum of unsegmented invertebrates with a large muscular foot beneath a soft body. The soft body is covered with a layer of tissue called a ⋆mantle which usually secretes a hard shell. Most molluscs feed using a rasp-like ⋆radula. e.g. Chitons, bivalves such as oysters and clams, univalves (Gastropods) such as snails, whelks, limpets and periwinkles, squids, octopuses, and nautili. See Appendix 1.

**Monera** Any member of the kingdom that includes the ⋆prokaryotic organisms. e.g. Bacteria, blue-green bacteria. Compare plant, animal, fungi and protista.

**Mongolian eyefold** Also called the **epicanthic fold**. A fold of skin which covers the points where the upper and lower eyelids meet. Characteristic of people in the Mongoloid race. e.g. Chinese.

**mongolism** See **Down's syndrome**.

**monitor** Any of the lizards in the genus *Varanus*, often called ⋆goannas, such as the common tree-climbing Lace Monitor of eastern Australia, *V. varius*. See Appendix 1.

**monocot** Short for **Monocotyledonae**. See below.

**Monocotyledonae** Called **monocot** for short. One of the two major divisions of the flowering plants where the plant's seeds have only one ⋆cotyledon. The flower parts are usually in threes, the vascular bundles are scattered in the stem, and the leaves have parallel veins. e.g. Grasses, orchids, lilies, irises, palms. See Appendix 1.

**monoculture** The growing of only one type of crop over large areas of farmland or forests. This can sometimes be ecologically unstable as insect pests and plant diseases can spread readily throughout the whole area.

**monocyte** Also called ⋆**macrophages**.

**monoecious** A plant species that has separate male and female flowers on the same plant. e.g. Hazel, sweet corn. Compare ⋆dioecious.

**monogamy** Having only one mate. e.g. Swans.

**monohybrid** An organism which is ⋆heterozygous (hybrid) in at least one genetic feature. e.g. With Mendel's peas the alleles controlling height are different (i.e. Tt). Compare dihybrid.

**monomer** A simpler molecular unit that can be ⋆covalently bonded with other units to form a ⋆polymer. e.g. Glucose is the monomer in starch. The approximately 20 naturally occurring ⋆amino acids are the monomers that make up proteins.

**monophyllous** Possessing or composed of a single leaf.

**monosaccharide** The simple ⋆carbohydrate sugars with general formula $C_nH_{2n}O_n$, the most common having six (⋆hexose) or five (⋆pentose) carbon atoms. e.g. Glucose, fructose, galactose, mannose are hexose sugars; ribose and deoxyribose are pentose sugars. See Fig. C2.

**Monotremata** An order of egg-laying mammals, including the ⋆echidna and the ⋆platypus, both of which are found only in Australia and nearby regions. See Appendix 1.

**montane** 1 Relating to mountains, and the fauna and flora that live there. 2 The region of a mountain extending below the timberline.

**moor** An area of poorly drained land, where acid-loving plants are dominant on the ⋆peaty soils.

**mopoke** An owl of the genus *Ninox*, especially the boobook found throughout Australia and New Zealand, and nearby areas. It has a call resembling the word mopoke.

**morbillivirus** A genus of ⋆virus with a single strand of RNA including those that cause measles, mumps and some types of influenza. See also equine morbillivirus.

**morphine** A crystalline compound, the most active part of the ⋆narcotic ⋆opium. It is used by doctors to treat extreme pain.

**morphogenesis** The development of the form and structure of an organism, or a part of an organism.

**morphology** The study of the form and structure of living things, especially the shape, arrangement, and interrelationships of the internal parts.

**mortality** The death of individuals of a population, especially death in large numbers. e.g. Mortality rate measures that rate at which members of a population die, usually expressed as 'per 100 000 per year' for humans.

**mosquito** Any member of the two-winged (order ⋆Diptera) insects in the family Culicidae, the females of which need to feed on blood for their eggs to mature correctly. The males feed on juices found in fruits. e.g. *Anopheles*, the malaria mosquito; *Aedes*, the yellow-fever mosquito. See Appendix 1.

**moss** 1 A small, leafy, nonvascular land plant that often grows in clumps. Belonging to Class Musci, Phylum (Division) Bryophyta they reproduce through ⋆alternation of generations.

The ★gametophyte is the most visible, with small leaf-like structures arranged around a short stem. ★The sporophyte grows from the top of the gametophyte, with a ★capsule containing spores at the end. When these spores are released they grow into a gametophyte plant. 2 A clump or mass of these plants, usually growing in crevices on rocks, wood or hard soil, as they are one of the first plant colonisers in an area. See Appendix 1.

**moth** Any of the large group of insects, order Lepidoptera, which can be told apart from ★butterflies by their antennae not being in a club shape (the males have a feathery-shaped antenna), and by their nocturnal habit. They feed using a ★proboscis. See Appendix 1.

**motile** Being able to move without apparent external stimulus (i.e. spontaneously).

**motivation** That which prompts an organism to behave in a certain manner.

**motor neurone** A ★neurone which carries ★impulses from the central nervous system (*efferent) to motor organs such as muscles and glands. The cell body lies within the ★grey matter of the spinal cord, and the ★axon extends out through the ventral ★root. Inside the spinal cord, the motor neurone may connect directly to a ★sensory neurone, or to an interneurone. See Fig. R2.

**mould** 1 Any growth of a surface-hugging ★fungus such as ★mildew. 2 The hollow part of a shape formed by the burial of a fossil.

**moulting** 1 Also called **ecdysis**. **a** Shedding of the ★exoskeleton of arthropods such as insects, crustaceans, spiders, etc., to allow growth in body size. See also ★moulting hormone. **b** Also applied to the shedding of skin by reptiles. e.g. Snakes shed their skin in basically one piece while lizards shed it in patches. 2 The periodic shedding of hair or feathers by mammals or birds.

**moulting hormone** Also called **ecdysone**. Produced by insects and crustaceans, it stimulates ★moulting and the growth that this then allows. See also juvenile hormone and metamorphosis.

**mount** 1 Male animal climbing onto the rear of a female animal for ★copulation. 2 To place a ★specimen into a microscope slide, and then prepare it for examination through a microscope.

**mountain ash** Any of several ★eucalypts, especially the tall and straight *Eucalyptus regnans* found in Victoria and Tasmania.

**mountain devil** *Lambertia formosa*, a shrub in family Proteaceae, with a hard fruit that has the appearance of a devil's head.

**Mousterian** A cultural period in the stone age (middle Palaeolithic) in which ★Neanderthal Man flourished with the characteristic stone tools. See also culture.

**mouth** 1 The opening through which food enters the body (★ingestion). 2 In vertebrates, the first part of the ★alimentary canal where food is ground and crushed (★masticated) and saliva added. Also called the **oral** or **buccal cavity**.

**mouth parts** Paired structures that surround the mouth of arthropods, and which are used for securing, manipulating, and chewing food. e.g. Insects (★labrum, ★mandibles, ★maxillae, ★labium). Arachnids (★chelicerae). Crustaceans (★mandibles, *maxillae, *maxillipeds). See Fig. M4.

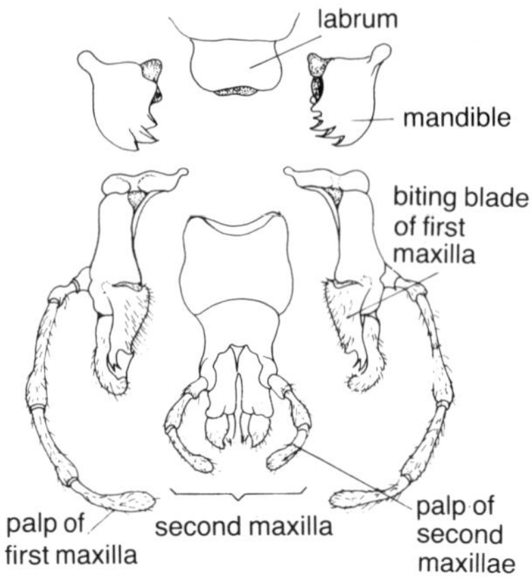


Front view: mouthparts of a cockroach

**Fig. M4 Mouth parts**

**MRI** Short for **magnetic resonance imaging**, a method for producing images of internal body tissues using magnetic fields and radiowaves, rather than dangerous X-rays. At one time called **nuclear magnetic resonance tomography** (NMR). Compare CT, PET and ultrasound.

**mRNA** Short for **messenger RNA**. The RNA (ribonucleic acid) molecule that travels from the DNA in the nucleus to the ribosomes in the cytoplasm carrying the information needed to synthesise proteins.

RNA differs from DNA in being made of only one strand, having ribose sugar in its 'backbone' rather than deoxyribose, and using the bases uracil (DNA uses thymine instead), cytosine, adenine and guanine. The mRNA is ⋆transcribed from the DNA in the nucleus by enzymes. As it travels out of the nucleus it is 'edited' by other enzymes: Non-coding parts called ⋆introns are removed, and the coding ⋆exons joined back together. The ends are also modified ('⋆capped' and '⋆tailed'). On reaching the ribosomes, the mRNA is ⋆translated into proteins. As the ribosome moves along the mRNA, it accurately locates (in an order specified by the ⋆codons) the ⋆tRNA with their attached amino acids, so that enzymes can join together the amino acids in their correct order.

**mucilage** A gelatinous substance produced by many plants which readily absorbs water and swells. See ⋆gums.

**mucosa** A ⋆mucous membrane. See below.

**mucous membrane** A surface membrane (epithelial) that is moistened with a slippery ⋆mucus. Thus the organs inside the body cavity are covered by mucous membranes to allow easy movement between them. The linings of the respiratory tract are covered with mucous membrane so that the mucus can trap dust and smoke particles before they reach the ⋆alveoli. The alimentary canal is also covered with mucous membrane, the mucus it produces protecting the cells (made of protein) from being eaten away by digestive juices.

**mucus** The slimy and viscous sticky substance produced by special goblet cells in ⋆mucous membranes. The main components of mucus are ⋆glycoproteins which contain large amounts of carbohydrate to make this lubricating and protective material slippery.

**mudlark** Also called **pee-wee**, **pee-wit**, and **magpie lark**. This common Australian black-and-white bird, *Grallina cyanoleuca*, builds a mud nest high on the branches of trees.

**mugga** One of the ironbarks of eastern Australia, *Eucalyptus sideroxylon*.

**mule** 1 The sterile offspring from the mating of a male donkey and a female horse. 2 Any infertile offspring from a genetic cross.

**mulga** 1 Common name for any of several species of *Acacia*, especially *A.*

*aneura* which grows in the drier parts of Australia. 2 A name also used for plant communities where *Acacia* is the dominant tree.

**mulgara** The carnivorous crested-tailed marsupial mouse, *Dasycercus cristicauda*, Family ⋆Dasyuridae, which inhabits the arid interior of Australia. See Appendix 1.

**multicellular** Consisting of many cells.

**multiple alleles** A gene with three or more alleles. As there are only two chromosomes of each type in a cell, only two of the alleles can be present at any one time. e.g. The main blood groups in humans are produced by two of the following three alleles being present in any one person: A, B or O.

**multiple sclerosis** Called MS for short. A disease of the nervous system where the ⋆neurones in the white matter gradually lose patches of their ⋆myelin sheaths. This results in a 'short-circuiting' of the nervous system, and sufferers gradually become paralysed, numb, blind, unable to speak, etc., depending on which areas of the nervous system are affected.

**mumps** A viral infection which causes inflammation of the salivary glands, and in some cases can result in inflammation of the testes or ovaries.

**muricate** Covered with small hard bumps.

**Musci** Also called **Bryopsida**. The Class of bryophytes also called ⋆mosses. See Appendix 1.

**muscle** A specialised organ or tissue which can contract, thus causing a change in shape (e.g. lens in the eye), or movement of parts (e.g. muscles which move the skeleton). Muscles are usually classified as ⋆striated (or voluntary e.g. arm muscles), ⋆smooth (or involuntary e.g. lining the arteries), and ⋆cardiac (e.g. heart) depending on the appearance of the ⋆muscle fibres when viewed under a microscope, and on their mechanical behaviour. Most of the animal flesh that we eat is striated muscle.

**muscle fibre** One of the many long thin 'cells' that lie along the length of a muscle. They are held together by connective tissue through which also pass blood vessels and nerves. See Fig. M5.

Formed from the fusion of many cells, a muscle fibre contains many nuclei (see acellular) and has diameter of 10 μm to 100 μm, and can be several centimetres long. Each muscle fibre is packed by many ⋆myofibrils which run the length of the cell. The striated pattern results from the bands present on these myofibrils, the bands being due to the arrangement of many long thin filaments of the proteins ⋆actin and ⋆myosin. Each segment between the bands is called a ⋆sarcomere. It is relative movement between these filaments that causes striated muscles to contract. There are two broad functional types of muscle fibres. Slow twitch fibres work slowly, but can contract repeatedly or for extended periods. Fast twitch fibres operate much more quickly, but soon tire.

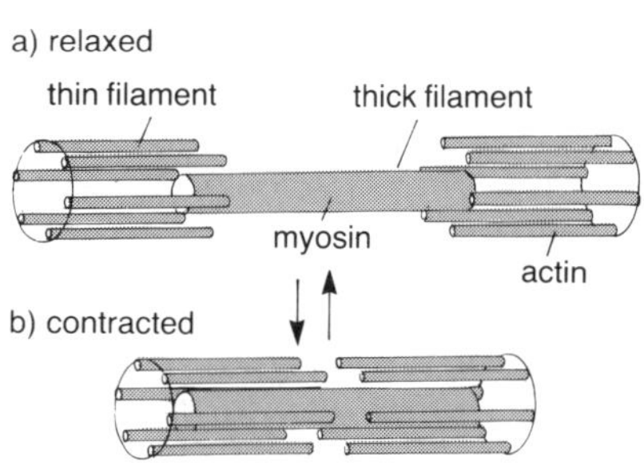


**Fig. M5 Muscle fibre**

**muscle spindle** A stretch ⋆receptor in vertebrate muscle. The muscle spindle measures both the static and movement components of the muscle's function, thus allowing fine control of its actions. See also mechanoreceptor.

**muscular dystrophy** A disease which causes the muscles to progressively deteriorate until the person is completely helpless.

**mushroom** The popular name for edible gill ⋆fungi of the genus *Agaricus*, especially the field mushroom *A. campestris*, and *A. bisporus*, the commercially grown mushroom. See also toadstool.

**mussel** Any ⋆bivalve of the class Pelecypoda, especially those marine and freshwater forms commonly eaten. See Appendix 1.

**mutagen** A radiation or chemical substance which can cause ⋆mutations.

**mutant** An organism or gene which has undergone ⋆mutation.

**mutation** A sudden change in the characteristics of an organism which may be transmitted to offspring if the mutation takes place in a sex cell.

Mutations arise from a change in the chemical structure of the DNA which makes up the genes on the chromosome (⋆gene or point mutation), or changes in the number or structure of the chromosomes (⋆chromosomal mutation). If these mutations occur in the gametes (sperm, ova), then they can be inherited by offspring. If mutation occurs in other cells, then it is not inherited (although it may be harmful e.g. cause a cancerous growth) and is called a ⋆somatic mutation. In nature, mutations are purely random events, there being no way of predicting which gene will mutate, and when. Only the statistical probability of a mutation occurring can be measured. This probability can be artificially increased by using chemical (e.g. mustard gas, some pesticides, LSD) or radiation (e.g. X-rays, radioactivity) ⋆mutagens. However, a vast majority of mutations are harmful to the organism. The action of ⋆natural selection on the very few beneficial mutations is believed to be the source of variation for ⋆evolution.

**mutton bird** Common name for the Short-tailed Shearwater, *Puffinus tenuirostris*, which migrates from islands in Bass Straight, between Tasmania and Victoria, to the northern Pacific off Alaska, and back again, each year. A number of other shearwaters are also called mutton birds, especially the wedge-tailed shearwater (*P. pacificus*) of coastal Queensland, northern New South Wales, and Western Australia.

**mutualism** A form of ⋆symbiosis where two organisms of different species live in close association with each other to the advantage of both. e.g. Cleaner shrimp and fish. The cleaner shrimp set up 'cleaning stations' for fish on the Great Barrier Reef. The shrimp are allowed to move over the body, into the gills, and even into the mouths of fish, including large predatory fish such as the Barracuda. The shrimps clean off parasites and dead tissues to the benefit of the fish, and receive a meal in return. See also ⋆mycorrhiza below. There are many other examples.

**myall** A species of *Acacia* (wattle) found in semi-arid areas of New South Wales and southern Queensland.

**mycelium** A tangled mass of filaments (⋆hyphae) which makes up the main body (vegetative body, ⋆thallus) of a ⋆fungus. e.g. The edible part of a mushroom is made of mycelium.

**mycology** The study of fungi.

**mycorrhiza** 1 The ⋆symbiotic association of fungi with the root system of green plants. Usually both parties benefit (⋆mutualism), as the plant will often not grow very well without its associated fungus. e.g. ⋆Radiata pine (*Pinus radiata*). All *Eucalyptus* appear to have their associated mycorrhizal fungi, which may help in absorbing water and nutrients from the soil, and assist in resisting disease. The fungus benefits, receiving food made by the plant. 2 The structure formed when a fungus associates with a root.

**myelin** A complex mixture of fats (⋆lipids) and protein that forms an insulating layer (⋆myelin sheath) around most vertebrate, and some invertebrate ⋆neurones. It is closely associated with the ⋆Schwann and ⋆glial cells. The length of neurone covered by myelin sheath is called the internode, while the gaps left between are called the ⋆nodes of Ranvier. Nerve impulses can travel more quickly along myelinated neurones than along unmyelinated neurones. See also ⋆multiple sclerosis, and Fig. N1.

**Müllerian duct** The ⋆oviduct of vertebrates. In most vertebrates the ducts are paired (single in birds), and in humans are called Fallopian tubes or oviducts. See Fig. R5.

**myocardiograph** An apparatus which records the movements of heart muscles (their contractions), producing a record called a myocardiogram.

**myocardium** The muscular walls of the heart.

**myofibril** The contractile elements of a ⋆striated muscle fibre, made up of thick (⋆myosin strand) and thin elements (⋆actin strand) arranged in units called ⋆sarcomeres. See also muscle fibre and Fig. M5.

**myoglobin** An oxygen-binding molecule, similar to ⋆haemoglobin, found in muscular tissue. It stores oxygen for use during strenuous muscle activity, a time when the blood supply may not provide sufficient oxygen until breathing rate and heart rate have time to respond.

**myopia** Also called ⋆**shortsightedness**. See also lens.

**myosin** A protein, that with ⋆actin makes up the ⋆myofibrils, the main contracting apparatus of the muscles. The molecules of actin and myosin slide past each other as the muscle contracts, and relaxes. See also muscle fibre.

**myriapod** A group of arthropods, with many body segments and numerous walking legs, the main members being the classes Chilopoda (⋆centipedes with one pair of legs per body segment) and the Diplopoda (⋆millipedes with two pairs of legs per body segment). See Appendix 1.

**Myrtaceae** A large family of dicot flowering plants with Australia as one of its main areas of distribution. It includes genera such as ⋆*Eucalyptus*, ⋆*Angophora*, ⋆*Leptospermum*, and ⋆*Melaleuca*. See also Appendix 1.

**myxomatosis** A highly infectious viral disease of rabbits, introduced into Australia to reduce the rabbit population. See biological control. Compare calicivirus.

**Myxomycota** The Phylum of simple fungi-like organisms called ⋆slime moulds. They possess both plant and animal characteristics. See also acellular and Appendix 1.

**myxovirus** An RNA-containing ⋆virus which produces diseases in humans such as ⋆influenza, ⋆mumps and ⋆measles.

# Nn

**nabarlek** Often called the Little Rock Wallaby, *Peradorcas concinna*, in the family Macropodidae. See Appendix 1.

**NAD** Short for **nicotinamide adenine dinucleotide** which is a ⋆co-enzyme that functions as a hydrogen carrier in oxidation-reduction reactions. e.g. The ⋆electron transport chain of aerobic ⋆respiration. NAD is reduced to NADH when hydrogen atoms are transferred to it from the chemical being oxidised. A specific enzyme (called a ⋆dehydrogenase) acts on the chemical being oxidised (the substrate) which has been obtained from food (the ⋆substrate). The reduced co-enzyme NADH can then provide energy elsewhere in the cell where it is oxidised back to NAD, producing ATP. Compare NADPH.

**NADH** See NAD above.

**NADP** Short for **nicotinamide adenine dinucleotide phosphate**. The ⋆co-enzyme NADP functions as a hydrogen carrier in ⋆oxidation-reduction reactions. NADP is reduced to NADPH. Although it is similar to NAD, it is mainly involved in synthesis reactions rather than breakdown.

**NADPH** See NADP above.

**naiads** A ⋆nymph of an aquatic insect that undergoes incomplete ⋆metamorphosis. e.g. Dragon fly.

**nano-** 1 Used with other words to mean very small. e.g. nanoplankton. 2 When used in measurement, one billionth or $10^{-9}$. See nanometre below.

**nanometre** Written **nm** for short. One billionth or $10^{-9}$ metres. One thousandth of a micrometre. Used as a unit of measurement in microscopy, and for some ⋆electromagnetic radiation.

**narcosis** A state of deep sleep, unconsciousness, and inability to feel pain, as produced by a drug. e.g. ⋆Narcotics are drugs which produce such effects.

**narcotic** A drug which, in moderate doses, relieves pain, acts as a sedative, and induces profound sleep. In excess, such drugs produce a coma or total loss of control of body functions. Narcotics such as opium, morphine, heroin, methadone and alcohol are all psychologically and physically addictive. See dependence.

**nardoo** Aquatic ferns of the genus *Marsilea*. The spores are produced in a hard casing and are rich in stored starch. Used by Australian Aborigines as food.

**nares** ⋆Nostrils of a vertebrate.

**nasal cavity** A cavity or chamber in the head of four-limbed (quadruped) vertebrates that contains the smell (⋆olfactory) receptors. It opens to the back of the mouth (⋆pharynx) in one direction, and to the outside through the nostrils (⋆nares) in the other direction. In humans, hairs in the nostrils and a lining of sticky ⋆mucous membranes help to trap dust particles. Extensive blood vessels around the cavity help warm the incoming air.

**nastic movement** In plants, a response to a stimulus that is independent of the direction of the stimulus. Nastic movements include slow growth movements (e.g. twisting of the stems of climbing plants in seeking objects to entwine), or much faster responses (e.g. closing and opening of flowers in response to changes in light intensity; closing of leaves of the touch-sensitive plant, *Mimosa pudica*).

**natal** Of or relating to birth.

**native** A plant or animal found naturally in a particular region. See indigenous.

**native cat** More correctly called *Quoll.

**natural killer cells** See NK cells.

**natural selection** A mechanism proposed by Charles Darwin (1809 –1882) in 1859 to account for the origin of species, and often summarised as 'the survival of the fittest'. If environmental conditions remain constant, then those who are sick, injured, or who carry a harmful *mutation are removed by predators, leaving the 'fitter' organisms to reproduce and pass on their characteristics. If, however, environmental conditions should change, then a few organisms may have a mutation that gives them some advantage in survival, such as in finding food, mates or shelter, or in escaping predators. These organisms are thus 'fitter', leaving more offspring, so that their new features gradually spread throughout the whole population. See industrial melanisms and resistance for examples.

**nature** 1 All the processes and activities which take place in the universe. 2 In a more limited sense, all the processes and activities which take place around us that are not due to humans.

**nauplius** A tiny oval and unsegmented free-swimming *larvae, the first stage in the development of certain crustaceans. e.g. Prawns, shrimps.

**nausea** A sick feeling in the stomach. A feeling that one is about to vomit.

**nautilus** Any *cephalopod mollusc of the genus *Nautilus*, with a spiral chambered shell, the animal living in the outermost (and biggest) chamber. See Appendix 1.

**navel** Also called the **umbilicus**. A depression in the middle of the belly being the remains of the *umbilical cord attaching us to our mother when we were in the womb.

**Neanderthal Man** Common name for a type of fossil ***Homo neanderthalensis*** whose remains were originally found in the Neander Valley in Germany. Since the original study, fossil finds have shown Neanderthal Man to have lived widely throughout Europe and nearby areas during the *Palaeolithic period, some 40 000 to 100 000 years before the present (bp). Roughly the same size as modern humans, but more stockily built with a slightly larger brain, on average. Skull has a more sloping forehead, more prominent eyebrow ridges, and less of a chin than present day mankind. Fossil remains associated with mainly the *Mousterian culture. Compare Cro-magnon Man.

**Nearctic** A *biogeographical realm which includes the Americas north of Mexico. Characteristic animals include the mountain goat, caribou, muskrat, raccoon, mountain lion and pronghorn antelope. Compare *Neotropical realm. (Fig. B5.)

**neck** 1 A relatively narrow region that connects the head to the main part of the body. 2 Any relatively narrow region connecting two parts. e.g. Neck of a tooth or bone.

**necrosis** Death of cells or of tissues, which are still surrounded by living tissues. Caused by damaging external factors such as poisons, excessive heat or lack of oxygen. Compare apoptosis.

**necrotizing fasciitus** Means 'killing of the connective tissue'. The effects of a ⋆*Streptococcus A.* infection where the ⋆toxins produced by the bacteria destroy the ⋆connective tissue. The bacteria normally only cause a sore throat or impetigo but have also earned the name 'flesh-eating bacteria' because of this disease.

**nectar** The sugary fluid secreted by special glands of ⋆flowers, or more rarely by leaves, to attract animals which bring about ⋆pollination. Pollinators include birds, bats, mammals (e.g. Honey possum), and butterflies, moths and other insects.

**nectary** The glands of a flower which secrete ⋆nectar.

**negative feedback** See feedback and Fig. F1.

**negro-head beech** Also called the ⋆**Antarctic Beech**.

**nekton** The animals that actively swim in seas and lakes. Compare ⋆benthos and ⋆plankton. See Fig. O1 and pelagic.

**nematocyst** Part of the stinging cell of cnidarians (or coelenterates and some other groups) consisting of a sac full of fluid in which lies a relatively long coiled thread with an arrow-like head. This arrow and thread can be shot out (discharged) on stimulation. Its function is to help gather food, and for protection. Some stings are very painful to humans (e.g. bluebottle), and others deadly (e.g. sea 'wasp' or box jellyfish). See also cnidoblast.

**Nematoda** The roundworms. This large phylum contains the long, cylindrical and unsegmented worms, pointed at both ends and often ⋆parasitic. e.g. Hookworms and ⋆filaria in humans, roundworms in pigs, eel worms in daffodil bulbs. See Appendix 1.

**Nemertea** Also spelt **Nemertina**. A small phylum of mostly marine worms called ribbon worms. Nemertea are flat and bilaterally symmetrical like Platyhelminthes, but have both mouth and anus, and a ⋆proboscis for defence and capturing food. See Appendix 1.

**Neo-Darwinism** The modification of Darwin's Theory of ⋆Evolution by natural selection to include the ideas of modern genetics. Modern ⋆genetics explains the origins of ⋆variation (e.g. ⋆mutation, ⋆independent assortment of ⋆alleles during meiosis). Studies of the genetics of whole ⋆populations has emphasised the importance of ⋆genetic drift and the ⋆isolation of populations. Compare punctuated equilibrium and Darwinism.

**Neocene** The Miocene and Pliocene epochs combined. See Appendix 2.

**Neolithic** Of or relating to the New Stone Age cultures which are believed to have started about 10 000 years before the present (bp). Plants and animals were domesticated. Compare ⋆Palaeolithic.

**neoplasm** Any abnormal new growth such as a tumour (⋆malignant or ⋆benign).

**neoteny** The retaining of characters and structures of the ⋆larval stage in an organism that should have undergone ⋆metamorphosis to become an adult. e.g. Amphibians such as the salamanders and axolotl undergo little change as they mature (compare frogs). One type of axolotl retains the gills of the larval stage.

**Neotropical** A ⋆biogeographic realm which includes the Americas south of Mexico, and the West Indies. Characteristic animals include guinea pigs, sloths, rheas and hummingbirds. Compare Nearctic realm. (Fig. B5.)

**nephridium** Plural is **nephridia**. In various invertebrates, an internal single or branched tubular structure which acts as an ⋆excretory organ, opening to the outside through holes. Found in flatworms, ribbon worms, rotifers, segmented worms (annelids), and the larvae of some molluscs.

**nephron** The unit of ⋆excretion found in the vertebrate ⋆kidney. It is made up from the ⋆Malpighian body: a tangled knot of capillaries called the ⋆glomerulus enclosed in a cup-shaped structure called ⋆Bowman's capsule. The urinary ⋆tubule (surrounded by capillaries) carries the urine away from the Malpighian body to where it is collected in the ⋆pelvis of the kidney. See Fig. K1.

**neritic** Inhabiting only the coastal waters to the edge of the continental shelf (often a depth of about 200 m), as compared with oceanic. See Fig. O1.

**nerve** A bundle of ⋆nerve fibres (⋆neurones) with accompanying connective tissue and blood vessels.

**nerve cell** A **neurone**. See below.

**nerve ending** A structure at the far end of a ⋆nerve fibre which is outside the central nervous system (i.e. part of the ⋆peripheral nervous system). In ⋆sensory nerves, impulses start there, while in ⋆motor nerves they finish at the nerve ending. The end may be made up of fine fibres (free nerve endings e.g. heat receptors), or may have special structures present (e.g. Meissner's corpuscle which detects touch in the skin). Motor neurones end at an end plate, an area of muscle cell membrane under the ends of the nerve fibres. See neuromuscular junction.

**nerve fibre** An ⋆axon of a ⋆neurone (nerve cell), together with its insulating ⋆myelin sheath (if present). The nerve fibre conducts the impulse, travelling faster in fibres that have larger diameters and have a myelin sheath.

**nerve impulse** The signal or 'message' sent along a neurone in the form of an electro-chemical change (NOT electricity). See impulse.

**nerve net** A relatively simple type of nervous system that consists of a decentralised net-like distribution of neurones that connect to various parts of the animal's body. Impulses pass in all directions, producing a general response. Found in cnidarians (coelenterates) and echinoderms.

**nervous system** A system which allows rapid coordination of the body (compare endocrine system), so that the various internal functions and external activities of an animal work together. Signals are sent along the long thin ⋆neurones of the ⋆peripheral nervous system, to-and-from the controlling ⋆central nervous system. All multicellular animals except sponges have a nervous system, ranging from simple ⋆nerve nets to the complex nervous systems of humans. The peripheral nervous system is divided into ⋆somatic nervous system (controlling voluntary movements), and the ⋆autonomic nervous system (controlling involuntary functions).

**nest** 1 A place or structure prepared by a bird for receiving, incubating and hatching eggs, and for the care of immature young. Some species also build nests for sleeping (roosting). 2 In general, any place where an animal deposits its eggs, rears or conceals its young. e.g. Crocodile and platypus nests.

**neural** Of or relating to a nervous system. e.g. **a** A neural arch (vertebral arch) forms a bony hollow in the vertebrae through which the spinal cord passes. **b** A neural plate is a flat area of ⋆ectoderm in a vertebrate ⋆embryo that folds during growth to form the neural groove, which then closes up to form the neural tube, from which the spinal cord and brain later develop.

**neuroglia** See glial cells.

**neurohormone** Any chemical substance released from special ★neurosecretory cells and that act as hormones, producing a specific effect at another site in the body.

Neurohormones are usually made in the cell body of a neurone, and then transported along the axon to be stored in ★vesicles in the nerve endings, until they are released in response to impulses that travel along the axon. The neurohormones then travel through the bloodstream to reach their site of action. Many neurohormones are produced in the ★hypothalamus, and act on the ★pituitary gland to control the production and release of its hormones (e.g. Gonadotropin releasing factor). Other neurohormones including ★oxytocin and ★vasopressin are produced in the ★hypothalamus, but transferred and released from the pituitary; nerves from the hypothalamus control the release of ★noradrenaline from the ★adrenal medulla. Chemically, most neurohormones are small ★peptide molecules. Compare neurotransmitter.

**neurohumour** See neurotransmitter.

**neurology** The scientific study of the ★nervous system, especially its structure, functions, and abnormalities.

**neuromuscular junction** The special ★synapse between the fibres on the ends of the ★motor neurone and a muscle cell membrane, allowing finely controlled muscle contractions to be made. The area of muscle cell membrane below the nerve endings is called the end plate.

**neuron** See neurone below.

**neurone** Also called **neuron** or **nerve cell**. A single nerve cell, the main structural cell of the nervous system. See Fig. N1.

Associated with it are support cells e.g. ★Glial cells, ★Schwann cells. There are three main parts, the relative size and shapes of which can vary greatly. The cell body contains the nucleus and is mainly concerned with the functions common to all cells. The ★dendrites are a group of fibres that carry nerve ★impulses to the cell body from the previous neurone or sense organ. The ★axon is the nerve fibre that carries the impulses away from the cell body, and to the next neurone, or to a motor organ. The signals are transmitted from one neurone to the next across the ★synapse using chemical ★neurotransmitters.

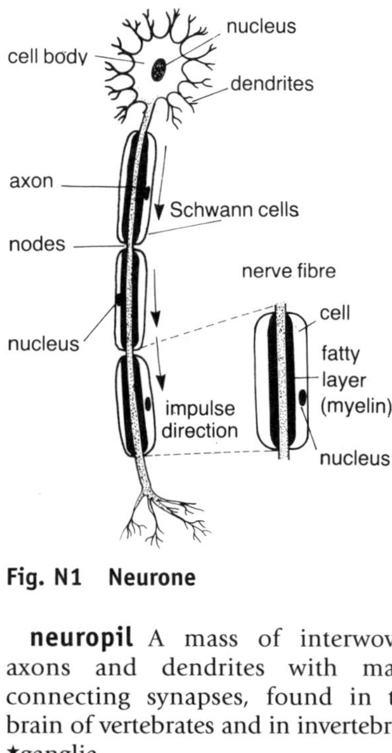


**Fig. N1 Neurone**

**neuropil** A mass of interwoven axons and dendrites with many connecting synapses, found in the brain of vertebrates and in invertebrate ★ganglia.

**neurosecretory cell** A nerve-like cell that releases one or more ★neurohormones into the blood stream.

**neurosis** Any emotional disorder where fear, anxiety, compulsive acts, and physical complaints are experienced without any obvious cause.

**neurotransmitter** A locally acting

chemical which transmits the nerve impulse across the ⋆synapse from one ⋆neurone to another, or from a neurone to a motor organ or gland. The neurotransmitter is made in the cell body and transferred to the ends of the axon where it is stored in ⋆vesicles until an impulse stimulates its release. Substances known to act as a neurotransmitter in vertebrates include ⋆acetylcholine, ⋆adrenaline and ⋆noradrenaline, ⋆dopamine, ⋆serotonin, and ⋆glycine. Compare neurohormone.

**neuter** Sexless. Lacking reproductive organs, or having imperfectly developed or functionless sex organs. e.g. The ⋆caste called worker bees is neuter.

**neutron** An elementary atomic particle found in the ⋆nucleus of atoms that has almost the same mass as a ⋆proton, but has no electrical charge.

**neutrophil** A white blood cell (⋆leucocyte) which appears in large numbers during infections, making up a large proportion of the pus that is produced. About 55 to 66% of white blood cells are neutrophils, which move like an amoeba, and readily ⋆engulf invading bacteria, toxins, etc. (i.e. they are ⋆phagocytic).

**newt** Any of a number of small tailed amphibians such as those in the genus *Triturus* that look like lizards but have a slimy skin. They are not native to Australia. See Appendix 1.

**niacin** A vitamin of the B group. See nicotinic acid.

**niche** Also called an **ecological niche**. **1** The role played by a particular organism in its total ⋆environment. This includes the way that it influences the living and non-living things around it, and the way in which they in turn influence the organism. See also ⋆Gause's principle. **2** The particular locality that an organism occupies within its habitat. e.g. Periwinkles that live in moist crevices on rock platforms.

**nicotinamide adenine dinucleotide** See **NAD** above.

**nicotinamide adenine dinucleotide phosphate** See **NADP** above.

**nicotine** A poisonous organic compound obtained mainly from the tobacco plant. It has been used to make insecticides, and is the ⋆addictive component of cigarette smoke.

**nicotinic acid** A vitamin of the B group that humans cannot make themselves, and thus must be included in their diet (e.g. milk, yeast, fresh meat, etc). Too little of this vitamin in the diet results in the disease pellagra. Nicotinic acid is needed to make the co-enzymes ⋆NAD and ⋆NADP.

**nictitating membrane** A transparent membrane which forms a third eyelid in some amphibians, reptiles, birds, and a few mammals. It is normally located beneath the lower eyelid, but can be drawn across the eyeball independently of the upper and lower eyelids.

**nictitation** Blinking.

**nidation** The ⋆implanting of a fertilised egg (zygote) into the walls of the womb (⋆uterus).

**nipple** **1** The conical structure on the female breast through which milk is discharged. **2** Any similar looking structure, especially on the male.

**nit** **1** The egg of a human body ⋆louse, usually left attached to fibres of hair or in clothing. **2** The young hatched from these eggs.

**nitrate** A highly soluble salt of nitric acid, such as sodium nitrate. The nitrate ion, $NO_3^-$, is an important soil nutrient that can be easily absorbed by plants.

**nitrification** The conversion of

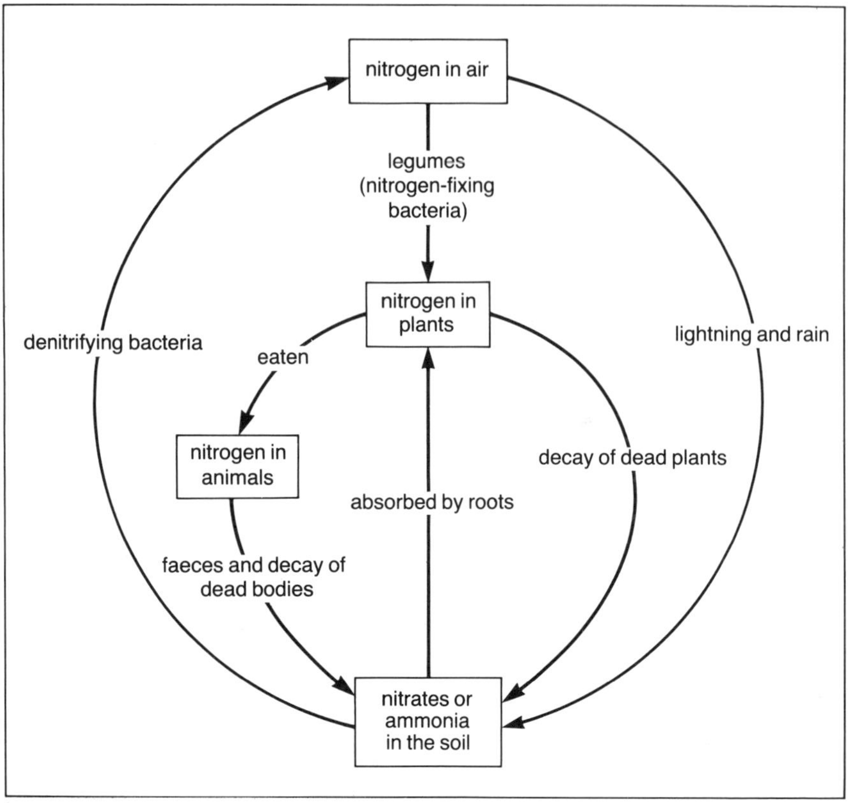


**Fig. N2 Nitrogen cycle**

complex nitrogen-containing organic compounds (from plants and animals) into nitrates which are then easily absorbed by plants through the roots. Nitrifying bacteria help convert ammonia ($NH_3$) produced from decaying organic matter into nitrites ($NO_2^-$) and then into ⋆nitrates ($NO_3^-$).

**nitrobacteria** Soil bacteria involved in ⋆nitrification. e.g. *Nitrosomonas* bacteria convert ammonia into nitrites, and *Nitrobacter* bacteria convert nitrites into nitrates.

**nitrogen** The colourless, odourless gaseous element that makes up 78% of the Earth's atmosphere. Nitrogen is an essential element in the body to make proteins. It is so chemically unreactive that animals and plants cannot use it directly from the air. It is converted into soluble form by ⋆nitrogen fixation, lightning, etc. See nitrogen cycle.

**nitrogen cycle** The cycle through which nitrogen atoms pass from the atmosphere into soluble nitrogen compounds in the soil which, in turn, are used by plants for growth. The nitrogen in the plants then passes to animals when the plants are eaten, and eventually the nitrogen is returned to the air. See Fig. N2. Lightning and ⋆nitrogen fixation are important processes by which nitrogen enters the soil. When animals and plants die, nitrifying bacteria (★nitrobacteria) may convert the proteins back into soluble nitrates. Other bacteria in the soil (called denitrifying bacteria), convert some of the nitrates etc. back into nitrogen. Compare carbon cycle.

**nitrogen fixation** The process by which atmospheric nitrogen is converted into soluble ★nitrates etc., and hence into complex compounds such as protein. This is achieved by some blue-green bacteria in the sea, but mainly by nitrogen-fixing bacteria on land. Some live in the soil (e.g. *Azotobacter*, *Clostridium*), while others (e.g. *Rhizobium*) live in nodules on the roots of ★legumes such as peas, beans, clover and wattles (a type of ★symbiosis). A small amount of nitrogen is fixed during lightning, while much larger amounts are synthetically fixed by humans in the manufacture of nitrate fertilisers.

**nitrogenous base** Organic ★bases containing nitrogen, especially the five bases that make up the 'letters' of the genetic code in DNA and RNA. e.g. ★Cytosine, ★thymine, ★guanine and ★adenine in DNA, with ★uracil replacing the thymine in RNA. See Fig. B3.

**NK cells** Short for **natural killer cells**. A type of ★lymphocyte that acts much like killer ★T-cells but is not specific. NK cells do not have to recognise its targets before attacking as do killer T-cells.

**nm** Short for ★**nanometre**.

**NMR** Short for **nuclear magnetic resonance tomography**, now called **magnetic resonance imaging (MRI)**.

**nocturnal** Animals that move around and seek their food mainly at night. e.g. Owls, bandicoots. Compare diurnal and crepuscular.

**node** 1 In plants, a joint or a region on a stem where leaves or branches are attached. Compare internode. 2 In animals, a rounded structure (e.g. ★lymph nodes) or a region that is thinner (e.g. ★node of Ranvier).

**node of Ranvier** One of a number of narrow areas along a myelinated nerve fibre, where there is no insulating myelin sheath, thus exposing the ★axon. Movement of sodium ($Na^+$) and potassium ($K^+$) ions during the passage of a nerve ★impulse takes place through special pores in the axon's membrane at these points. See Fig. N1.

**nodule** A small rounded mass, such as the root nodules of ★legumes, inside which ★nitrogen fixation occurs.

**nomadic** An organism without a fixed territory, but which wanders about seeking food supplies etc. e.g. Many birds such as honeyeaters move between different regions as flowering occurs. Some early humans and the Australian Aborigines were also nomadic.

**nomenclature** The system of distinguishing names used to classify living things. e.g. ★Binomial system.

**non-disjunction** The failure of a normal (★homologous) pair of chromosomes to separate during the first division of ★meiosis. Thus some daughter nuclei have one more than the normal haploid number of chromosomes for that species, while others have one less. In women, non-disjunction seems to increase with age and results in extra chromosomal material in the egg. Thus non-disjunction is one of the main causes of ★Down's syndrome, where three copies of chromosome 21 result.

**nonpolar** A molecule that has an even distribution of electrical charge over its surface so that no parts are positive or negative. e.g. Oxygen, petroleum. Compare polar.

**nonspecific immunity** See immunity.

**Noogoora burr** A tall herbaceous plant *Xanthium occidentale* that carries large ⋆burrs, and is poisonous to young livestock if eaten.

**noradrenaline** Also called **norepinephrine**. 1 A ⋆neurotransmitter in the ⋆sympathetic nervous system. 2 A ⋆neurohormone released by the ⋆adrenal medulla into the bloodstream. In both cases, it causes narrowing of the arteries and veins, and more rapid beating of the heart etc. It differs from ⋆adrenaline in having one less ⋆methyl group. See Fig. A5.

**norepinephrine** Also called **noradrenaline**. See above.

**Norfolk Island Pine** See pine.

**normal curve** A bell-shaped curve which illustrates the normal distribution of a quantity. In such a distribution, the mean, mode, and median are all equal. See Fig. N3.

**Fig. N3 Normal curve**

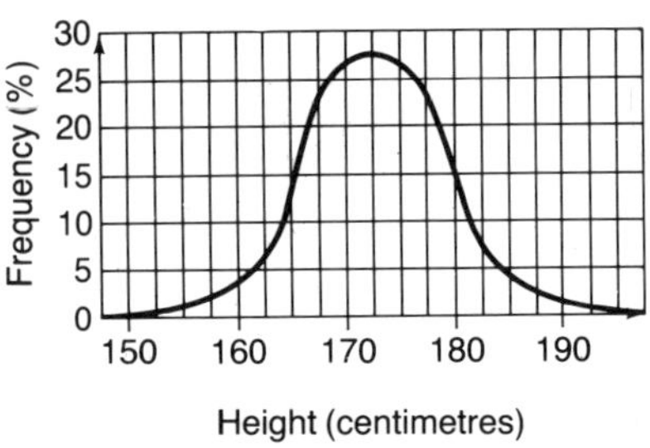


**normal flora** The bacteria and fungi that live on the surfaces of the body but do not cause disease. These normal inhabitants help protect the body against ⋆pathogens by competing for food and living space. They sometimes produce chemicals that are toxic to pathogens.

**nose** The usually protruding part of the face that contains the nostrils (⋆nares), allowing air to enter and leave the nasal cavity.

**nostril** One of the external openings of the nose. In humans, hairs are found inside to help filter dust etc. from the incoming air.

**notochord** A longitudinal, flexible rod located between the main nerve cord and the alimentary canal, at some stage in the development of all ⋆chordates. In vertebrates the notochord only exists in the embryo, and is replaced by the ⋆vertebral column. In chordates such as ⋆lancelets, and larval stages of the ⋆tunicates, the notochord forms the main skeleton.

**Notoryctidae** The family of marsupial mammals which contains the ⋆marsupial mole. This almost blind marsupial burrows underground in central Australian sand-dune country. See Appendix 1.

**nourishment** The food needed for an organism to survive.

**nucellus** The tissue surrounding the embryo sac, and within the protective layers called the ⋆integument, in the ⋆ovule of seedplants.

**nuclear envelope** Also called **nuclear membrane**. See below.

**nuclear magnetic resonance tomography** See MRI.

**nuclear membrane** Also called the **nuclear envelope**. A double membrane (with space between them) that surrounds the nucleus. Pores in this membrane connect the interior of the nucleus to the ⋆endoplasmic reticulum in the cytoplasm outside. See Figs C4 and C5.

**nuclease** One of a group of ⋆enzymes that breaks down (by ⋆hydrolysis) ⋆nucleic acids to ⋆nucleotides and ⋆nucleosides.

**nuclei** Singular is ⋆**nucleus**.

**nucleic acid** A complex long-chain organic compound found in the cytoplasm of cells. Nucleic acids are polymers of ⋆nucleotides, and especially refer to ⋆DNA and ⋆RNA.

**nucleo-protein** A complex of protein and ⋆nucleic acid. In the nucleus, the DNA forms a complex with proteins called ⋆histones and protamines. In the cytoplasm, ribosomal RNA forms a complex with proteins to form the ribosomes.

**nucleolus** A small dark body found in the resting (interphase) nucleus of cells. It contains DNA, RNA and protein, and is the site for the production of ⋆ribosomal RNA. The ⋆ribosomes, however, are assembled in the cytoplasm.

**nucleoside** A ⋆nucleotide without the phosphates.

**nucleosome** Bead-shaped structures around which DNA is wound when not being ⋆transcribed. Made of proteins called ⋆histones, nucleosomes allow the very long DNA molecule (about 2 metres in humans) to fit into a microscopic cell (about 20 µm long).

**nucleotide** A molecule formed from a purine or pyrimidine base (e.g. adenine, guanine and cytosine, thymine, uracil respectively), linked to a 5-carbon (pentose) sugar (e.g. ribose or deoxyribose), which is in turn linked to phosphoric acid. An example of a mononucleotide is ATP, while two nucleotides joined together (dinucleotides) include NAD and NADP. Polynucleotides, or a polymer of nucleotide monomers, include ⋆RNA and ⋆DNA. Compare nucleoside and see Fig. N4.

**nucleus** Plural is **nuclei**. **1** That part of a cell where the genetic material is enclosed in a ⋆nuclear membrane. Thus a nucleus can be called the 'control center' of a cell. Cells with a nucleus are called ⋆eukaryotic, while those without (e.g. bacteria) are called ⋆prokaryotic. The resting phase (interphase) of a cell nucleus contains a tangle of chromatin strands, which

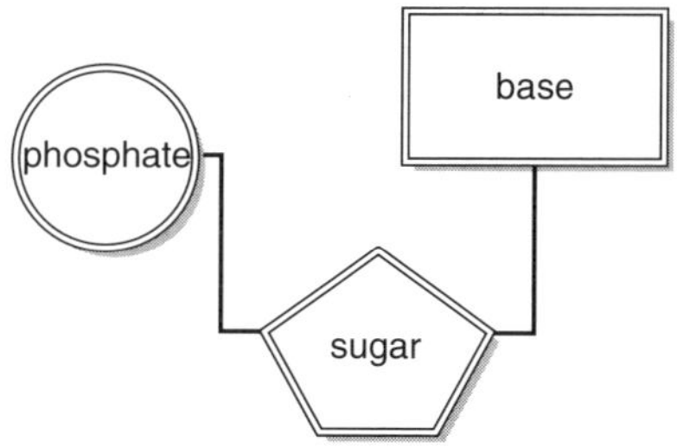


**Fig. N4 Nucleotide**

become organised into chromosomes before mitosis or meiosis. Also present in the nucleus are the ⋆nucleolus, and various enzymes needed for replication of DNA, and transcription of RNA. 2 A distinct part of grey matter deep inside the vertebrate ⋆brain. See Fig. B7. 3 The center of an atom, composed of ⋆protons and ⋆neutrons, and around which the ⋆electrons orbit.

**numbat** Also called **banded anteater**. A small marsupial, *Myrmecobius fasciatus*, which feeds on ants and termites. Its distribution is restricted to south-western Western Australia and it is the animal emblem of Western Australia. See Appendix 1.

**nut** 1 A dry fruit that remains closed when it matures. It forms from more than one carpel, but only one seed develops. Nuts usually have a dry outer wall. e.g. Hazelnut, chestnut, acorn. 2 Often more loosely applied to any woody nut or fruit. e.g. Walnut (which is a drupe), Brazil nut (which is a seed).

**nutation** A growth movement in plants, especially young climbers, whereby the growing tip spirals around, thus helping it to wrap around its supports.

**nutrient** Any food substance which helps supply living things with energy, and materials for growth and maintenance. See macronutrients and micronutrients.

**nutrition** 1 In animals and fungi, the process of taking in and using nutrients (e.g. ingestion, digestion, assimilation). 2 In plants, the production of organic compounds from inorganic materials.

**nyctinasty** The ★nastic movements of a plant in response to daily changes in the intensity of light and/or temperature. e.g. Leaves that fold, and flowers that close overnight.

**nymph** The immature form of an insect that only undergoes partial change at ★metamorphosis. The nymph resembles the adult, but has no wings or reproductive organs. e.g. Green vegetable bug. See also instar. Compare larva.

# Oo

**obcordate** A term used to describe leaf shape. See Fig. L1.

**obese** Excessively fat. Usually when a person is more than 20% above normal weight for their height, frame and age.

**objective** The microscope ⋆lens closest to the ⋆specimen. See Fig. L2.

**oblanceolate** A term used to describe leaf shape. See Fig. L1.

**obligate** Essential, limited. Describes an organism that is limited to living in one type of environment. e.g. Parasites can live only on their host and not as independent organisms. Compare faculative.

**oblique** Slanting. **1** In plants, the veins of a leaf at an angle to the midrib. **2** A muscle that does not run along (longitudinally) or across (transversely) the body.

**obovate** A term used to describe leaf shape. See Fig. L1.

**occipital** Relating to the back of the head. e.g. Occipital bone is at the rear base of the skull, and has a large hole (⋆foramen magnum) where it sits on the top of the uppermost vertebrae, and through which the spinal cord connects with the brain.

**occlusion** Closing or shutting off. e.g. A coronary occlusion is the blockage of the coronary artery by a blood clot.

**oceanic** Of or relating to the open oceans. Compare neritic. See Fig. O1.

**ocellus** Plural **ocelli**. A simple eye-spot that is found in some invertebrates, and that can detect the intensity of light, but cannot form an

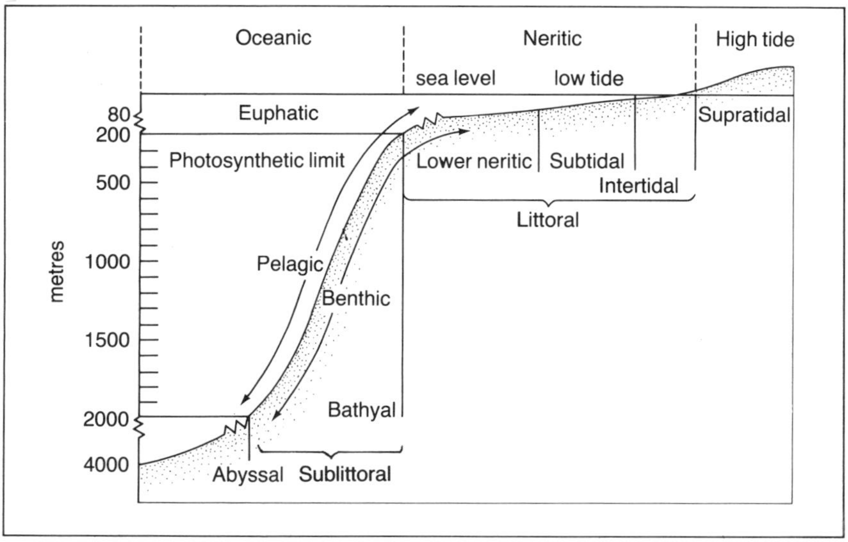


**Fig. O1 Ocean**

image. e.g. Many insects have two ★compound eyes and three ocelli.

**octopus** A cephalopod mollusc in the order Octopoda with a soft oval body and eight sucker carrying arms arranged around the hard bony 'beak' in the mouth. See Appendix 1.

**Odonata** The order of insects that includes the dragonflies and damselflies. They are long and slender with two pairs of similar and equal-length wings. Odonata have chewing mouthparts, and their juvenile forms are aquatic. See Appendix 1.

**oedema** Once called **dropsy**. An excessive accumulation of fluid in a body cavity and in the spaces around cells. e.g. The swelling around a ★bruise or a splinter (a result of the inflammatory response).

**oesophagus** The muscular food-pipe connecting the back of the mouth (pharynx) to the stomach. The chewed food, moistened when mixed with saliva, is formed into a ball called a ★bolus, and forced down the food pipe by ★peristalsis. See Fig. D2.

**oestradiol** One of the group of hormones called ★oestrogens. See below.

**oestrogen** A group of hormones (e.g. oestradiol, above) produced mainly in the female placenta and ovaries, but also in the ★adrenal cortex and the male ★testes. See Fig. O2.

They are found in all vertebrates, and even some species of plants. The best known actions of oestrogens are as female sex hormones, where they stimulate the growth and continued functioning of the sex organs, especially the womb (uterus) and birth canal (vagina). They also help to produce the ★female secondary sexual characteristics (e.g. breast development), and influence behaviour through their action on the hypothalamus. Release of oestrogen from the mature ★Graafian follicles is controlled by hormones released by the ★pituitary (e.g. ★FSH and ★LH).

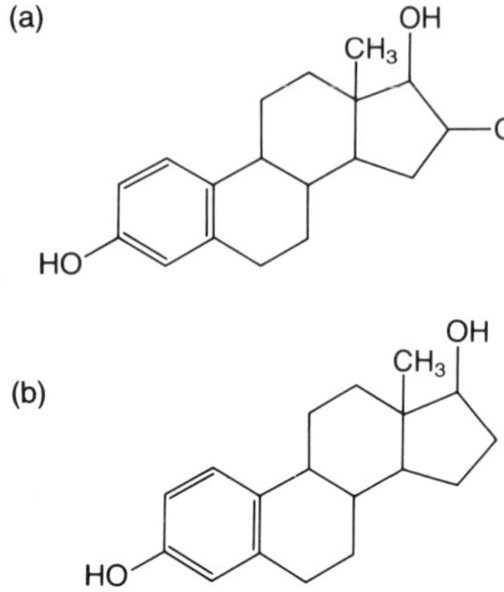


**Fig. O2 Oestrogen (a) Oestriol (b) β-oestradiol**

**oestrous cycle** The reproductive cycle that occurs in most female mammals, either more or less continuously one after the other (e.g. rats each few days, cows once a month), or once a year during the breeding season (e.g. wolves, deer), as long as pregnancy does not occur. The period when the female is receptive to mating (★oestrus) may last a few hours to three of four weeks, depending on the species. The whole process is under the control of hormones secreted from the ★pituitary, which is in turn under the control of the ★hypothalamus. The reproductive organs undergo corresponding changes (somewhat similar to the ★menstrual cycle, but without menstruation). Humans and apes have no oestrus, and mating can occur at any time.

**oestrus** An animal in oestrus is also said to be 'on heat'. A short period in the oestrus cycle of most mammals when the female is physically and behaviourally receptive to mating (★copulation). The ovum is released at this time, sometimes spontaneously (e.g. rats), but in other cases by the stimulation of mating (e.g. rabbits). Primates, including humans and apes, often do not have oestrus as they are able to mate at any time (see menstrual cycle).

**oil-immersion** A technique used with ⋆light microscopes to increase the ⋆resolving power of the lens, especially when high magnifications are being used. A drop of clear oil is placed on the ⋆coverslip of a microscope slide, and a special oil-immersion lens is lowered so that the drop is caught between the coverslip and the lens. Since the oil and the lens have the same refractive index, there are less interfaces where the light is bent, and the amount of light reaching the lens is increased. See Fig. L4.

**oils** Vegetable oils obtained from oil-seeds such as castor bean, sunflower seeds, etc. These simple ⋆lipids are compounds of fatty acids (e.g. oleic, linoleic) and glycerol, and function mainly as energy storage compounds for the ⋆seeds.

**Oldowan** A cultural period in the stone age (early ⋆Palaeolithic) in which ⋆*Australopithecus* flourished, with some very basic tools being attributed to this hominid. See also culture.

**olfactory** Of or relating to the sense of smell. e.g. Olfactory cells line the top of the nasal cavity, and connect to the olfactory lobe of the brain via the olfactory nerves.

**Oligocene** An epoch of the Tertiary period following the Eocene epoch. See Appendix 2.

**Oligochaeta** The class of annelids that includes the earthworms and freshwater bloodworms. These segmented worms have only a few bristles (⋆chaeta), are ⋆hermaphrodites, and lack the head 'appendages' or body 'projections' found on the ⋆Polychaetae (e.g. beachworms). See Appendix 1.

**oligotrophic** Low in nutrients. Bodies of water (e.g. lakes) that contain little organic matter and are thus nutrient poor.

**ommatidium** Plural is **ommatidia**. The single visual unit found in the ⋆compound eyes of arthropods. A lens on the outer end focuses light through a transparent material onto light-sensitive cells at the other end. Each compound eye may have from a few ommatidia (e.g. ants) to 10 000 or more (e.g. dragonflies). See Fig. O3.

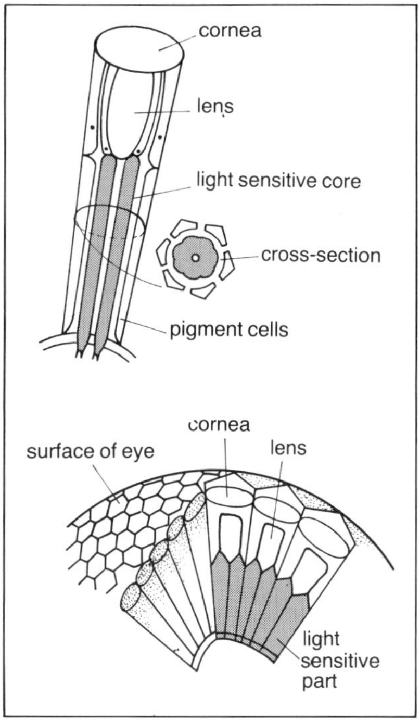


**Fig. 03 Ommatidium**

**ommatophore** A moveable structure with an eye on the end. e.g. Eye stalk of snails and crabs.

**omnivorous** Eating both plants and animals. Compare herbivorous and carnivorous.

**oncogene** Cancer-causing ★genes either found naturally in cells, or introduced to cells by certain viruses.

**oncology** The study of tumours (cancer).

**'one gene, one enzyme' hypothesis** The idea that each ★gene is concerned with directing the synthesis of one ★enzyme (or better, one protein or polypeptide).

**ontogeny** The development of an individual organism during its lifetime.

**Onychophora** A small phylum of invertebrates also called ★Peripatus. See Appendix 1.

**oocyte** The cell (primary oocyte) in an ovary that undergoes ★meiosis to produce an ★ovum. During the first meiotic division, a new large cell (secondary oocyte) is formed plus a ★polar body. The large cell may then divide to form the ovum and a second polar body. In some species, the primary or secondary oocyte may be released at ★ovulation without the formation of an ovum.

**oogamy** Sexual reproduction involving unlike ★gametes. Usually one is a large female gamete (the egg) that is unable to move by itself, and the often smaller male gamete that can move itself. e.g. Human egg and sperm.

**oogonium** 1 The female sexual organ of many algae and fungi, containing one or more female gametes (eggs). 2 A cell in the ovary of an animal that undergoes mitosis to produce the ★oocyte.

**ooze** A soft mud and slime, especially the mud rich in calcium carbonate (mainly from small organisms called ★globigerina), that covers part of the ocean bottom.

**open forest** Also called **dry sclerophyll forest**. A forest dominated by eucalypts, with shorter and more widely spaced trees than in ★tall open forest, and an understorey of shrubs with small hard leaves (★xeromorphic). See also forest and sclerophyll.

**operant conditioning** Also called **instrumental conditioning**. One of the many possible behaviours available to an organism is ★reinforced, thus increasing the probability of that behaviour occurring again. e.g. A pigeon in a cage pecks indiscriminately at everything in the cage. However, only if a particular coloured spot is pecked will food be produced. The pigeon soon learns the connection (i.e. the behaviour is reinforced), and soon pecks at the coloured spot whenever it needs food. Compare classical conditioning.

**operation** The procedures used by a doctor (surgeon) to correct disease or injury to the body.

**operator gene** A gene which is part of an ★operon, and acts as an on/off switch for the structural gene that is next to it along the DNA molecule. The operator gene is controlled by ★repressor molecules. See operon below.

**operculum** A lid or covering. For example **1** The cap on the flower bud of *Eucalyptus*. See Fig. G5. **2** The cap or lid on a moss capsule. **3** The cap on certain pollen grains pushed aside by the emerging pollen tube. **4** The cover for the gill opening in fishes. **5** The plate which covers the opening to a gastropod shell (e.g. snail, periwinkle) when the animal retreats inside. **6** The lid on the eggs of certain insects.

**operon** A group of related genes next to each other along the DNA in the nucleus, which produce one or more enzymes. See Fig. O4.

The operon consists of a ★promoter, then ★operator, and finally one or more ★structural genes. The whole sequence is controlled by a ★regulator gene, producing a regulator molecule.

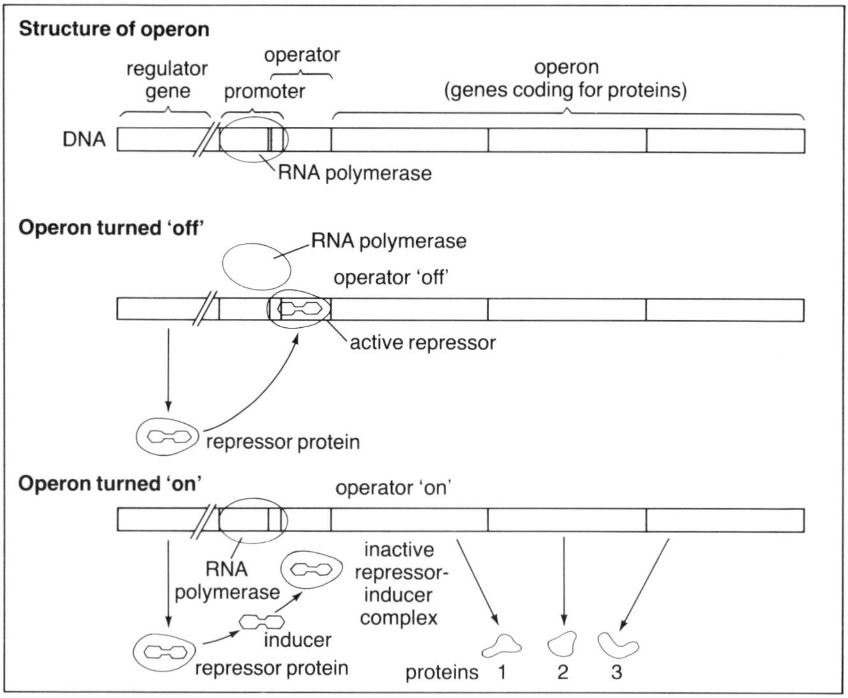


**Fig. 04 Operon**

The regulator gene may be located elsewhere amongst the chromosomes. For the structural genes to be *transcribed the enzyme catalyst called mRNA *polymerase must attach itself to the promoter, and then proceed along the DNA strand. If a repressor molecule is attached to the operator, between the promoter and structural gene, then transcription cannot proceed. The repressor molecule becomes detached when another molecule, called an *inducer, combines with the repressor molecule, thus changing its shape. Without the repressor molecule present, the mRNA polymerase can proceed along the DNA, transcribing it into mRNA. Compare cistron.

**Ophiuroidea** A class of echinoderms which includes the *brittle-stars and basket-stars. Their long flexible arms easily break off, and are attached to a disc-like body. See Appendix 1.

**ophthalmic** Of or relating to the eye.

**opiate** 1 Narcotics derived from or similar to opium. e.g. Morphine, heroin, codeine. 2 Opiate-like peptides which occur naturally in the body, helping to control pain etc. e.g. *Enkephalin and β-*endorphin. 3 In a broad sense the word applies to any sense-dulling, stupor-inducing drug.

**opium** A narcotic drug prepared from the dried juice of the unripe buds of the opium poppy, and used to make the other *opiates. See also dependence.

**opossum** Possum-like *marsupials found in North and South America. See Appendix 1.

**opposable thumb** The ability of the thumb of most primates to be brought into contact with any of the fingers of the hand. This opposability in humans gives a 'precision grip'. Compare the human hand (opposable) to the foot (big toe is not opposable).

**opposite** A term used to describe how the leaves are arranged in pairs on opposite sides of the stem of a plant. Compare alternate. See Fig. L1.

**opsin** The protein part of the visual pigments of the eye. See rhodopsin.

**optic chiasma** The point where the ★optic nerves partially cross-over, with the ★neurones from the left side of each eye going to the right half of the brain, while the neurones from the right side of each eye go to the left half of the brain.

**optic nerve** The nerve from the ★retina of the eye, which leaves the eyeball at the ★blind-spot, and connects to the visual area at the rear of the ★brain. See Fig. B7.

**oral** Of or relating to the mouth or region of the mouth.

**orang-utan** Also spelt **orang-outang**. A large tree-living, long-armed ape living in Borneo and Sumatra, *Pongo pygmaeus*. See Appendix 1.

**orbicular** A term used to describe leaf shape. See Fig. L1.

**orbit** The eyesocket. The cavity in the vertebrate skull that contains the eye.

**Orchidaceae** Also called **orchids**. A family of monocot flowering plants, with both terrestrial and ★epiphytic species. Renowned for their beautiful flowers. The floral emblem of Queensland is the Cooktown Orchid, *Dendrobium bigibbum*. See Appendix 1.

**Order** The taxonomic group above the level of Family, and thus made up of many similar Families. Many similar Orders make up a Class. See classification and Appendix 1.

**Ordovician** The second period of the Palaeozoic era, following the Cambrian, and before the Silurian. See Appendix 2.

**organ** A part of the body of an organism which forms a structural and functional unit, and is made up of one or more ★tissues. e.g. In plants, a leaf; in animals, a kidney.

**organ of Corti** A structure within the ear's spiral ★cochlea which contains the 'hair' cells that detect the vibrations carried from the eardrum to the cochlea by the ★ossicles. See Fig. E1.

**organelle** A 'little organ'. A structure within a cell that performs a definite function. e.g. ★Golgi bodies, ★mitochondria, ★plastid.

**organic** 1 Related to, or originating in, or derived from living or once-living organisms. 2 Chemical compounds containing carbon, with the exception of certain simpler compounds such as carbon dioxide $CO_2$, carbon monoxide CO, and carbonates such as calcium carbonate $CaCO_3$.

**organic base** In biology, the five chemical 'letters' of the genetic code: ★adenine (A); ★cytosine (C); ★guanine (G); ★thymine (T); and ★uracil (U). See purine and pyrimidine.

**organism** A living plant, animal or protist that is capable of carrying out metabolic activities, reproducing, etc. See also living thing.

**organiser** 1 The part of the DNA molecule which is transcribed to produce ribosomal RNA. See nucleolus. 2 A region of cells in an animal ★embryo which, by the release of certain chemicals, can cause cells to ★differentiate in another part of the embryo.

**orgasm** The complex series of pleasurable physiological responses in the genital organs and skin that is the climax of sexual intercourse.

**Oriental realm** A ⋆biogeographical region extending from Asia south of the Himalayas, through Indonesia as far as ⋆Wallace's Line. Wallace's Line passes between Borneo and the Celebes, south past Bali. Characteristic animals include the water buffalo, orang-utan, tiger and mongoose. See Fig. B5.

**orifice** A mouth-like opening, as on the end of a tube.

**origin of life** An explanation for the origin of living things from non-living matter.

The Earth is believed to be about 4.5 billion years old, while life as we know it is thought to have evolved 2 to 3 billion years ago. Fossils, however, did not become abundant until about 600 million years ago. Although there is much disagreement amongst biologists as to how living cells developed from non-living matter (often called chemical evolution), the following steps appear essential, even if not in this order. **a** The formation of the simpler molecules presently found in living things. e.g. Sugars, amino acids. **b** The combination of these simpler molecules to form more complex ones such as polysaccharides, proteins, DNA, RNA. **c** The formation of simple cells which could obtain energy from some type of respiration, and that could reproduce. **d** The evolution of cells which could make their own food by photosynthesis (autotrophs), and the evolution of cells that obtain food by feeding on other cells (heterotrophs). **e** The evolution of multicellular organisms.

Although numerous theories have been proposed for all these steps, only those relating to the first stage are based on much experimental evidence. Assuming that the Earth's atmosphere had much the same composition as that found on Jupiter and Saturn today, a mixture of methane ($CH_4$), ammonia ($NH_3$), hydrogen ($H_2$), and water ($H_2O$) was subjected to a week's bombardment with electric sparks. Other similar experiments used ultraviolet light. In each case a complex mixture of compounds was formed, including urea, lactic acid, and four different amino acids. Thus scientists have proposed a similar environment for the origin of the simpler organic molecules on Earth.

**ornithology** The study of birds and bird life.

**Orthoptera** An order of insects which includes the grasshoppers, locusts, katydids, crickets, cockroaches, mantids and stick insects. They have relatively large hind legs adapted for jumping, leathery front wings, membrane-like rear wings, and chewing mouthparts. See Appendix 1.

**osmoregulation** The means by which animals maintain the correct concentrations of water, salts and small molecules in their bodies.

Freshwater animals have body fluids which are more concentrated than the freshwater around them. Thus they have to pump out the water that diffuses into their bodies. Protozoans use a ⋆contractile vacuole. Other animals use their excretory systems, including ⋆kidneys, ⋆Malpighian tubules, and ⋆nephridia. Marine animals must conserve water, and remove excess salts. As their body fluids are less concentrated than the ocean around, water tends to diffuse outwards. Many reduce water loss by having an impermeable covering. e.g. Mucus layer that covers most fish. They also have efficient kidneys, able to extract water from what they drink, leaving a concentrated urine in which to excrete the salts. Terrestrial animals which live in dry regions avoid excessive loss of water by having impermeable skins (e.g. dry scales), few sweat glands, and efficient kidneys that reabsorb most of the water, excreting often nearly solid urine. Such animals often hide in burrows etc. by day, and feed in the relative cool of the night.

**osmosis** 1 In living things, the net movement of a solvent, usually water, across a ⋆semi-permeable membrane (a membrane that allows relatively easy passage to water (or other solvent), but stops or retards the movement of larger chemicals). The net movement of water is from the side with a lower concentration of chemicals (and thus higher concentration of water) to the side with a higher concentration of chemicals (and thus lower concentration of water). 2 More generally, the movement of a chemical across a semi-permeable membrane from the side of higher concentration to the side with lower concentration for that chemical.

**osmotic pressure** The pressure due to ⋆osmosis, usually measured as the force per unit area that has to be applied to a solution to prevent osmosis taking place.

**ossicle** A small bone or bone-like structure. e.g. The three tiny bones within the middle ear (⋆malleus or hammer, ⋆incus or anvil, and ⋆stapes or stirrup). See also ear and Fig. E1.

**ossification** The formation of bone. See also osteoblast.

**Osteichthyes** A class of fish with a bony skeleton (at least in part). Flaps help cover the gills, and they have two-chambered hearts. Compare Chondrichthyes. See Appendix 1.

**osteoblast** A cell found in growing ⋆bones that deposits layers of bone around itself. See also osteocyte below.

**osteocyte** A mature ⋆bone cell which has formed from an ⋆osteoblast after the bone has stopped growing.

**ostium** 1 One of the tiny holes in the body wall of sponges (⋆Porifera), through which currents of water enter, bringing food and oxygen. The current of water is produced by waving ⋆flagella inside the sponge. 2 One of the small openings in the heart of an arthropod, through which the blood enters from the body cavity.

**outbreeding** The mating of individuals of the same species which are not closely related (don't have the same ancestors during the last few generations). This process introduces new genes, and hence increases the genetic variation in the next generation. Compare inbreeding.

**outer ear** The parts of the ⋆ear outside the eardrum (⋆tympanum). Reptiles and amphibians do not have an outer ear, the eardrum being at the skin's surface. Birds and mammals have a short tube connecting to the surface, and mammals also have external 'ears' (⋆pinna) to help direct sound into the ear (e.g. bats). See Fig. E1.

**oval window** The plate covering the opening to the ⋆cochlea, and connected to the ⋆ossicles of the ⋆middle ear. Vibrations transmitted by the ossicles reach the liquids of the inner ear through the oval window. See Fig. E1.

**ovarian follicle** The layer of cells that forms a sac around an ⋆oocyte in an ovary. In mammals, the ovarian follicle contains a cavity, and is called a ⋆Graafian follicle.

**ovary** 1 In animals, the usually paired organs which produce eggs (ova). In vertebrates, they usually also produce sex hormones. See oestrogen, progesterone, Graafian follicle, oocyte and Fig. R4. 2 In flowers, the swollen base of a ⋆carpel which contains one or more ⋆ovules. Sometimes several carpels are joined together, forming a compound ovary. See Fig. F2.

**ovate** A term used to describe leaf shape. See Fig. L1.

**oviduct** A tube which carries eggs (ova) from the ovary to the womb (uterus) in mammals, or to the outside of the body in other vertebrates. See also Müllerian tubes, Fallopian tubes and Fig. R4.

**oviparous** The type of reproduction where the mother lays eggs which hatch outside the body. The embryo is poorly developed, and there is a large supply of food in the large yolksack. e.g. Birds, monotremes, reptiles, amphibians, fish, and most invertebrates.

**ovipositor** A specialised organ or structure for the depositing of eggs, especially those of female insects. See Fig. I5.

**ovoid** Shaped like an egg with the broader end at the base.

**ovoviviparous** The type of reproduction where the mother produces shelled eggs which, however, remain to develop within her body. There is no placental connection to the mother, and the young are born live. e.g. Some reptiles, sharks.

**ovulation** The release of one or more mature egg (ova), or ⋆oocytes, from the ovary into an oviduct. Ovulation appears to be controlled by the production of ⋆LH in the pituitary, under the control of the hypothalamus. See reproduction.

**ovule** 1 In flowering plants (angiosperms), one or more bodies found in the ⋆ovary at the base of a ⋆carpel, each containing an egg (ovum). After fertilisation, the ovum develops into the ⋆embryo, whilst the ovule becomes the seed that surrounds it. The ovary wall develops into the fruit. 2 In ⋆gymnosperms the ovules are unprotected, and are carried in the scales of the female ⋆cones.

**ovum** Also called the **egg cell**. The unfertilised female gamete or reproductive cell. Often produced in an ⋆ovary. Compare sperm.

**oxidation** A chemical reaction in which one or more of the following occurs: **1** Oxygen combines with another substance, often producing heat. e.g. Burning. **2** Hydrogen is removed from a compound. **3** A substance loses electrons. Compare reduction.

**oxidation-reduction** Also called **redox** reactions. Whenever an ⋆oxidation reaction occurs, there is an accompanying ⋆reduction reaction.

**oxidative phosphorylation** The combining of phosphate and ADP to produce ⋆ATP. This energy-storing reaction occurs at a number of points along the ⋆electron transport chain (Fig. E3) during aerobic ⋆respiration. The enzymes in this 'chain' are arranged in order along the inner membrane of ⋆mitochondria. The energy is supplied by NADH and $FADH_2$ produced in the ⋆Krebs' cycle. As NADH is converted to NAD and $FADH_2$ is converted to FAD, the energy they contain is transferred and stored as ADP is converted to ATP.

**oxygen debt** A lack of sufficient oxygen for ⋆respiration if the body, or part of the body, has been working too hard. In animals, a form of ⋆anaerobic respiration provides some energy, the wastes being stored as ⋆lactic acid until sufficient oxygen can be supplied for it to be removed. A build-up of lactic acid in muscles results in cramps. e.g. 'stitch'. In plants, the anaerobic respiration produces ⋆alcohol (ethanol).

**oxygen (O)** A tasteless, odourless and colourless gas that makes up about 21% of the Earth's atmosphere. It readily undergoes reaction with most elements, as in burning, rusting and corrosion. Oxygen is needed by the body for cell ⋆respiration.

**oxygen quotient** The rate of oxygen consumption of an organism or tissue. Measured in microlitres of oxygen per milligram of dryweight per hour.

**oxyhaemoglobin** Oxygenated *haemoglobin, formed when haemoglobin in red blood cells passes through gills or lungs and combines with oxygen. See respiratory pigments.

**oxytocin** A peptide *hormone produced in the *hypothalamus, transported to and stored in the rear (posterior) *pituitary before release into the blood stream. Causes the cells around the milk-producing glands to contract, ejecting milk. Causes the muscles of the womb (uterus) to contract, both during sexual intercourse to help sperm reach the oviducts, and during labour to help push the baby out through the birth canal (vagina).

**oyster** A marine *bivalve of the Family Ostreidae which grows attached to rocks and other solid objects. Although widely sought as a food delicacy, its function in nature is as a scavenger, removing dead and dying plant and animal material from the water it filters through its body. See Appendix 1.

**ozone** A strong smelling form of *oxygen, $O_3$, instead of the more common $O_2$. It is produced naturally by the action of ultra-violet light on normal oxygen high in the *stratosphere. It is also produced by electrical sparks, especially lightning flashes. See also ozone layer.

**ozone hole** Areas of the *ozone layer near the north and south poles where the percentage of ozone decreases in the spring and continues at this level for several months before increasing again. When *CFCs and other chemicals are released into the atmosphere, the chlorine-containing molecules rise and are broken down by sunlight, after which the chlorine reacts with and destroys ozone molecules. For this reason, the use of CFCs is being banned in many countries.

**ozone layer** A band in the atmosphere, 10 to 30 km above the Earth, where the concentration of ozone is relatively high. The ozone layer is vitally important to life on Earth, as ozone strongly absorbs life-damaging ultraviolet (UV) radiation produced by the Sun (e.g. UV helps produce sunburn). The percentage of ozone in the atmosphere is decreasing, leading to concern that the increasing amounts of ultra-violet light will lead to more skin cancer. See ozone hole.

# Pp

**pacemaker** 1 A small piece of tissue in the walls of the right ⋆atrium of the heart which causes the atria to contract. Called the sinoatrial node or ⋆S-A node for short. The signal passes to another pacemaker, called the ⋆A-V node (atrio-ventricular node), which causes the ⋆ventricles to contract about 0.1 seconds later. 2 A battery-powered device connected to the heart to take over the role of the S-A node.

**pachytene** A stage during ⋆prophase I of the first division of ⋆meiosis. During this stage the two sets of paired chromosomes (⋆tetrad) thicken and the chromatids become visible.

**pademelon** Any of the several species of small wallabies of the genus *Thylogale,* mostly found in regions of thick scrub or moist, dense undergrowth. See Appendix 1.

**pain** 1 An unpleasant, hurtful and distressing physical feeling or sensation from some part of the body. Usually associated with disease or bodily damage. 2 The sensory perception of signals sent from pain receptor cells found under the surface of the skin.

**pairing** Also called **synapsis**. The lining up side-by-side of homologous chromosomes during prophase I of meiosis.

**Palaearctic realm** A ⋆biogeographical region which includes the continent of Europe, Asia north of the Himalayas, and north Africa. Characteristic animals include reindeer, polecats, chamois, hedgehogs and European bison. See Fig. B5.

**Palaeocene** The first epoch of the Tertiary period, after the Cretaceous. See Appendix 2.

**palaeoecology** A study of the ecology that existed on Earth when the fossil plants and animals were alive.

**Palaeolithic** The Old Stone Age period of human history, extending from the end of the Pleistocene (about 1.8 million years ago) to the start of the ⋆Neolithic (about 10 000 years ago). Both Neanderthal Man and Cro-magnon Man are believed to have existed during this time, with stone tools being perfected, and the introduction of tools made from ivory and bone.

**palaeontology** The study of life in the past, based on fossil remains.

**Palaeozoic** The first era of the geological time scale that contains considerable fossil remains. It begins with the Cambrian period about 600 million years ago, and extends to the Permian period that ended 225 million years ago. See Appendix 2.

**palate** The roof of the vertebrate mouth. The front (anterior) part of the palate is hard (membrane-covered bone), while the rear (posterior) part is a soft flap of skin. The soft palate is raised during swallowing to prevent food entering the nasal cavity.

**palea** The inner or upper scale or ⋆bract, usually smaller than the ⋆lemma. Together they enclose a single grass flower.

**palisade** The layer of chlorophyll-rich cells under the surface (epidermis) on the upper side of most leaves. See Fig. S9. The column-like cells are often arranged in neat rows, and are one of the main sites for photosynthesis. Together with the ⋆spongy layer beneath, they make up the ⋆mesophyll of the leaf. Leaves of some plants, e.g. *Eucalyptus*, have palisade on both sides, with the spongy layer between.

**palmate** Parts arising from a common centre. e.g. **1** A compound leaf divided into lobes that radiate from a common centre. See Fig. L1. **2** A system of leaf veins that radiate from a common centre.

**palmitic acid** One of the most common fatty acids in plants and animals, making up a high proportion of ⋆lipids (vegetable oils and animal fats). See also stearic and oleic acids.

**palp** A small and usually blunt appendage in many invertebrates, situated near the mouth. It has many functions. e.g. In polychaete worms it is sensory. In bivalve molluscs it helps produce feeding currents of water. In insects and crustaceans the palps are associated with the mouthparts.

**palpate** To touch or feel during a medical examination for the purpose of diagnosis.

**palpitation** An unusual throbbing, pounding, rapid, or fluttering heartbeat.

**palsy** Paralysis.

**pampas** The grass-steppes of South America.

**pancreas** Also called **sweetbread**. In vertebrates, a long gland that lies along the base of the stomach, releasing hormones into the bloodstream, and digestive juices into the beginning of the small intestine. See Fig. D2.

⋆Exocrine tissues make up most of the pancreas. These are composed of groups of special cells (acini) which produce the ⋆pancreatic juices which enter the start of the small intestine through the ⋆pancreatic duct. ⋆Endocrine tissues, called ⋆Islets of Langerhans, are scattered throughout the pancreas. They produce the hormones ⋆insulin and ⋆glucagon, which enter the bloodstream and help control the concentration of sugars there.

**pancreatic duct** The tube connecting the pancreas to the start of the small intestine (duodenum). The ⋆bile duct connects to the pancreatic duct just before its enters the duodenum. See Fig. D2.

**pancreatic juice** Digestive enzymes such as ⋆trypsinogen, ⋆lipase, ⋆amylase and ⋆maltase produced by special ⋆exocrine tissues (acini) in the ⋆pancreas. The pancreatic juice produced is alkaline, and drains into the duodenum through the ⋆pancreatic duct.

**pandemic** The spread of a disease throughout a whole country, continent, or the whole world.

**Pangaea** The crescent-shaped supercontinent made up of all the present continents fitted together to form one large landmass. Continental drift (⋆plate tectonics) is believed to have split Pangaea into ⋆Laurasia and ⋆Gondwana and later into the present-day continents. See Fig. P7.

**pangenesis** A now-rejected theory of inheritance proposed by Charles Darwin to explain how new varieties of organisms could arise. He felt that the environment could indirectly influence the gametes, thus producing new varieties upon which natural selection could act. Compare Lamarckism.

**panicle** A flower cluster (⋆inflorescence) consisting of numerous ⋆racemes. See Fig. F2.

**pantothenic acid** A water-soluble vitamin of the B group, required in the diet of humans. Sources include yeast, liver, peas, etc. Pantothenic acid is converted by the body into ⋆coenzyme A.

**pap smear** A test for cervical cancer in women. A sample of cells is scraped from the neck (*cervix) of the womb (*uterus), and spread onto a microscope slide (a *smear), where the cells are examined for abnormalities.

**paper chromatography** A method of separating the components in a mixture of chemicals. The basic technique involves placing a sample of the mixture onto some absorbent paper, and dipping this into a solvent such as water or alcohol. As the liquid solvent soaks past the sample, some components travel faster than others so they are separated and can then be identified. See also chromatography.

**paperbark** Trees with paper-like bark, especially the broad-leaved tea-trees of the genus **Melaleuca*, in the Family Myrtaceae.

**Papilionaceae** Also called ***Fabaceae**. See also Appendix 1.

**papilla** Plural is **papillae**. A small conical, nipple-like projection above a surface. e.g. Tongue papillae which mark the entrance to cavities lined with taste buds. Orderly rows of papillae under the skin surface cause it to bulge, producing our fingerprints.

**pappus** A tuft of fine hairs or scales on the tops of fruits of many plants in the daisy family (*Asteraceae). These tufts remain attached to the seeds, thus aiding in wind *dispersal of the seeds. e.g. Dandelion, thistle.

**parallel evolution** Independent evolution of similar features by related organisms (with a common ancestor) due to adaptation to similar environments. Compare convergent evolution.

***Paramecium*** A genus of microscopic protozoans, whose oval-shaped bodies are covered by movement-producing *cilia. See also Ciliata.

**paranoia** A form of mental illness where the patient is suspicious, suffers delusions, and feelings of being persecuted, spied upon or endangered.

**paraplegic** A person who is paralysed in both legs, usually due to injury or disease of the spinal cord. Compare quadriplegic.

**parasite** An organism which, at some stage in its life history, lives in or on another organism (its *host), and from which it obtains food, shelter, etc. In this form of *symbiosis the host is usually harmed. e.g. Malaria protozoans, plant 'rusts', tapeworms, body lice.

**parasympathetic nervous system** A part of the *autonomic nervous system of vertebrates which generally relaxes the body, thus balancing the tension-producing *sympathetic system. It stimulates digestion, but generally slows down other functions and restores the body to normal after emergencies. The nerves leave the spinal cord at the top (anterior) and lower (posterior) ends. The nerve endings release the *neurotransmitter acetylcholine at the organs and glands that they connect with.

**parathyroid glands** Four pea-sized glands embedded in the back and side surfaces of the *thyroid gland. The parathyroid glands secrete hormones (*calcitonin and *parathyroid hormone (PTH)) which help keep the concentrations of calcium and phosphate in the blood at a constant level.

**parathyroid hormone** Called **PTH** for short. Secreted from the parathyroid glands, PTH raises the levels of calcium in the blood by causing bones to release calcium, the kidneys to reabsorb it more efficiently from the urine, and the stomach to absorb it from the diet (which requires the help of vitamin-D). Compare calcitonin. See also rickets.

**parenchyma** Plant tissue made up of thin-walled cells, often loosely packed, that forms the *ground tissue (packing tissue) of a plant. Such tissues occur in the pith and cortex of stems, and the spongy tissues in the middle of leaves. They can be sites for photosynthesis and/or storage.

**parietal** Relating to the wall of an organ, or of a cavity. e.g. Parietal bones form part of the roof and sides of the cranial cavity (skull).

**Parkinson's disease** A ⋆chronic (long lasting) disease that gets progressively worse, where the person suffers tremors, stiffness, and slowness of movement. At least partly due to a lack of the ⋆neurotransmitter ⋆dopamine in the ⋆basal ganglia of the brain.

**parotid gland** One of the ⋆saliva-producing glands, located just in front of and below the ear. It is connected to the mouth cavity by a duct.

**parthenogenesis** The development of an egg-cell (ovum) without fertilisation by a sperm cell (male gamete). e.g. Aphids, rotifers. honey bees, and some wasps. The offspring have the same chromosomes, and are thus genetically identical, to the parent.

**parturition** Childbirth. Labour, and giving birth to young.

**passerine** A bird in the order Passeriformes, the largest order in the class Aves. Passerines are called perching birds because their large first toe points backward, the other three forwards as they perch. They also hop rather than run/walk.

**passive immunity** See **acquired immunity**.

**passive transport** See **diffusion**.

**pasteurisation** The method of killing disease-causing (⋆pathogenic) bacteria in milk (and some fruit juices, beer and wine) by special heating techniques.

**patella** 1 In mammals, a bone protecting the joint, and its tendons, of the hind limb. In humans it forms the kneecap. See Fig. S3. **2** In arachnids, a segment of the leg between the femur and the tibia.

**pathogen** Disease-causing organism. e.g. Certain bacteria, virus, fungus, protozoan or animal ⋆parasite.

**pathology** The study of the nature of disease; its causes, effects, etc.

**PCB** Short for **polychlorinated biphenyl**. Very long-lived substances, chemically similar to ⋆DDT and ⋆dioxin. For years these compounds have been used in making plastics, de-inking fluids for recycling newspaper and for producing non-carbon copy paper. Because of their extreme toxicity, use of PCBs is now restricted to insulators in electrical transformers and capacitors.

**peat** Partly decomposed plant material, especially that waterlogged in swamps and bogs, where there is insufficient oxygen for complete decay (i.e. anaerobic conditions).

**peck order** The social hierarchy among groups of animals where the more dominant members 'rule over' the more submissive ones. Widespread, but especially obvious in birds (e.g. chickens) which exert their dominance by pecking. Any bird higher up the 'peck-order' is aggressive towards birds lower down, especially at feeding time. Submissive birds usually give way during disputes.

**pectic substances** Complex ⋆polysaccharides found in the middle ⋆lamellae of plant cell walls, and found in large quantities in some fruits. e.g. Apples, pears, citrus fruit. Some of the pectic substances such as pectin and pectic acid form ⋆gels with sugar solution, and are thus used to produce jams.

**pectin** The ⋆monomer which forms ⋆pectic substances.

**pectoral** Of or relating to the breast or chest. e.g. Pectoral fins of fish.

**pectoral girdle** Also called the **pectoral arch** or **shoulder girdle**. The skeletal structure of vertebrates to which the forelimbs or fins are attached. In humans, it comprises the two ★scapula (shoulder blades) and ★clavicles (collar bones). Compare pelvic girdle.

**pectoral muscles** Muscles of the chest. **1** Pectoralis major connects the clavicle, sternum and second to sixth ribs, to the humerus. It helps move the arm. **2** Pectoralis minor connects the third to fifth ribs to the scapula, and helps move the shoulder joint, or the ribs during forced inhaling.

**pedicel** A short stalk, e.g. the stalk supporting a single flower. Compare peduncle.

**pedigree** A diagram or list to show the relationships in a family over several generations. Commonly called a family tree.

**pedipalp** The second pair of appendages on the head of arachnids. They have various functions. e.g. In scorpions, they are claw-like and used to eat prey. In other arachnids they are sensory organs. Male spiders use them to place sperm sacs into the reproductive organs of the females.

**peduncle** A stem or stalk which supports many flowers (an ★inflorescence). Compare pedicel.

**peewee** See mudlark.

**Peking Man** Fossils of ★*Homo erectus* found near present day Beijing (once called Peking). See also Java Man.

**pelagic** Living in the open waters of lakes or the oceans, but not on the bottom. Those that drift with the currents are called the ★plankton. Those that swim actively are called the ★nekton. See Fig. O1.

**Pelecypoda** A class of molluscs, also called ★bivalves due to the two shells that are hinged on one side. e.g. Clams, oysters, mussels, scallops, cockles and 'shipworm'. See Appendix 1.

**pellagra** A disease due to lack of ★nicotinic acid, a vitamin of the B group. Symptoms include skin and nerve problems, and ★diarrhoea.

**pellicle** A protective membrane layer that surrounds some single-celled organisms. e.g. Protozoans such as ★Ciliata.

**peltate** Of a leaf, having the stalk attached to the lower surface instead of at the edge. e.g. *Nasturtium*. See Fig. L1.

**pelvic bone** See pelvic girdle, below.

**pelvic girdle** Also called the **hip girdle**. That part of the skeleton to which are attached the hind limbs or pelvic fins. In humans, it is made of three pairs of fused bones—the ilium, ischium, and pubis. It is also attached to the sacrum, which is in turn attached to the sacral vertebrae. Compare pectoral girdle. See Fig. S3.

**pelvis** **1** The basin-like cavity formed by the pelvic girdle, the sacrum, and the coccyx. See Fig. S3. **2** The bones which surround this cavity. See pelvic girdle. **3** In the kidney, the funnel-shaped top end of the ureter into which drain the urine-collecting ducts. See Fig. K1.

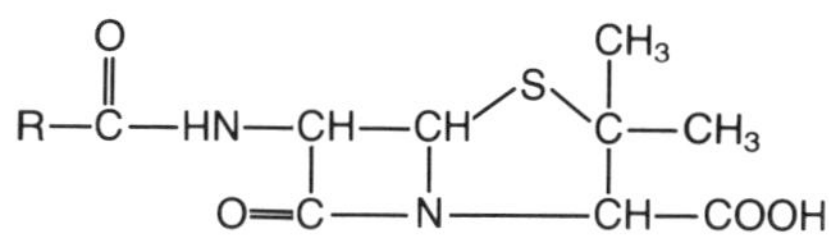


R = functional group which varies for different types of penicillin

**Fig. P1 Penicillin**

**penicillin** An ★antibiotic, extracted from the fungus *Penicillium*, used to control the growth of bacteria. Penicillin interferes with the making of the cell wall, causing the bacterium to split open and die. See Fig. P1.

***Penicillium*** A genus of moulds in the class Ascomycetes, which naturally produces a chemical called ★penicillin that controls the growth of bacteria. Special varieties of *Penicillium* have been bred to increase the amount of penicillin produced.

**penis** The male reproductive organ, used to transfer sperm from the male to the female. To achieve this the penis becomes erect as it fills with blood during sexual stimulation. See ★erectile tissue. In mammals, the penis is also used to remove urine from the body.

**penniveined** Of a leaf, pinnately ★veined, where the numerous lateral veins diverge at a fairly regular angle from both sides of the midvein. See Fig. L1.

**pentadactyl limb** The five-fingered, five-toed limbs found in most vertebrate classes: amphibians, reptiles, birds and mammals. See Fig. P2.

There are three main parts. The upper arm or upper leg (thigh) contains one long bone (humerus in arm, femur in leg). The forearm or foreleg (shank) contains two more-or-less parallel bones (radius and ulna in arm, tibia and fibula in leg). The hand or foot contains many small bones: the wrist or ankle (carpels in wrist, tarsals in ankle); palm or sole (metacarpals in palm, metatarsals in sole); fingers or toes (phalanges for both). There are many modifications of this basic pattern. e.g. Horses' hooves, whales' fins and birds' wings. This limb arrangement provides evidence of evolution from common ancestors.

**pentose** A ★monosaccharide sugar with five-carbon atoms. e.g. Ribose, deoxyribose. An important part of ★nucleic acids. See carbohydrates and Fig. R6.

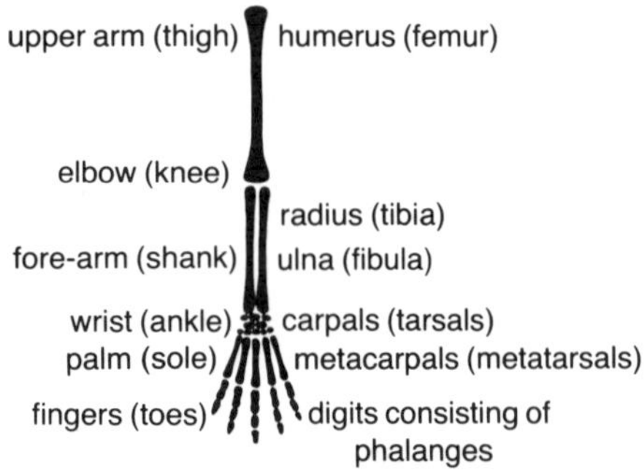


**Fig. P2 Pentadactyl limb**

**peppermint** 1 A herb, *Mentha piperita*, widely used to produce peppermint oil for flavouring. 2 Any one of a group of *Eucalyptus* species with a characteristic smell from their crushed leaves, and a fibrous but not stringy bark.

**pepsin** An enzyme which splits up (by ★hydrolysis) proteins into short lengths of amino acids (★peptides) when in acidic conditions (i.e. a ★peptidase). In vertebrates, pepsin (in the form of ★pepsinogen) and hydrochloric acid are secreted by the walls of the stomach.

**pepsinogen** The chemical secreted by the walls of the stomach, and converted to the enzyme ★pepsin by hydrochloric acid.

**peptic ulcer** An ulcer (raw and open flesh) associated with the digestive action of acidic stomach juices. The protective ★mucus coating is overcome, and the digestive juices attack the protein walls of the stomach (gastric ulcer) or upper parts of the small intestine (duodenal ulcer). There is some evidence that a bacterium may be partly the cause of some of these ulcers.

**peptidase** A group of enzymes that splits up polypeptides (protein fragments) by attacking certain peptide bonds (by hydrolysis) between specific amino acids. Some split off only the amino acids on the ends of a chain (exopeptidases e.g. ⋆pepsin). Others only attack peptide bonds between amino acids that are away from the ends (endopeptidases e.g. ⋆trypsin). See also proteases.

**peptide** A compound formed when two or more amino acids are joined together by ⋆peptide bonds to form a chain-like molecule. Long chains of amino acids are also called ⋆polypeptides.

**peptide bond** The bonding found between amino acids in all peptides, polypeptides, structural proteins and enzymes. The amino group ($-NH_2$) of one amino acid joins to the carboxyl group ($-COOH$) of another, thus forming a peptide bond (Fig. P3), with a molecule of water ($H_2O$) being set free.

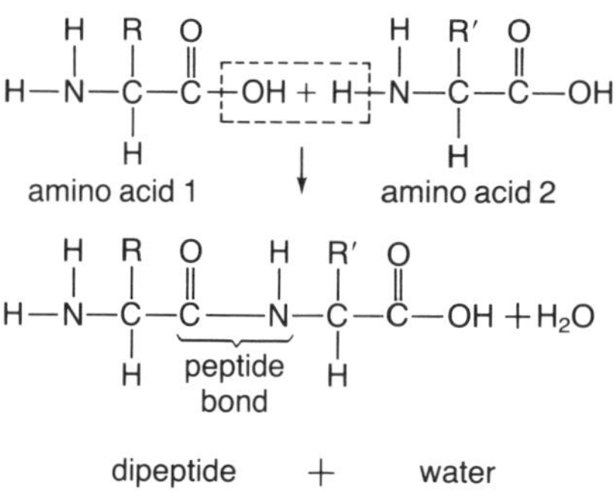


**Fig. P3 Peptide bond**

**peptone** A mixture of large protein fragments (polypeptides) produced by the early stages of protein digestion.

**Peramelidae** The family of marsupial mammals that includes the ⋆bandicoots. These small rat-like marsupials with a rather pointed snout, are ⋆omnivorous, living and feeding on the ground. See Appendix 1.

**perennial** A plant which continues to grow from year to year and does not die after flowering and fruiting only once. In some herbaceous perennials the above-ground stems die down each year, but the plant survives using underground parts. e.g. roots, corms, tubers, bulbs, etc. In woody perennials (trees and shrubs), the stems survive and grow each year, thus allowing them to reach very large sizes. Compare annual, biennial, ephemeral.

**perfect** Of a flower, one that has both male and female organs. i.e. Stamens and carpels.

**perforation plate** The remains of the end walls between two ⋆vessels of a plant's ⋆xylem. The lack of end walls allows nearly continuous xylem tubes to be created for the movement of water and dissolved minerals from the roots to the leaves. See transpiration. Compare tracheids and sieve plates.

**perianth** The parts of a flower which surround the reproductive structures. It usually consists of two whorls (or rings). The petals are collectively called the ⋆corolla. Outside of the corolla are the sepals, collectively called the ⋆calyx. In insect-pollinated flowers the corolla, and sometimes the calyx as well, are brightly coloured to attract insects. Wind pollinated flowers usually have little or no calyx or corolla.

**pericardial cavity** The space surrounding the heart.

**pericardium** A smooth membrane that lines the ⋆pericardial cavity. See also mesothelium.

**pericarp** The fruit wall which develops from the ovary wall. In some fruits the pericarp is fleshy, when it usually has three layers: a tough outer ⋆epicarp, a fleshy middle ⋆mesocarp, and an ⋆endocarp on the inside. In dry fruits the pericarp becomes hard, leathery or papery. If the pericarp splits open at maturity, the fruit is said to be ⋆dehiscent, but if it remains closed it is said to be *indehiscent. See Fig. F4.

**pericycle** The outermost layer of the *stele (vascular cylinder), lying just inside the *endodermis. In stems it is only a few cells thick, but is more easy to see in roots since lateral roots arise from it.

**periderm** The outer protective layers or *bark of older roots and stems, consisting of the inner cork *cambium and the outer layers of cork. The cork cambium is the site of cell division, producing new cork layers to protect the outside of the root or stem (trunk). The surface is often broken by *lenticels, passageways for air to enter through the otherwise impervious surface.

**perigyny** Flowers in which the sepals, petals, and stamens are attached above the ovary to the edges of a flat or cup-like receptacle. See also hypogyny and epigyny. See Fig. F2.

**period** *Menstruation.

***Peripatus*** Soft-bodied, 'caterpillar'-like animal of damp forests. Appears to have features of both annelid worms and arthropods. See also Onychophora and Appendix 1.

**peripheral nervous system** All parts of the nervous system that are not a part of the *central nervous system. It includes the *sensory and *motor nerves that run to and from the central nervous system. See also autonomic and somatic nervous systems.

**Perissodactyla** An order of herbivorous placental mammals which includes the horses, donkeys, tapirs, rhinoceroses. They are hoofed, but with an odd number of toes (1, 3, 5), and are sometimes called the odd-toed *ungulates. See Appendix 1.

**peristalsis** A wave of muscular contraction passing along a tubular organ. e.g. *Oesophagus (food pipe) moving the *bolus (lump) of food from the mouth to the stomach. Intestines, mixing the food and moving it along as it is digested and absorbed.

**peritoneum** The smooth membrane that lines the abdominal cavity and the abdominal organs (i.e. contains stomach and intestines), thus allowing easy slippage between them. See also mesothelium.

**peritonitis** An inflammation of the *peritoneum, especially after rupture of the *appendix which releases bacteria into the abdominal cavity.

**periwinkle** Any of a number of small univalve (gastropod) molluscs, especially the smaller dark varieties on the upper zones of rock platforms. e.g. Genus *Littorina*. See Appendix 1.

**permafrost** Ground which is permanently frozen, even if a few centimetres on the surface thaws in summer. e.g. Artic regions.

**permanent teeth** The second and final of the two successive sets of teeth in mammals, replacing the *milk or deciduous teeth. In humans, there are 4 incisors, 2 canines, 8 molars and 2 wisdom teeth in each jaw.

**permeable** Able to be passed through. Compare semi-permeable and impermeable.

**Permian** The last period of the Palaeozoic era, after the Carboniferous but before the Triassic. See Appendix 2.

**peroxisome** A *microbody which contains a number of enzymes. e.g. Enzyme that can break down damaging hydrogen peroxide produced as a byproduct of other cell reactions. In plant cells, enzymes involved in *photorespiration may be present.

**persistent** Remaining attached for some time, especially until mature. e.g. Cones on a pine tree. Bark on a tree-trunk.

**perspiration** Sweat. Produced by ⋆sweat glands in the skin. As the sweat evaporates, it absorbs energy from the body, thus cooling it and helping to maintain constant body temperature. Also used to excrete some wastes, as about 0.5% is made up of urea, lactate, and salt.

**pervious** ⋆Permeable.

**pest** An organism directly harmful to humans, or to their agriculture. e.g. Mosquitoes, grasshoppers, fungal 'rusts'.

**pesticide** A chemical substance for destroying pests. e.g. Insecticides; weedkillers; fungicides.

**PET** Short for **positron emission tomography**. A short-lived radioactive isotope is injected into the blood stream. This isotope emits positrons that soon hit nearby electrons and so produce gamma rays. Instruments placed around the body detect these gamma rays and construct a picture of what is happening inside. Compare CT, MRI and ultrasound.

**petal** One of the usually coloured leaf-like structures of a ⋆flower (the inner whorl of the ⋆perianth), the colour, size, shape, patterning and movements acting to attract pollinators. Petals often have 'honey-guides' to direct pollinators to nectar. In wind-pollinated flowers, petals are absent or very small and inconspicuous. See Fig. F2.

**Petauridae** The family of marsupial mammals which includes the ringtail possum and larger gliders. The ringtail possums are mainly leaf-eaters, and have ⋆prehensile tails. The gliders are famous for their ability to glide between trees using flaps of skin between their legs to form a 'parachute'. See Appendix 1.

**petiolate** Having a ⋆petiole.

**petiole** Of a leaf, the slender stalk that attaches the blade to the plant stem. Leaves which lack a petiole, such as grasses, are said to be ⋆sessile. See also abscission.

**petrified** Used to describe a ⋆fossil that has been turned to stone. Usually, percolating waters replace most of the original material with silica, but leave the original shape.

**petroglyph** A drawing or carving in rock made by prehistoric people.

**pH** A measure of how acidic or alkaline (basic) an aqueous solution is. It is a measure of the hydrogen ion concentration $[H^+]$. Solutions with pH of 7 are neutral. If the pH is above 7, the solution is alkaline (basic), and if below 7 the solution is acidic.

**Phaeophyta** The brown algae, a phylum of mostly marine plants including seaweeds such as kelp and neptune's necklace. The green chlorophyll cannot be seen due to a brown pigment. See Appendix 1.

**phage** Also called **bacteriophage**. A ⋆virus which infects ⋆bacteria. The virus enters a bacterium and takes over the cellular machinery, directing it to make more phages. Eventually the bacterium dies, splitting open to release more phages. See Fig. V4.

**phagocyte** A cell that can ingest into its cytoplasm particles from its surroundings, a process called ⋆phagocytosis. See Fig. E6. Although many protozoans are phagocytes (e.g. *Amoeba*), the term is more usually applied to white blood cells in the body. e.g. ⋆Polymorphs and ⋆macrophages. Phagocytes may not destroy all invaders, but often reduce the numbers of pathogens enough for other parts of the immune system to finish the job.

**phagocytosis** A type of ⋆endocytosis where a cell ⋆engulfs a particle (e.g. bacterium) by flowing around it and forming a ⋆vacuole within its cytoplasm. ★Lysosomes fuse with the vacuole, adding enzymes to digest the particle. See Fig. E6.

**phalanger** A tree-living marsupial of the Family Phalangeridae, including the cuscus, and the brushtail and scaly-tailed possums. Their tails are not completely ★prehensile, and they mainly live on a diet of fruit and leaves. See Appendix 1.

**Phalangeridae** See phalanger above.

**phalanges** Singular is **phalanx**. Any of the rod-shaped bones in the fingers and toes. See Fig. P2.

**phalanx** See phalanges above.

**pharyngeal** Of or relating to the ★pharynx.

**pharmacist** A person who prepares and dispenses medicines.

**pharynx** In vertebrates, the portion of the digestive tract which connects the mouth (buccal cavity) to the ★oesophagus. When food touches the walls of the pharynx, it stimulates the reflex muscular actions involved in swallowing. In fish and aquatic amphibians, the gill slits connect the pharynx to the outside. In other vertebrates the nasal cavity and larynx connect to the pharynx, as do the paired ★Eustachian tubes that connect to the middle ear.

***Phascogale*** A genus of small carnivorous marsupials in family Dasyuridae. e.g. ★Tuan and ★Wambenger. See Appendix 1.

**Phascolarctidae** The family of marsupial mammals which includes the ★koala. These tailless animals live on a diet of *Eucalyptus* leaves. See Appendix 1.

**phase** 1 The steps in prophase I of ★meiosis: leptotene, zygotene, pachytene, diplotene, diakinesis. Compare stage. 2 Any part of the development of an animal which results in a different coat or skin colour during its yearly breeding cycle.

**phase contrast microscopy** The use of a ★microscope which allows living, and thus normally transparent, material to be viewed in shades of grey. This is achieved by recombining light that has taken different paths through the specimen. Because the refractive index varies from one part of the specimen to another, the light travels at many different speeds. Special optics recombine the light to produce a much clearer image than that seen with normal light microscopes.

**Phasmatidea** An order of insects which includes the stick-insects and leaf-insects. These vegetation eaters are long and thin, or long and flat. When wings are present, there are four, but the front ones are small and thickened. See Appendix 1.

**phellem** The ★cork found in bark. Part of the ★periderm.

**phenols** Also called **phenolics**. Compounds containing a benzene ring and made from phenol (carbolic acid) $C_6H_5OH$. Naturally occurring phenols, or compounds containing a phenol ring, include lignin, the amino acid tyrosine, and ★tannins.

**phenotype** The physical make-up and appearance of an organism, resulting from the genes present (★genotype), and due to the influences of the environment. Thus organisms with the same genotype can have quite different phenotypes.

**phenylalanine (phe)** One of the about 20 naturally occurring ★amino acids. The lack of a certain enzyme, due to the inheritance of a recessive gene, means that phenylalanine cannot be used by the body, resulting in the disease phenylketonuria. See below.

**phenylketonuria** Called **PKU** for short. An inherited disease that results in mental retardation. The body is unable to use the amino acid ⋆phenylalanine.

**pheromone** A chemical substance released by organisms into their surroundings in very small quantities to influence the behaviour or growth and development of other organisms of the same type. e.g. Sex attractants released by female moths to attract the males for mating. Queen-bee substance, produced by the queen, and fed to the workers so that their ovaries do not develop, preventing the making of queen-bee cells in the hive. See also chemotaxis.

**phlegm** Mucus, especially the thick mucus discharged through the mouth. This mucus is produced in the respiratory passages to trap dust and other particles, and is carried to the rear of the mouth by cilia, and by coughing.

**phloem** The main food-carrying tissue in vascular plants, which together with the ⋆xylem makes up the ⋆vascular tissue. Foods, such as sugars and proteins are carried from the leaves where they were made during ⋆photosynthesis, by a process called ⋆translocation. The phloem is a tissue composed of ⋆sieve tubes, associated ⋆companion cells, ⋆fibres and ⋆sclereids. The cytoplasm in the sieve tubes has no nucleus, and is continuous with the cytoplasm of adjacent cells. The companion cell appears to control its functioning. See Fig. P4.

**phobia** An abnormal, excessive dread or fear. e.g. Claustrophobia is a fear of confined spaces. Xenophobia is a fear of strangers.

**Pholidota** The order of placental mammals which includes the pangolins. These anteaters lack teeth, and their body is covered by ⋆horny scales. See Appendix 1.

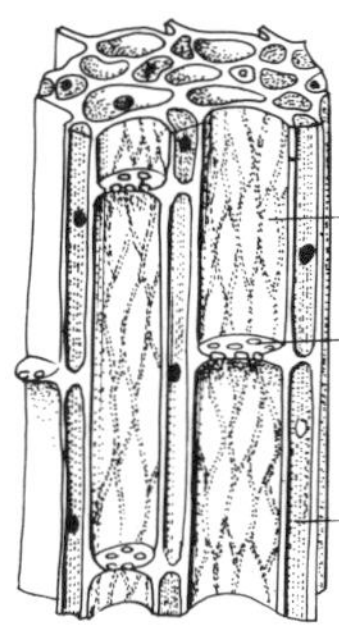


**Fig. P4 Phloem**

**phoneme** One of the basic sounds that make up speech.

**phosphate** A salt containing the phosphate ion ($PO_4^{3-}$). Important component of ⋆ATP, ⋆NADP, and ⋆nucleotides. See also ⋆phosphorylation.

**phospholipid** Complex ⋆lipids that are important in cell membranes. e.g. ⋆Lecithin. One of the three fatty-acids normally found joined to glycerol in lipids is replaced by phosphate which in turn is joined to some other small molecule, often an alcohol.

**phosphoprotein** A group of proteins where one or more phosphate groups are bound to protein. e.g. ⋆Casein.

**phosphorescence** A type of ⋆luminescence that lasts for a period of time after the stimulating radiation is removed.

**phosphoric acid** $H_3PO_4$, an important chemical in cell metabolism. See phosphate.

**phosphorus** (P) A non-metallic element that comes in various forms (allotropes). e.g. Red and yellow phosphorus. Phosphorus is ⋆essential for living things, being important in ATP, DNA, RNA, and many enzymes and proteins.

**phosphorylation** Occurs when a

phosphate group is added to a compound, a reaction catalysed by an enzyme. e.g. The adding of a phosphate to ADP to produce ATP, a process that stores chemical energy; adding phosphates to a protein, which often changes its shape thus making it more or less chemically active.

**photomicrograph** The photograph of a specimen taken through a microscope, and thus usually magnified.

**photon** The elementary particle of light. Light acts like a particle when it interacts with certain molecules. e.g. During photosynthesis, photons of light knock electrons from magnesium atoms, found in chlorophyll molecules.

**photoperiodism** The response of plants to the relative length of day and night. Photoperiodism usually controls flowering, and is important in tuber and bulb formation, leaf fall, and bud dormancy. See also ★long-day, ★short-day, and ★day-neutral plants.

**photophosphorylation** The synthesis of ATP from ADP and phosphate during ★photosynthesis. The reactions involved are part of the ★electron transport chain.

**photoreceptor** A sensory organ which is sensitive to light. e.g. ★Eye; ★ocellus.

**photorespiration** Respiration which occurs in many ★$C_3$ plants during active photosynthesis and the presence of light. A two-carbon molecule produced in the Calvin cycle is converted back to $CO_2$. This reduces the efficiency of $C_3$ plants as it wastes energy and carbon dioxide. Little photorespiration appears to occur in $C_4$ plants. The benefit of photorespiration to $C_3$ plants is unknown.

**photosynthesis** The complex series of chemical reactions which occur in the ★chloroplasts of green plants where carbon dioxide and water combine in the presence of light and chlorophyll to produce sugars and oxygen.

$$6CO_2 + 6H_2O \xrightarrow[\text{chlorophyll}]{\text{light}} C_6H_{12}O_6 + 6O_2$$

Thus light energy from the Sun is converted into chemical energy, and used to produce carbohydrates, which in turn are used to make all the other materials of the plant. See Fig. P5. The above reaction is misleading as all the $O_2$ released actually comes from the $H_2O$. A better representation would be:

$$6CO_2 + 12H_2O^{\star} \xrightarrow[\text{chlorophyll}]{\text{light}} C_6H_{12}O_6 + 6O_2^{\star} + 6H_2O$$

Photosynthesis can be divided into two phases.

**a Light reaction**: Takes place in the ★grana of the chloroplasts (Fig. P5). Chlorophyll molecules absorb light energy (see action spectrum, Fig. A3 and photosystems 1 and 2), emitting high-energy electrons that pass along the ★electron transport chain, producing energy-rich ★ATP molecules. Meanwhile the electrons at the chlorophyll are replaced when water is split into oxygen gas and hydrogen ions, the hydrogen ions being used to convert $NADP^+$ to NADPH. Thus the overall reactions are:

$$2NADP^+ + 2H_2O + ADP + P \xrightarrow[\text{chlorophyll}]{\text{light}} 2NADPH + O_2 + 2H^+ + ATP$$

**b Dark reaction**: The dark reaction takes place in the ★stroma of the chloroplast, and continues in the cell cytoplasm outside. The ATP and NADPH made in the light reaction is used to convert carbon dioxide ($CO_2$) into carbohydrate, a process called ★carbon fixation, the main part of which is called the ★Calvin cycle (Fig. C1). There are two types of dark reactions. In the ★$C_3$-carbon-fixation cycle:

$$6CO_2 + 12NADPH + 18ATP + 12H^+ + 12H_2O \longrightarrow C_6H_{12}O_6 + 12NADP^+ + 18ADP + 18P$$

Not all dark reactions occur in the same way. The $C_3$-cycle described above is replaced in various tropical grasses (e.g. sugar cane, corn, millet, sorghum) and certain other plants (e.g. succulents) by what is called the ⋆$C_4$-carbon-fixation cycle:

$6CO_2 + 12NADPH + 30ATP + 12H^+ + 24H_2O \longrightarrow C_6H_{12}O_6 + 12NADP^+ + 30ADP + 30P$

Although $C_4$ plants use more ATP than $C_3$ plants, this is no problem as there is plenty of sunshine and (periodically) water for the light reaction. However, levels of $CO_2$ are often low in $C_4$ plants as the stomata are closed to stop excess loss of water by transpiration in extreme heat. The $C_4$ plants can still photosynthesise with very low concentrations of $CO_2$, and have a method of moving the $CO_2$ to increase its concentration in certain parts of the leaf. This increased concentration has a second advantage in reducing ⋆photorespiration which in $C_3$ plants can use up nearly 50% of the products of photosynthesis. The overall effect is to allow $C_4$ plants to grow much more rapidly than $C_3$ plants. $C_3$ plants, however, can carry out photosynthesis at much lower light intensities than can $C_4$ plants.

**photosynthetic pigments** The light-absorbing pigments of photosynthesis. In most plants (not ⋆prokaryotes),they are located in the ⋆grana of the ⋆chloroplasts. In green plants the main pigment is chlorophyll-*a*, which absorbs mostly red and blue light (see action spectrum and Fig. A3). Green light is reflected, making most leaves appear green. Also present are accessory or 'helping' pigments whose job it is to collect light and pass it on to the main pigment. Accessory pigments include chlorophyll-*b*,

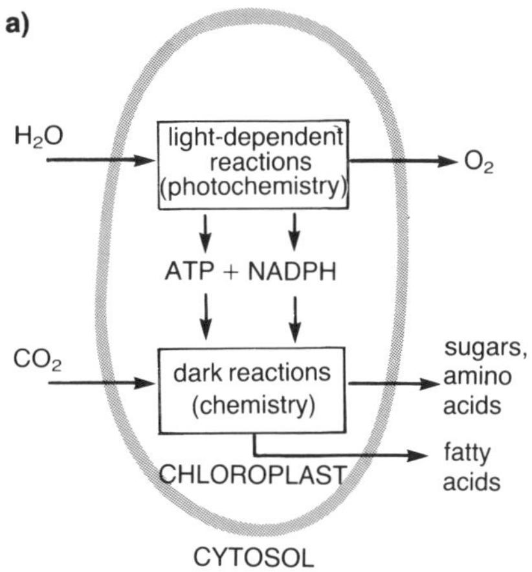


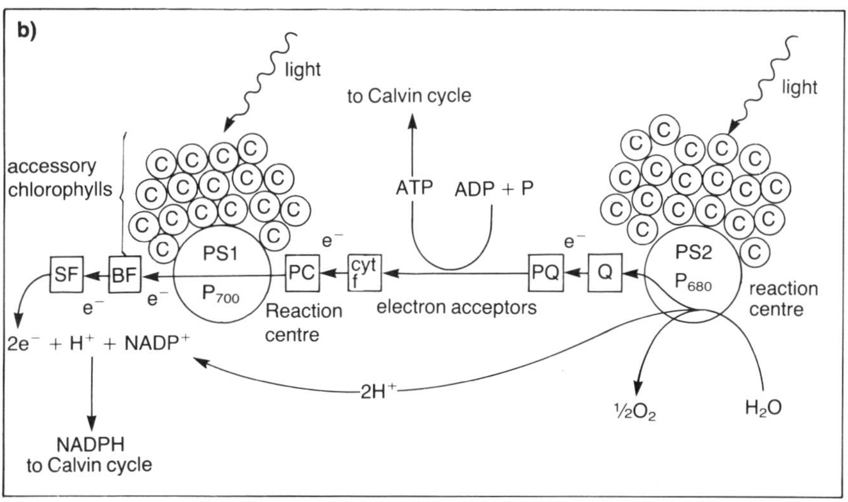


**Fig. P5 Photosynthesis (a) Light and dark reactions (b) Photosystems 1 and 2 of the light reaction**

carotenes and xanthophylls. The carotenes and xanthophylls (called collectively the carotenoids) are yellow, orange, red, or brown pigments seen in many flowers and fruits. In green leaves there is enough chlorophyll to mask the other pigments. Other photosynthetic pigments exist, and are found in diatoms, blue-green algae, etc.

**photosystems 1 and 2** Two photochemical systems which operate one after the other to perform the light reactions of ★photosynthesis (Fig. P5). Each of the photosystems has a form of chlorophyll-*a*, plus various accessory (helping) pigments. Photosystem 2 contains a form of chlorophyll-*a* called P680 which most efficiently absorbs light of 680 nm wavelength. When light energy is absorbed, water is broken down: $2H_2O \rightarrow 4H^+ + O_2 + 4e^-$ The electrons pass along the ★electron transport chain to photosystem 1. As they do so, energy is stored in ATP: ADP+P Æ ATP

Photosystem 1 contains a form of chlorophyll-*a* called P700 as it absorbs light of wavelength 700 nm most efficiently. At photosystem 1 $NADP^+$ is reduced to NADPH, using absorbed light energy and the electrons that have passed along the electron transport chain: $NADP^+ + H^+ + 2e^- \rightarrow NADPH$

**phototaxis** The movement (★taxis) of an animal or plant in response to light. e.g. Slaters (woodlice) move away from bright light (negative phototaxis). Certain single-celled algae will move towards light (positive phototaxis) as long as it is not too bright. Some chloroplasts within cells will align themselves to receive as much light as possible.

**phototropism** The growth (★tropism) of a plant in response to light. Most shoots grow towards light (positive phototropism), under the control of plant growth substances (e.g. ★auxins). Some roots (e.g. sunflower) grow away from light (negative phototropism).

**Phycomycetes** A class of fungi made mainly of thread-like ★hyphae with no conspicuous fruiting body. e.g. Bread mould, potato ★blight, downy ★mildew, ★rusts. See Appendix 1.

**phyllode** A flattened ★petiole (leaf stalk) that looks like a leaf, and provides a surface area for photosynthesis. e.g. The leaves of some wattles (*Acacia*) are phyllodes.

**Phylloxera** A genus of plant lice, especially *Phylloxera vastatrix*, which has caused very serious losses to the grape growing industry.

**phylogeny** The evolutionary history of an organism. e.g. The phylogenetic system of ★classification attempts to classify organisms according to their ancestry.

**phylum** Plural is **phyla**. The largest classification units of plants, animals, etc. Thus a kingdom contains many phyla, while each phylum in turn contains many classes. In plant classification, the word ★Division is often used instead of Phylum. See also classification.

**physician** A person who is legally qualified to practise medicine. Compare surgeon.

**physiological adaptations** 1 Any ★evolutionary change in the physiology of an organism that makes it better able to survive in its environment. e.g. The production of concentrated urine by the kidneys of desert-dwelling organisms; the swelling and shrinking of guard cells. 2 The process by which the organ of an organism ceases to respond to a constant ★stimulus. e.g. sense organs such as smell and sight.

**physiology** The study of the functioning and activities of living organisms and their parts. e.g. Transmission of nerve impulses; contraction of muscles.

**phytochrome** A protein pigment that helps govern plant responses called ⋆photoperiodism. e.g. Flowering, germination, growth of roots, stems and leaves, etc. Phytochromes occur in very low concentrations in all plants, and respond to light at the red end of the spectrum. There are two forms, readily changed from one to the one. Red phytochrome (PR or P660) absorbs red light (wavelength about 660 nm) which converts it quickly into far-red phytochrome (PFR or P730), which absorbs light in the far-red (nearly infra-red) end of the spectrum (wavelength 730 nm) converting it back to red phytochrome. In the absence of light, far-red phytochrome slowly changes back to red phytochrome of its own accord. Thus phytochrome is a kind of switch used by plants to control many of its processes.

**phytoplankton** The plant members of the ⋆plankton, microscopic organisms which float freely in the surface waters of oceans, lakes, etc., being carried about by currents and wind. Phytoplankton include microscopic algae, especially ⋆diatoms and ⋆dinoflagellates, and are thus the start of aquatic food chains. It has been estimated that more that 50% of total world photosynthesis is due to the oceanic phytoplankton. Compare zooplankton.

**pigment** 1 Any colouring matter. 2 A substance which absorbs light, usually of only a small band (group) of wavelengths. e.g. Chlorophylls, carotenoids.

**piles** Also called ⋆**haemorrhoids**.

**pili** Singular is **pilus**. 1 A hair or hair-like structure. 2 One of the thin hair-like projections from the wall of a ⋆bacterium.

**pimple** An inflamed swelling or elevation on the skin, filled with pus. See acne and black heads.

**pine** 1 Any member of the gymnosperm genus *Pinus*, evergreen trees with needle-like leaves, and reproducing by ⋆cones. e.g. *Pinus radiata* is an introduced pine widely grown for softwood timber. There are no species of *Pinus* native to Australia. See also Coniferopsida and Appendix 1. 2 More generally, any conifer native to Australia. The cypress pine (genus *Callitris*) grows over much of the drier parts of Australia. In the more northern parts of Australia, Hoop Pine (*Araucaria cunninghamii*) and Bunya Pine (*A. bidwillii*) are large trees, while the famous Norfolk Island Pine (*A. heterophylla*) grows on that island. The genus *Agathis* contains the famous Kauri pine of New Zealand (*A. australis*) and there are several native pines belonging to this genus in Australia. The Huon Pine of Tasmania, *Lagarostrobos franklinii*, is the most famous of the many conifers native to that state.

**pineal eye** Also called the **median eye**. An organ containing photoreceptors that forms a third 'eye' on the top of the head of lampreys, tadpoles and reptiles. Thought to help control ⋆photoperiodic responses.

**pineal gland** A pea-sized structure in vertebrate ⋆brains located just above the ⋆cerebellum. It produces the hormone melatonin, which in various animals lightens the skin, causes gonads to become inactive, etc. Since production of melatonin decreases in sunlight, the pineal gland may be involved in the control of reproductive cycles. The pineal gland is thought to be an endocrine organ in humans, but its exact function is uncertain.

**pinna** 1 In plants, a leaflet which is part of a compound leaf. See Fig. L1. 2 In animals, the external ⋆ear, or outside flaps of skin and cartilage that funnel sound waves into the ear canal and eardrum. See Fig. E1.

**pinnate** Leaves which carry paired leaflets (*pinnae) along opposite sides of an axis (midrib). See Fig. L1.

**Pinnipedia** An order of marine placental mammals that includes the sea-lions, walruses, and seals. All four limbs are flippers, and these carnivorous mammals have teeth like carnivores. See Appendix 1.

**pinnule** The paired leaflets along opposite sides of a *pinna, itself a leaflet along the sides of the midrib of a leaf. e.g. Some fern fronds. See Fig. L1.

**pinocytosis** The taking in or engulfing of small globules of fluid by a cell. Compare with phagocytosis. See also endocytosis.

**pioneers** In ecology, the first organisms to become established during any one stage of a *succession. e.g. spinifex grasses on a sand dune.

**Pisces** A once-used grouping of vertebrates which includes the cartilaginous and bony fishes.

**pistil** A term used in a number of different ways. e.g. With *flowers, it describes a single *carpel (simple pistil) or a group of fused carpels (compound pistil).

**pistillate** Of a flower that has female parts (pistil) but no male parts. See also staminate, monoecious, dioecious.

**pit** A thin part of the cell walls between two adjacent plant cells. The cytoplasm extends from one cell to another through the pits. See plasmodesmata.

**pit organ** Pits on the heads of certain snakes that contain sensory cells that detect sources of heat. Pit organs are located between the eyes and nose in pit vipers, and along the edges of the jaw in pythons and boas.

**pitcher plant** *Insectivorous plants with pitcher (jug) shaped leaves to help catch and digest insects and other small animals.

**pith** Also called a **medulla**. The soft central core of plant stems where the vascular tissue is arranged in a ring around the inside of the edge.

**pituitary** A major *endocrine gland attached to the base of the vertebrate *brain, and connected by a short stalk to the *hypothalamus. Popularly called the 'master gland' of the body. The pituitary is divided into three regions. **a** The front or anterior lobe produces a number of hormones that control growth and nutrition (*trophic hormones), including *growth hormone, *ACTH, *LH, *FSH, and *prolactin. **b** The intermediate (middle) lobe (absent in birds and some mammals) secretes a melanocyte-stimulating hormone. **c** The rear (posterior) lobe stores and releases a number of hormones made in the hypothalamus, and then passed to the pituitary along the connecting stalk. These hormones include *vasopressin (ADH) and *oxytocin. All these hormones are released by the pituitary into the bloodstream.

**placebo** An inert substance such as a sugar pill, which has no drug effect, but which is given to some patients to allow a comparison with a drug being tested. This comparison is necessary as some people seem to get better from just taking any pill, whether it contains a drug or not (called the placebo effect). See also randomised trial.

**placenta** 1 In primates, the organ attached to the wall of the womb (uterus) through which the growing baby (*foetus) receives nourishment and eliminates wastes. The placenta grows into the wall of the womb where its capillaries become closely associated with the capillaries of the mother. This allows food and oxygen to diffuse from the mother's blood into the blood of the baby, and for carbon dioxide and other wastes to be passed back again. The baby's blood then travels through the ⋆umbilical cord, connecting the placenta and foetus. In other mammals the attachment may not be so close. The placenta often produces hormones which help suppress ⋆ovulation and ⋆menstruation (to prevent a second pregnancy while one embryo is already growing), to stop the uterus from contracting (and thus expelling the growing embryo), and to prepare the breasts to produce milk. At birth, the umbilical cord is cut, and the placenta is expelled as the ⋆afterbirth a short time later. See also Fig. E4 and Appendix 1. **2** In plants, the part of the ovary which carries the ovules. See placentation below.

**placental mammals** The sub-class (also called Eutheria) that contains the majority of living ⋆mammals. Placental mammals give birth to young that are relatively well developed after a long period in the mother's womb (uterus), where they have been nourished through a ⋆placenta. Examples native to Australia include whales, bats, bush rats and mice, and the dingo. Introduced placentals include foxes, cats, rabbits, house rats and mice, dogs, and the various mammals used in agriculture. Compare marsupials and monotremes.

**placentation** In plants, the way (or pattern) in which the ovules are attached to the walls of the ovary. e.g. Axile (on the axis), marginal (attached to the edge), basal (attached to the base).

**plague** Also called **black death**. An acute disease, highly ⋆contagious and producing a fever, caused by the bacterium *Pasteurella pestis*. It is primarily a disease of rats and other rodents transmitted by fleas. Under unhygienic conditions, the fleas can pass the disease to humans. Bubonic plague is characterised by swellings under the armpits and in the groin. Pneumonic plague affects the lungs and has a high fatality rate.

**planarian** A free-swimming ⋆flatworm in the order Tricladida. It has only one opening to the intestines. See Appendix 1.

**planigale** Any of the flat-skulled carnivorous marsupial 'mice' of the genus *Planigale*, Family Dasyuridae, which includes the smallest known marsupials. They are renowned for their ferociousness, considering their small size. See Appendix 1.

**plankton** Microscopic organisms which float freely in the surface waters of oceans, lakes, etc., being carried about by currents and wind. ⋆Phytoplankton are minute plants and are the beginning of the various aquatic food chains. They are fed upon by minute animals called ⋆zooplankton.

**plant** A living organism that can be generally told apart from animals in the following ways: ability to carry out photosynthesis; cell walls containing cellulose; lack of ability to move about readily. Other features often present include ⋆alternation of generations, the ability for continued growth throughout life from cells that constantly undergo mitosis (⋆meristems), and lack of quick response to stimuli. Compare animal, protist, fungi, monera.

**Plantae** The ⋆**plant kingdom**. See Appendix 1.

**plantigrade** The type of walking (gait) where the whole foot (digits and metatarsals or metacarpals) remain flat on the ground. e.g. Primates, bears. Compare digitigrade and unguligrade.

**plaque** **1** A clear area on a ⋆culture plate growing ⋆bacteria, caused by destruction of bacterial cells by ⋆bacteriophages. **2** A similar clear area in a ⋆tissue culture of animal cells, caused by a ★virus. 3 A deposit of bacteria etc. on the teeth which may lead to tooth decay. 4 A patch which develops on the walls of ★arteries, usually made of deposited ★lipids and some scar or connective tissue. Plaque contributes towards stiffening blood vessel walls, narrowing of the arteries, and ruptured arteries, all associated with heart attacks.

**plasma** The yellowish liquid part of ★blood or ★lymph that remains after the various types of blood cells have been removed. It is mainly made of water in which the other non-cellular components of the blood are carried. e.g. Plasma proteins, inorganic ions, sugars, hormones. Compare serum.

**plasma cell** A type of white blood cell (★leucocyte) that develops from ★B-cells. The lymphocytes migrate to the connective tissue near a site of long term infection, and when stimulated by antigens, mature into plasma cells which produce large amounts of appropriate antibodies. See also immune system.

**plasma membrane** Generally called the **cell membrane**. The membrane that surrounds the protoplasm of a cell. In plants, it is located just inside the cell wall, and also surrounds the vacuoles. See Fig. P6.

Cell membranes are between 5 and 10 nm thick, and are made from two layers of ★lipids (mainly ★phospholipids). The polar ends of the lipids are exposed on the upper and lower surfaces of the membrane, while the non-polar ends are in the centre. Small protein-lined pores extend through the membrane, allowing water, ions and certain small molecules to pass by diffusion. Larger molecules require ★active transport, including ★endocytosis and ★exocytosis. ★Glycoproteins that are on the outer surface of cells are important in cell recognition by the immune system (see immunological tolerance).

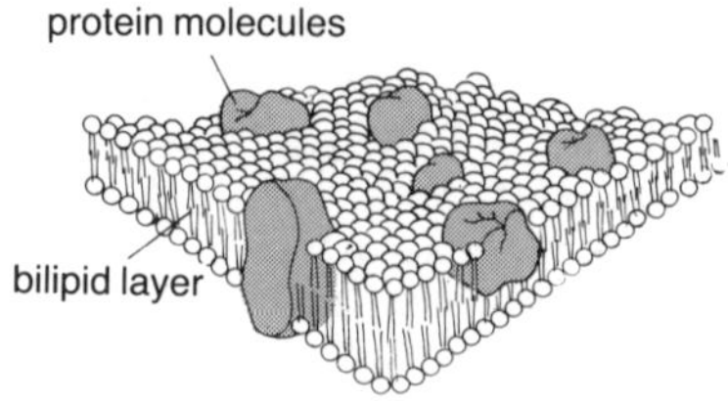


**Fig. P6 Plasma membrane**

**plasma proteins** Proteins carried in the blood ★plasma, including antibodies, hormones and blood clotting substances.

**plasmid** Any small piece of ★DNA which is not part of the chromosomes, exists in the cytoplasm of a bacterium, and is able to duplicate independently. Naturally occurring plasmids include the R factors which can carry genes giving the bacterium resistance to antibiotics. These R factors can be exchanged between bacteria, including bacteria in a different genus. Also present are F factors, the $F^+$ factor making the bacterium male, and the $F^-$ factor making it female. Artificially produced plasmids are widely used in ★genetic engineering to introduce genes into bacteria. See also episome and cloning.

**plasmodesmata** Strands of cytoplasm that pass through plant cell walls, usually at the ★pits, thus connecting the cytoplasm of adjacent cells.

**plasmodium** 1 An amoeba-like mass of protoplasm with many nuclei contained within one membrane, forming a stage in the life cycle of slime fungi (slime moulds—Myxomycetes) where reproduction does not occur. 2 A protozoan of the genus *Plasmodium*, which includes the agents that cause the disease malaria.

**plasmolysis** The loss of water from a plant, bacterial or blue-green algae cell to the point where the protoplasm shrinks, and starts to pull away from the cell wall, causing the cell to become *flaccid. See osmosis and hypertonic.

**plastid** A small membrane-enclosed body (organelle) found in the cytoplasm of cells, especially plant cells. e.g. *Chloroplasts contain green chlorophyll (and many other pigments and enzymes) needed for photosynthesis. Chromoplasts come in a variety of colours, and contain pigments that give flowers, fruits and (some) leaves their special colours. *Leucoplasts are colourless, and used for starch storage.

**plastron** The lower or ventral part of a tortoise shell.

**plate tectonics** Sometimes called **continental drift**. The theory that the surface of the Earth is divided into a number of constantly moving crustal-plates, the continents sitting on top of some of them. This movement of the crust can be used to explain the present positions of the continents, and hence the present distribution of living things (see biogeographical realms and Fig. B5). The original 'supercontinent' called *Pangea is believed to have begun to split up during the *Cretaceous (about 200 million years ago) to form the continents of *Laurasia and *Gondwana. These large continents are then believed to have split up to form the present continents (see Fig. P7). See also sea-floor spreading (Fig. S1).

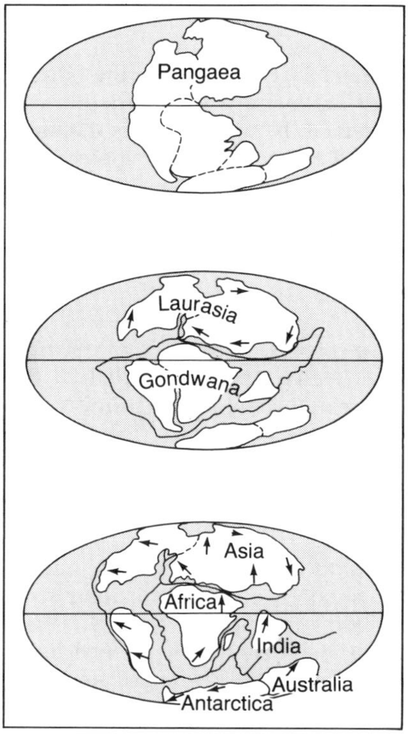


**Fig. P7 Plate tectonics**

**platelet** Minute bodies, fragments of cells formed in the *bone marrow, and found in the blood of mammals. There are about 250 000 per cubic millimetre of blood, and they play an important role in blood clotting by helping convert prothrombin into *thrombin. See blood cells.

**Platyhelminthes** Also called the **flatworms**. A phylum of flat unsegmented worms, the shape giving a large surface area for external respiration. They have no circulatory system, and only one opening for food to enter, and wastes to leave, the branched digestive system. Most flatworms are *hermaphrodites, and many are *parasitic. e.g. *Tapeworms, *flukes, *planarians.

**platypus** Also called the **duckbill platypus**, or *Ornithorhynchus anatinus*. One of the two types of egg-laying *monotreme mammals, both found only in eastern Australia and Tasmania. See also echidna.

**pleiotropy** A single gene that can have several different effects on an organism. e.g. In the tobacco plant, a single gene controls the length of anthers, capsules and petioles.

**Pleistocene** The first epoch of the Quaternary period in which the ice ages occurred. See Appendix 2.

**pleura** In mammals, the smooth membrane that lines the chest (thoracic) cavity, and the lungs, thus allowing easy friction-free movement between them as the lungs expand and contract during breathing. See also mesothelium.

**pleurisy** An inflammation of the ⋆pleura.

**plexus** An interwoven network of nerve fibres or blood vessels. e.g. **Solar plexus**, a network of ⋆sympathetic nerve fibres connected to the organs in the abdominal cavity.

**Pliocene** The last epoch of the Tertiary period during which time humans and most modern species of mammals evolved. See Appendix 2.

**plumule** In plants, the (top) bud on the shoot from a germinating seed, including the first true leaves. In dicots it sits between the two ⋆cotyledons. In monocots, it is wrapped up inside the cotyledon. Compare radicle.

**pneumatocyst** A gas-containing cavity e.g. **a** The ⋆swim bladder of a fish. **b** A gas-filled bladder found on certain brown algae.

**pneumatophore** Also called a **breathing root**. A root that protrudes some distance above the soil. The surface of the breathing root has many ⋆lenticels to help the entry of air. e.g. Characteristic of some species of ⋆mangroves.

**pneumonia** Inflammation of the lungs caused by various bacteria.

**Poaceae** Also called **Gramineae**, or the grass family of monocot flowering plants. Flowers are inconspicuous, and the leaves form a sheath around a jointed stem. See Appendix 1.

**pod** **1** A dry ⋆dehiscent fruit that splits along two cracks to form two halves. e.g. Pods of legumes such as peas, beans, wattles. **2** A group of animals together. e.g. A pod of whales.

**podzol** Also spelt **podsol**. A type of ⋆soil that is common in high rainfall areas of eastern Australia. Podsols are acidic soils with most of the colloids, and iron and aluminium compounds leached out of the uppermost layers. This leaves such layers a light greyish colour, while deeper down the soils are brownish due to the accumulation of leached materials. Such soils are difficult to cultivate.

**poikilothermic** Also called **ectothermic** or '**cold blooded**'. Used to describe animals whose body temperatures rise and fall with the temperature of the surroundings. Such animals may maintain their body temperature by behavioural means. e.g. Lizards bask in the sun in winter, but burrow underground during the heat of a summer's day. This is a feature of all animals except birds and mammals. Compare homoiothermic.

**point mutation** See **gene mutation**.

**polar body** A minute cell, one of two produced during the first and second division of the ⋆meiosis which produces the egg (ovum). Each contains a nuclei, but almost no cytoplasm. See also oocyte.

**polarity** **1** The property of many molecules where one end is slightly negative, and the other end slightly positive. Gives rise to ⋆hydrogen bonding. e.g. In water molecules, the oxygen 'end' is slightly negative, and the hydrogen 'end' slightly positive. **2** The existence in individual cells or in tissues of a head and tail end. e.g. If a ⋆planarian worm has both head and tail removed, a new head and tail grow at the same ends as the ones removed.

**polarising microscope** A ⋆microscope that uses polarised light to make visible certain parts of transparent cells, especially areas that are crystallised. It has been used to study membranes, fibrils and cell division.

**pollen** Small dust-like sacs that carry a number of ⋆haploid male 'sperm' of conifers and flowering plants.

Pollen is produced in male cones of gymnosperms, and when released, is carried by the wind to the female cones. In flowering plants, the pollen is produced in the ⋆anthers of the flower's ⋆stamens. The pollen grains vary, depending on how they are transferred between flowers (⋆pollination). The pollen carried by birds and insects is often sticky or patterned; pollen carried by the wind is generally smooth, dry, and comparatively small. When a pollen grain reaches the stigma of a flower, it ⋆germinates to produce a ⋆pollen tube. The 'sperm' pass along the pollen tube to fertilise the eggs in the ovules.

**pollen tube** In flowering plants, a tube that grows from a germinating ⋆pollen grain down the style of a ⋆carpel, to the ⋆ovule. The male gametes move down this tube to fertilise the ova within the ovule.

**pollination** The transfer of pollen from an ⋆anther to a ⋆stigma in flowering plants, and from male to female ⋆cones in conifers.

Pollination is carried out by animals (certain nectar-feeding marsupials and birds; insects such as bees and moths), the wind, and occasionally by water. Self-pollination is the transfer of pollen from the anther to stigma of the same flower, a form of ⋆inbreeding. Cross-pollination is the transfer of pollen from the anther of one flower to the stigma of another flower, a form of ⋆crossbreeding. Some flowers totally prevent self-pollination as pollen that falls on the stigma of their own flower will not germinate. In other cases the anthers and stigma mature at different times, thus preventing self-pollination. In some flowers when cross-pollination does not occur, self pollination is possible. Some flowers nearly always self-pollinate, often before the flower bud even opens.

**pollution** Addition of materials to the environment (air, water, soil, etc.) making it unfavourable to organisms that normally live there under natural conditions. See also acid rain, biodegradable, recycling, temperature inversion.

**Polychaeta** Also called **bristle worms**. A class of mostly marine segmented annelid worms, including 'beach worms' used as bait for fishing. The head usually has appendages, and each body segment has limb-like 'projections' with bristle-like chaetae on the ends. Compare Oligochaeta, and see Appendix 1.

**polygenic inheritance** The development of a given characteristic, such as height or weight, under the directions of a number of genes. Such features usually have a pattern of ⋆continuous variation in the characteristic concerned.

**polymer** A compound whose molecule often consists of many repeating units (⋆monomers) linked together. e.g. Starch, protein, nucleic acids, and human-made plastics.

**polymerase chain reaction** Called **PCR** for short. A technique for rapidly producing billions of copies of a length of DNA. Automated PCR machines are widely used in laboratories, and have been used to produce large amounts of DNA from very small samples such as those found in frozen specimens from the ice age.

**polymorph** A group of ⋆leucocytes (white blood cells), including ⋆neutrophils, basophils and eosinophils. Polymorphs make up about 70% of all white blood cells. They are ⋆phagocytes, and move freely throughout the bloodstream and tissues. Since their life is short they are continually being made in the bone marrow.

**polymorphism** The existence of two or more different forms of the same species. e.g. Visible examples include grey and white forms of the reef heron and black and spotted forms of the South American jaguar. Humans can also be called polymorphic due to differences in skin colour, blood types, hair type, body shape etc. Other examples of polymorphism are only revealed by chemical analysis of animal or plant tissues.

**polyp** 1 Also called a **hydranth**. A stationary form in the lifecycle of many cnidarians (coelenterates), especially corals, hydra and sea anemones. See Fig. P8. A cylinder-shaped body is attached by its base, while the other end has a ring of tentacles surrounding its mouth. The polyp of coral sits within a small cup that it has secreted. Sea-anemones only have a polyp stage, while hydra have a free-swimming ⋆medusa stage which settles on a solid object and develops into a polyp. 2 Cylindrical growths from mucous membranes, especially in the nose and colon.

**polypeptide** A molecule consisting of about 100 amino acids linked together in a single chain. Compare with ⋆peptide and ⋆protein.

**Polyplacophora** See Amphineura.

**polyploidy** The condition in which an organism has three or more complete sets of chromosomes in its nucleus, compared to the normal two sets (diploid). Polyploidy is rare in animals, but more common in plants. Commercial breeders use polyploidy to artificially produce new varieties (e.g. tomatoes, tobacco, sugar beets). Wheat used to make bread (*Triticum*) contains six sets of chromosomes, and scientists believe that polyploidy may have been important in the origin of this plant. See also chromosome mutation. Compare polysomy.

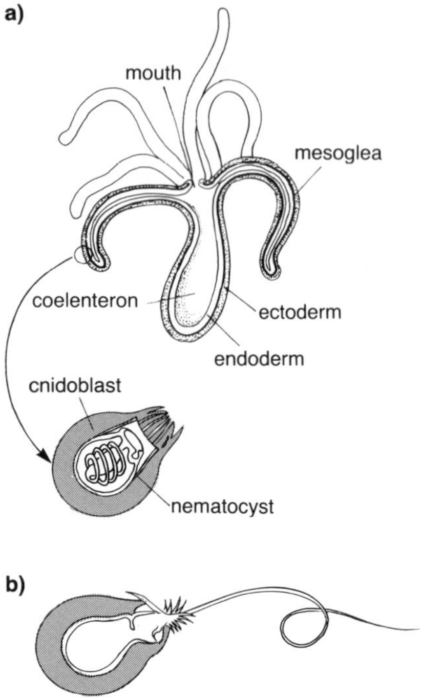


**Fig. P8 Polyp (a) Polyp and stinging cell (b) Discharged stinging cell**

**polysaccharide** One of a group of complex ⋆carbohydrates that contain more than two simple sugars (monosaccharides) linked together to form chain-like molecules. Polysaccharides are usually insoluble in water, and do not have a sweet taste. e.g. Cellulose, starch, glycogen, chitin. See Fig. C2.

**polysome** Also called a **polyribosome**. A string of ⋆ribosomes arranged along the same strand of messenger RNA, producing their proteins simultaneously. See translation.

**polysomy** The presence of one or more additional copies of one particular chromosome in a cell or an organism. e.g. Down's syndrome is due to three copies of chromosome 21. See ⋆trisomy. Compare polyploidy.

**polytene chromosome** A chromosome made up of many strands of DNA due to the repeated replication of DNA which then failed to separate in the normal way. The giant chromosomes that result are valuable for researches into chromosome structure and function. e.g. Giant chromosomes in the salivary glands of the fruit fly *Drosophila*.

**polyunsaturated** Used to describe fats and vegetable oils whose fatty acids contain one or more double covalent bonds between carbon atoms. Thus the molecule is more reactive, and does not contain as many hydrogen atoms as possible. For reasons of health, it is unwise to eat large amounts of saturated fats such as animal fat, and some vegetable oils. Compare ⋆saturated.

**pome** A type of fleshy false-fruit (⋆pseudocarp) where the succulent tissues develop from an enlarged receptacle rather than the ovary wall. e.g. Roses, ⋆apples and pears. See Fig. F4.

**Pongidae** The family of placental mammals in the Order Primates, that includes the ⋆anthropoid apes, gibbons, chimpanzees, orang-utans and gorillas. These tailless primates have forelimbs longer than their hindlimbs, binocular vision, relatively flat faces, and a well-developed brain. See Appendix 1.

**pons** Also called the **pons varolii**. The word pons means bridge, and is a region at the top of the spinal cord connecting it to the brain. Connections of various parts of the brain pass through the pons. See Fig. B7.

**population** All the members of a single species that are present in a defined area. Compare community.

**population genetics** The study of the number, variety, and distribution of ⋆genes in a ⋆population of organisms.

**Porifera** Also called **sponges**. A phylum of stationary animals containing many cells, but with little internal structure and no nervous system. Pores in the surface of the sponge (⋆ostia) allow food-carrying currents of water to enter, usually bought in by cells with waving flagella. The food particles are engulfed as they pass through, and the digested products shared amongst nearby cells. Between the inner and outer layers of cells there is often a body-supporting 'skeleton' of lime, silica, or a collagen-like protein. See Appendix 1.

**porphyrin** An important group of ⋆pigments that contain four complex pyrrole groups. e.g. ⋆Chlorophyll, ⋆cytochromes, ⋆haemoglobin, ⋆myoglobin.

**portal** An entryway. e.g. **1** Hepatic portal vein carries blood from the digestive organs and spleen to the liver. **2** In general, a vein which carries blood from one part of the body to another, but which does not go directly to the heart.

**Portuguese man-of-war** More commonly called a **bluebottle** in Australia. A cnidarian (coelenterate) of the genus *Physalia* found in warm oceans, and having a floating blue-coloured gas-filled bag from which long stinging tentacles hang. See Appendix 1.

**positron emission tomography** See **PET**.

**possum** Any of the many largely herbivorous tree-dwelling marsupials, Families Phalangeridae, Petauridae, Burramyidae, and Tarsipedidae. It includes the cat-sized ringtails and brushtails, and the small pygmy-possums and honey-possums. See Appendix 1.

**posterior** In animals, the rear end which is usually that part of the body furthest from the head or front end

(Fig. S11). With animals that walk upright (*bipedal, like humans), the *dorsal or surface nearest the backbone. Compare anterior.

**postpartum** After childbirth.

**postsynaptic** Applied to the membrane or nerve cell after the *synapse.

**potassium (K)** A soft silvery metal that is an *essential element for living things. It has roles in *co-enzymes, *osmoregulation, and especially the nerve *impulse.

**potoroo** Any of the several species of Australian rat-kangaroo in genus *Potorous*. These small kangaroo-like marsupials need dense ground vegetation in which to hide. See Appendix 1.

**pouch** The bag-like cavity or sac. e.g. The pouch or *marsupium found on female marsupial mammals, and in which the young complete their development. Cheek pouches of rodents. Brood pouch on a male seahorse where the eggs are kept until they hatch.

**poxvirus** A DNA-containing group of *viruses that includes the virus causing smallpox in humans.

**prairie** A large treeless area, with level or slightly undulating land covered with grass of north America. See also grassland.

**prawn** Any of the swimming decapod crustaceans, especially those in the genera *Palaemon* and *Penaeus*. Prawns are laterally compressed with 10 legs attached to the cephalothorax and many swimmerets attached to the abdomen. The first three pairs of legs have small claws, compared to shrimp which only have two pairs. See Appendix 1.

**Pre-Cambrian** Before the Cambrian, the first period of the Palaeozoic era. The Pre-Cambrian era is also sometimes called the Proterozoic era. The rocks of the Pre-Cambrian contain few fossils. See Appendix 2.

**preadapted** A feature of an organism that may be of no particular use in one environment, but which becomes an advantage in a different environment. See also evolution.

**prebiotic** Before life arose.

**precocial** Birds whose young, when hatched, are able to move about and help feed themselves. e.g. Stilt, Mallee-fowl. Compare altricial.

**precursor** A chemical substance from which another chemical substance is formed. e.g. *Nicotinic acid, one of the vitamin-B group, is a precursor for the co-enzymes *NAD and NADP.

**predation** The act of capturing and killing other animals for food.

**predator** An animal that hunts and kills other animals. Compare scavenger and parasite.

**pregnancy** The condition of being pregnant, or with some animals, *gravid. Carrying a baby (embryo or foetus) in the womb (uterus).

**prehensile** Able to grasp or clasp, especially by wrapping around. e.g. Tail of some monkeys and *possums.

**prehistory** The history of people before writing, especially that discovered by *archaeology. See also culture.

**premenstrual tension** Emotional and physical symptoms that appear in some women about a week before menstruation. e.g. Emotional upset, headache, abdominal cramps, and swelling and tenderness of the breasts.

**premolar** In mammals, a *tooth located between the canine and the first molar tooth. In humans there are four premolars in each jaw, used for crushing and grinding.

**prenatal** Before birth.

**prepuce** The ★**foreskin**.

**preservation** The act of keeping alive or in existence. Compare conservation.

**pressure** Force per unit area. Often measured in units of Newtons per square metre.

**pressure variation** In ★aquatic environments the ★pressure of water on an object increases with depth. In ★terrestrial environments the pressure of the air on an object decreases with height.

**prey** An animal hunted or seized for food.

**prickle** A short but sharp woody outgrowth from the surface (epidermis) of the plant, as in a rose. Compare spine and thorn.

**prickly pear** A cactus, *Opuntia stricta*, that was introduced to Australia from Mexico, and became a very serious agricultural pest early in the twentieth century. Introduction of a moth (★*Cactoblastis*) that feeds on prickly pear has controlled the plant. See biological control.

**primaries** The main flight ★feathers of a bird. Compare secondaries.

**primary growth** In plants, an increase in length due to cell division and subsequent cell expansion in the ★apical meristems of the shoots and roots. In gymnosperms and dicots, there is also ★secondary growth at the ★cambium, causing stems and roots to become thicker.

**primary host** The ★host normally occupied by a parasite when it is in the adult sexually mature stage of its lifecycle.

**primary structure** The linear sequence of amino acids that make up a protein molecule, or of nucleotides in DNA and RNA. Compare secondary and tertiary structure.

**primary transcript** The original RNA molecule transcribed from the DNA of the chromosomes. This is converted into messenger RNA by a process of 'editing' where ★introns are removed, and the ends chemically modified ('capped' and 'tailed'). See also transcription.

**Primates** An order of mammals which includes the prosimians, monkeys, apes, and humans. The main features of the primates include a relatively highly developed brain, binocular vision, opposable fingers and toes allowing a precision grip, often a prehensile tail. See Appendix 1.

**primitive** An organism, or feature of an organism, that is relatively simple and unspecialised, and thus thought to have appeared early in the evolution of the organism concerned.

**prion** An infectious particle made of protein but lacking DNA or RNA which may cause diseases such as ★kuru, ★scrapie and ★mad cow disease.

**probe** See gene probe.

**problem-solving** A type of thought where solutions are found to a problem that is new to the person concerned. There are a number of recognisable stages (preparation, incubation, illumination, verification), although they may overlap or be omitted in any one particular example of problem-solving.

**Proboscidea** The order of herbivorous placental mammals called the elephants. The upper incisors are very large, and are called tusks. The nose and upper-lip are modified as a trunk. See Appendix 1.

**proboscis** 1 The trunk of an elephant, or the elongated snout of animals such as the mole, tapir, shrew. 2 Any of the long and usually tubular structures used by insects and other invertebrates to feed on fluids. e.g. Coiled nectar-sipping tube of moths and butterflies.

**procaryotic** Also spelt ★**prokaryotic**.

**procumbent** Plants that spread out, lying along the ground without forming new roots. Prostrate.

**producer** An ★autotrophic organism in an ecosystem, which is able to make the complex materials that it needs from simpler inorganic materials, by ★photosynthesis (e.g. green plants) or by ★chemosynthesis (e.g. some bacteria). Producers begin all ★food chains, thus also forming the base of all food ★pyramids.

**productivity** A measure of the rate at which energy is fixed (e.g. by photosynthesis) by an organism or group of organisms. The primary or gross productivity measures the total amount of energy stored by ★producers, but some of this is used in respiration by the organisms concerned. The net primary productivity measures the energy stored in the tissues of the producer organisms, and is thus potentially available to ★consumers.

**profile** 1 In ★soil, a vertical section through all of its layers (★horizons), and extending down to the parent rock or material. See Fig. S5. 2 In vegetation, a diagram showing the type, numbers and size of plants present along a ★transect.

**progeny** Offspring or descendants.

**progesterone** A female sex hormone produced by the ★corpus luteum of the ovary, and during pregnancy, by the ★placenta. See Fig. P9.

With oestrogen, it promotes changes in the wall of the womb (uterus) to prepare it for implantation of the fertilised ovum. If fertilisation does not occur, the corpus luteum gradually stops production of progesterone, and ★menstruation occurs. If, however, fertilisation occurs and the foetus begins to grow, the placenta produces progesterone which suppresses production of more ova and the menstrual cycle. With ★oestrogen, the progesterone from the placenta also stimulates the development of the mammary glands.

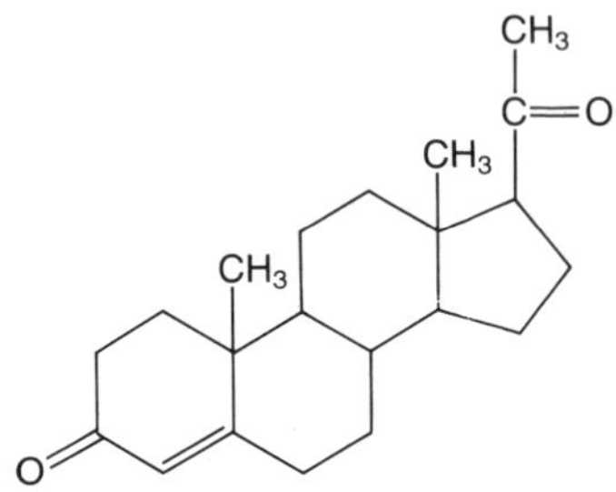


**Fig. P9 Progesterone**

**proglottid** One of the segments in the body of a ★tapeworm. The oldest and largest proglottids are found furthest from the tapeworm's neck and each proglottid is capable of producing many eggs.

**prognosis** A doctor's forecast of the course and duration of an illness based on the information available to him or her.

**prokaryotic** Also spelt **procaryotic**. Used to describe the cells of organisms where the nucleus is not separated from the cytoplasm by a nuclear membrane, where the DNA is not organised into chromosomes during cell division, and where there are no membrane-enclosed organelles such as mitochondria, lysosomes, etc. e.g. Bacteria, blue-green bacteria. Compare eukaryotic.

**prolactin** Also called **lactogenic hormone**, **luteotropic hormone**, or **LTH** for short. Produced by the front (anterior) lobe of the ★pituitary. In mammals it stimulates and maintains the production of milk by the mother after the birth of a baby. In some mammals it promotes the production of progesterone.

**proline (pro)** One of the 20 common ⋆amino acids found in proteins. Due to its ring structure, proline is an important amino acid in determining the shape of a protein. See Fig. A6.

**promoter** A part of the DNA molecule to which the enzyme ⋆RNA polymerase attaches itself, ready to start catalysing the ⋆transcription of the genes in an ⋆operon. Compare operator.

**prop root** Any of the aerial roots growing downwards from the stem and helping support the stem. e.g. Maize, mangroves. Compare buttress roots.

**propagation** Increasing the numbers of a species. Usually applied to plants, especially layering, growing cuttings, grafting and other vegetative (asexual) means of reproduction are used.

**prophase** The first stage of ⋆mitosis and ⋆meiosis during which the chromosomes become visible in the nucleus. In prophase I the homologous chromosomes pair up. Prophase I can also be divided into a number of phases: ⋆leptotene, ⋆zygotene, ⋆pachytene, ⋆diplotene, and ⋆diakinesis.

**prophylaxis** The prevention of the spread or development of disease, for example, by public health measures, vaccination, or inoculation against infection.

**proprioceptor** A sensory receptor found in muscles, tendons, joints. With the help of the balance organs in the ear, movement, change in position, muscle stretch or tension can be detected. See kinaesthetic senses, inner ear.

**prosimian** Any of a group of primates, suborder Prosimii, including lemurs, lorises, tarsiers, and tree-shrews. These primates are nocturnal, have rather pointed faces, and relatively small brains. Compare Anthropoid apes. See Appendix 1.

**Prosimii** See **prosimian** above.

**prostaglandin** One of a group of complex compounds, formed from fatty acids, that have been found in many human tissues. They act as local hormones, only influencing tissues near where they are released. They stimulate the contraction of smooth muscle, including that of blood vessels and the human womb (uterus). They are released during inflammation of damaged tissues. ⋆Aspirin is thought to work by reducing the production of prostaglandins that cause headache, fever, etc.

**prostate gland** A gland associated with the male reproductive system of mammals. In males, the ⋆urethra passes through the prostate gland, which adds an alkaline fluid to the ⋆semen to activate the sperm.

**prosthesis** A substitute for a missing or non-functioning part of the body. e.g. Artificial limb, hip-bone, or denture.

**prosthetic group** A non-protein substance that is usually bound by covalent bonds to a protein (⋆conjugated protein). e.g. Various metal ions, co-enzymes, lipids in ⋆lipoproteins, polysaccharides in ⋆glycoproteins.

**prostrate** Lying flat on the ground. Used to describe plants growing close to or along the ground.

**Proteaceae** The protea or banksia family of dicot flowering plants, including the waratah, banksia, grevillea, etc. The flowers are often clustered on large spikes or heads. See Appendix 1.

**protease** Also called **proteolytic**

**enzymes**. An enzyme that catalyses the hydrolysis of protein, thus helping to split it into peptide chains and amino acids. Most such enzymes attack only certain chemical bonds, and thus several different proteases are needed to complete digestion. e.g. Papain (obtained from paw-paw trees) used commercially to make meat tenderiser. ⋆Peptidases are proteases that work outside cells. e.g. Digestive enzymes. Other proteases work inside the cell. e.g. Cathepsins.

**protective colouration** The colouring of various organisms that help in ⋆camouflage, making it hard for predators to see them. e.g. Moths with a mottled pattern are hard to see on lichen-covered tree-trunks. Green-coloured snakes and insects are hard to see amongst green leaves. Compare mimicry.

**protein** A large group of macromolecules, all containing nitrogen, which are essential to the structure and function of all living things. See Fig. P10. Proteins are made up of one or more chains of ⋆amino acids chemically joined together by covalent ⋆peptide bonds (Fig. P3).

Short lengths of amino acids are called ⋆peptides, while those up to about 100 amino acids are called ⋆polypeptides. There are about 20 naturally occurring amino acids found in proteins (Fig. A6), the exact order of which in any particular protein is often crucial, and is genetically specified by messenger RNA (⋆mRNA) during protein synthesis (see translation). Proteins can be classified chemically or by shape. Chemically, the simple proteins are made only of amino acids. e.g. Albumin, globulins. ⋆Conjugated proteins are covalently bonded to non-protein substances. e.g. Glycoproteins to polysaccharides, lipoproteins to lipids. The two general shapes of proteins are called ⋆fibrous and ⋆globular. The fibrous proteins are linear, and are structural (collagen, keratin) or contractile (myosin, actin) in function. The globular proteins have a more complex, and often very specific, three-dimensional shape, and function as hormones, antibodies and enzymes. See also primary, secondary and tertiary structure.

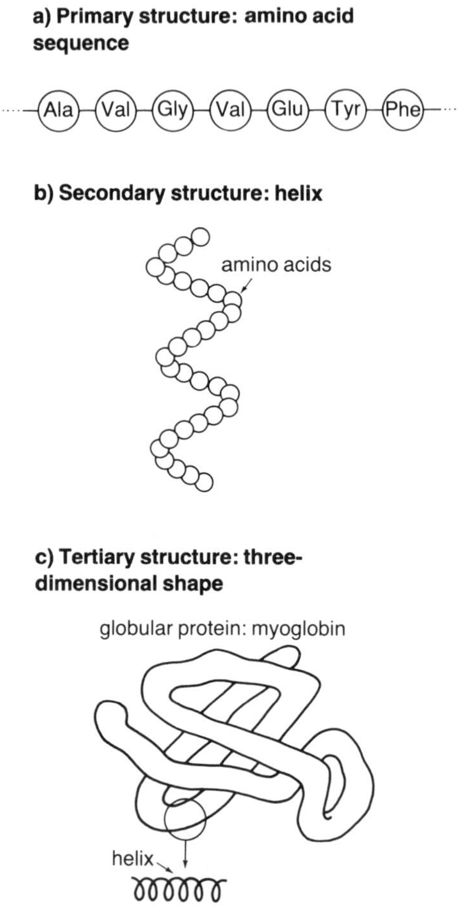


**Fig. P10 Protein**

**protein synthesis** Proteins are made (synthesised) from amino acids at the ribosomes in the cytoplasm of cells. The sequence of amino acids is ultimately determined by the genetic code of the DNA in the nucleus. This code is ⋆transcribed into primary RNA, which as it leaves the nucleus is 'edited' by various enzymes to convert it into messenger RNA (mRNA). Once the mRNA reaches the cytoplasm, one or more ribosomes begin to move along the molecule, 'translating' it into a sequence of amino acids. See translation for details.

**proteinase** Also called **endopeptidases**. See peptidase.

**proteolytic enzymes** See proteases above.

**Proterozoic** See pre-Cambrian above.

**prothallus** A small flattened, often heart-shaped, plant in the lifecycle of the ⋆pteridophytes (e.g. ferns, horsetails, clubmosses). The prothallus is the ⋆gametophyte stage, producing male and female gametes. In ferns the prothallus produces both male and female gametes. In other plants (e.g. *Selaginella*) there are separate male and female prothallia. In both cases, the sperm need droplets of water to swim through to reach the egg cell. See also alternation of generations.

**prothrombin** An inactive form (⋆precursor) of the hormone ⋆thrombin. Chemically it is a ⋆glycoprotein. Many co-factors, including calcium ions and platelets, are needed for the complex reaction that converts prothrombin to thrombin. See also blood clotting.

**protist** Any member of the kingdom called Protista, a group that includes the mostly single-celled ⋆eukaryotic organisms. e.g. *Amoeba*, *Paramecium*, seaweeds. Compare plant, animal, fungi and monera.

**proton** A positively charged particle found in the nuclei of all atoms. It has a single positive charge that just balances the negative charge of an ⋆electron, and has a mass of 1, almost the same as that of the uncharged particle called a neutron. A hydrogen ion, $H^+$, is a proton.

**protoplasm** All the substances within a cell, including the cell membrane.

The protoplasm is divided into the ⋆nucleus and ⋆cytoplasm. About 90% of protoplasm is water. Of the dry weight, approximately 65% is proteins, about 16% is lipids, and simple sugars make up around 12%. The remainder is various small compounds and mineral salts. The protoplasm is a highly ordered structure. The ⋆organelles act as separate structures. The ⋆endoplasmic reticulum divides the cell into compartments each with their own function. Within this membrane system the liquids are in a ⋆colloidal state. The whole cell is given a shape by a system of fine fibrils and tubules—the ⋆cytoskeleton.

**protoplast** All of a plant cell except the ⋆cell wall. Thus the protoplast is a plant's equivalent to an animal cell.

**Prototherian** Another name for the ⋆**monotremes**.

**Protozoa** Singular is **protozoan**. All the single-celled animals. Some form colonies of single cells, but there are no tissues present. Protozoans inhabit marine, freshwater, and damp terrestrial environments. Some are parasitic. See Appendix 1.

**proximal** The end nearest to an organism's body, or to the point of attachment. Compare distal.

**prune** 1 The selective cutting of branches or twigs from a tree or shrub, in order to improve the shape, fruiting power, etc. 2 The dried fruit of a plum.

**pseudocarp** 1 A false fruit. A ⋆fruit that develops from parts other than just the ovary. e.g. From a receptacle (apple, rose), or from a complete ⋆inflorescence. 2 A particular type of pseudocarp is confusingly also called a pseudocarp. Here the false fruit is also derived from a swollen receptacle, but the achenes (small single-seeded dry fruit) are embedded in its fleshy outside. e.g. Strawberry.

**pseudopodium** Called **pseudopod** for short. A 'false foot'. A temporary projection or extension formed from flowing cytoplasm, and used for cell movement and for ⋆engulfing food. Seen in *Amoeba*, and in ⋆leucocytes. See also phagocytes and Fig. E6.

**Psilopsida** A small class of vascular plants, also called **Fork-ferns**. These fern-like plants lack true roots or leaves, and have a simple vascular system. See Appendix 1.

**psychiatrist** A person who treats mental disease or emotional problems.

**psychologist** A person who studies the human mind and behaviour.

**psychosis** Serious mental illness, usually requiring hospital treatment.

**psychotherapy** The treatment of emotional and mental disorders by psychological methods rather than by physical or medical methods.

**psychosomatic** A physical sickness which is caused by, or strongly influenced by, the emotional state of the patient.

**Pteridophytes** Ferns. See Filicopsida and Appendix 1.

**pterodactyl** Any member of the Pterosauria, an order of extinct flying reptiles that lived during the Jurassic to Cretaceous. Their wings were large membranes supported by their forelimbs. See Appendix 2.

**ptyalin** An ⋆amylase enzyme present in the saliva of some mammals (including humans), which helps the breakdown of starch (by ⋆hydrolysis) into simple sugars. Also known as salivary amylase.

**puberty** The period of life when sexual maturity is reached.

In boys, sperm starts to be produced, and the male sex hormones such as ⋆testosterone begin to be released, resulting in the voice 'breaking' to become deeper. In girls, ⋆oestrogen and ⋆progesterone are produced, resulting in breast development and the beginning of menstruation. See secondary sexual characteristics for a more complete list of changes. Puberty can occur at many different ages, usually between 11 and 15, being slightly later, on average, for boys.

**pubescent** Covered in fine hairs.

**pubic bone** The pubis. See below.

**pubis** The forward projecting, and the smallest of the three bones making up one-half of the ⋆pelvic girdle or pelvis. See Fig. S3.

**puff** A swollen region of the ⋆giant chromosomes found in the salivary glands of *Drosophila* and similar insects, thought to represent a site of ⋆transcription of DNA into messenger RNA.

**pulmonary** Of or relating to the lungs. e.g. Pulmonary artery carries deoxygenated blood from the heart to the lungs. The pulmonary vein carries oxygenated blood from the lungs back to the heart. Compare systemic circulation.

**pulp** A soft spongy tissue. e.g. In animals, the tissue which fills the pulp cavity of the tooth. In plants, the soft and succulent part of a fruit, or the pith of certain stems.

**pulse** 1 The regular wave of pressure felt in the arteries due to the pumping action of the heart (heart beat). The high pressure results from the contraction of the muscular left ⋆ventricle of the heart, forcing blood into the ⋆aorta, and thence to other arteries. Under pressure, the artery expands (dilates), and this bulge passes along the arteries being felt as a pulse. The best locations to feel a pulse are the wrist, neck and temple. 2 The edible seeds of legumes such as peas and beans. Sometime the whole plant that produces these seeds is called a pulse.

**punctate** Marked with dots, depressions, or translucent glands.

**punctuated equilibrium** A theory of *evolution where new organisms evolve quickly (in geological time) rather than by gradual change. After the period of rapid change (say one hundred thousand years), there are long periods (say millions of years) of very little evolutionary change. Compare Neo-Darwinism.

**pungent** 1 Piercing. Ending in a sharp point. 2 A strong, acrid smell.

**Punnett square** A checkerboard diagram, devised by a geneticist named Punnett, and used to determine the percentages of different genotypes produced during any particular genetic cross. See Fig. P11.

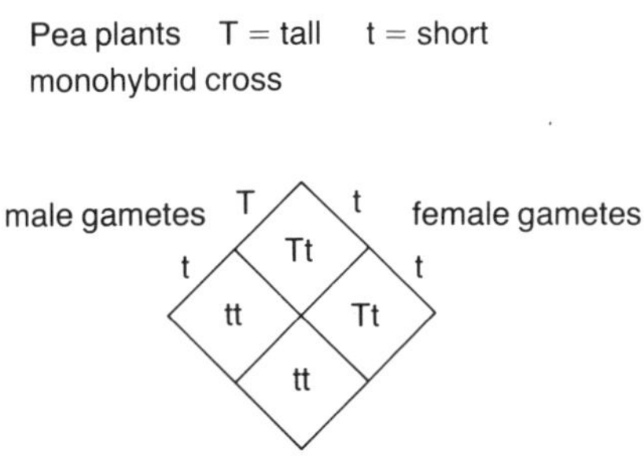


**Fig. P11 Punnett square**

**pupa** Plural is **pupae**. A stage in the lifecycle of some insects where they undergo complete *metamorphosis. It is usually a stage between larva and adult, spent in a protective *cocoon, *chrysalis or chamber in the ground. Inside, the larval body and organs are completely broken down, and a new adult structure is formed. e.g. Butterflies, moths, bees, beetles, true flies, etc.

**pupil** The opening in the *iris of the eye, through which light enters to reach the *retina. See Fig. E9. Since each retina is most sensitive to a particular intensity of light, the pupil size is varied to keep the light as close to the optimum intensity as possible. e.g. Mammals and cephalopods. In nocturnal animals, the pupil is only a slit during daylight, to reduce the intensity of the bright light reaching the very sensitive retina.

**pure line** 1 A succession of generations of an organism that show no variation (other than mutations) because they are completely *homozygous. Such organisms are said to breed true. Pure lines are usually produced by *self-fertilisation in plants, and *inbreeding in animals. 2 In practice the term is often applied to organisms that are homozygous for the genes producing a certain desirable character. They are said to breed true for the character under consideration. e.g. Height of stem in short-stemmed wheat or rice.

**purine** One of a group of nitrogen-containing bases built on a basic two-ring structure. e.g. *Adenine and *guanine that form part of the genetic code in nucleic acids such as DNA and RNA. Compare pyrimidine. See Fig. B3.

**pus** A yellowish fluid derived from blood, and containing many dead and dying white blood cells, and tissue debris. Pus is produced during *inflammation and the fight against invading bacteria, etc. See also abscess.

**putrefaction** To become putrid; the process of decaying and rotting, giving off unpleasant smells.

**pygmy glider** The smallest of the gliding possums, *Acrobatus pygmaeus*, Family Burramyidae, found in eastern Australia. Also called the feathertailed glider due to the appearance of its tail.

Feeds on nectar, sap and insects found in trees and fruits. See Appendix 1.

**pyloric sphincter** In vertebrates, a ring of muscle around the ⋆pylorus, the opening between the stomach and duodenum. This ring of muscle controls the rate at which food passes from the stomach to the intestines. See Fig. D2.

**pylorus** In vertebrates, the opening from the stomach to the start of the small intestine. See pyloric sphincter.

**pyramid** In ecology, a pictorial way of showing the relationships between the different ⋆trophic levels of ⋆food chains in a community. e.g. Pyramid of ⋆biomass; pyramid of numbers. See Fig. P12. In each case, at the base of the pyramid are the primary ⋆producers. Above are the ⋆consumers: the herbivores and various carnivores. Because the consumers on each trophic level use some of the food they eat to supply energy, the total biomass or energy (measured over a period of time) available to the next level decreases as one moves upwards, thus producing the pyramid effect. Also, the number of organisms tends to decrease as they become larger. See also amplification.

**pyrethrin** Either of two compounds obtained from the pyrethrum flowers, Family ⋆Asteraceae, and used as contact insecticides.

**pyrimidine** One of a group of nitrogen-containing bases built on a single ring structure. e.g. ⋆Cytosine, ⋆thymine, and ⋆uracil that form part of the genetic code in nucleic acids such as DNA and RNA. ⋆Thiamin (vitamin B1) is also constructed from pyrimidine. Compare purine. See Fig. B3.

**Pyrrophyta** A phylum of microscopic unicellular and colonial algae, including the dinoflagellates, unicellular algae with two flagella, which are prominent in the phytoplankton of tropical waters. See Appendix 1.

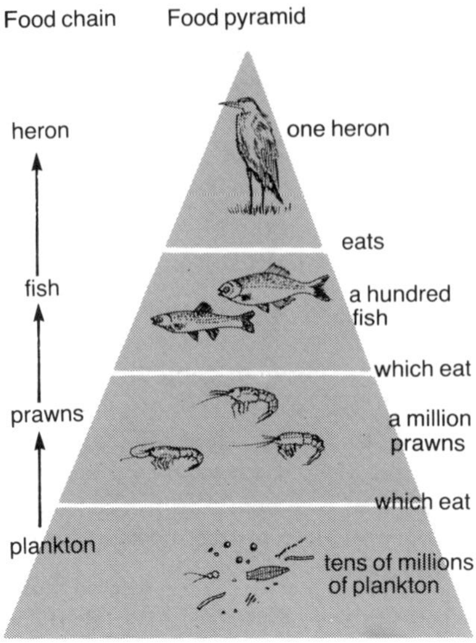


**Fig. P12 Food chain and food pyramid**

**pyruvate** The ion formed by the breakup of ⋆pyruvic acid in water:

$$\underset{\text{pyruvic acid}}{CH_3COCOOH} \rightleftharpoons \underset{\text{pyruvate}}{CH_3COCOO^-} + \underset{\text{hydrogen ion}}{H^+}$$

**pyruvic acid** A 3-carbon organic acid, $CH_3COCOOH$, which is important in respiration and photosynthesis. In respiration it is an intermediate in the ⋆glycolysis reaction, while in photosynthesis it is an intermediate in the ⋆$C_4$ reaction pathway.

**pythons** Any member of the snake family Boidae, which kill their prey by coiling their body around their victim and suffocating them, as breathing is prevented. See Appendix 1.

# Qq

**quadrangular** Of plant stems, having four corners when viewed in cross section.

**quadrat** An area of known size used in ecological studies to measure some property (e.g. distribution, number, frequency) of a population. For example, the numbers of each organism are counted in several randomly selected quadrats, allowing the population densities to be compared in different habitats. Compare transect.

**quadriceps femoris** A large muscle, which divides into four parts. It is attached to the kneecap (patella) and extends up the front of the leg to the pelvic girdle. It helps move the leg.

**quadriplegic** Also spelt **quadraplegic**. A person whose arms and legs are paralysed due to damage to the spinal cord in the neck region. Compare paraplegic.

**quadruped** An animal that has four feet or legs. See also ⋆quadrupedal and ⋆knuckle walking.

**quadrupedal** Walks primarily on four legs. Compare ⋆knuckle walking.

**qualitative** A purely descriptive study or record of an experiment without measurements or numerical analysis. Compare quantitative.

**quantitative** A study or experiment based on measurement and numerical analysis. Compare qualitative.

**quarantine** The isolation of organisms from one another to help prevent the spread of disease.

**Quaternary** A geological period that combines the ⋆Pleistocene and ⋆Recent periods. It makes up the second part of the Cainozoic era. See Appendix 2.

**queen** The fully mature, egg-laying female of a species of social insect. e.g. Queen bee.

**quickening** The time, about halfway through pregnancy, when kicks and other movements within the womb (uterus) give unmistakable evidence that life is present.

**quill** **1** A feather, especially one of the large feathers of a bird's wing or tail. **2** The hard and hollow end of a feather close to the body of a bird. **3** The stiff spine of a hedgehog, porcupine or ⋆echidna.

**quinine** A drug used to treat malaria, bitter tasting, and obtained from the bark of a tree of genus *Cinchona*.

**quinsy** An ⋆abscess in and around a ⋆tonsil, causing a very sore throat.

**quokka** A small wallaby, *Setonix brachyurus*, Family Macropodidae, found mainly on islands off the coast of Western Australia, and to a lesser extent, on the nearby mainland. It differs from members of genus *Macropus* in arrangement of its teeth, skull structure, chromosomes and blood proteins. See Appendix 1.

**quolls** A group of relatively larger carnivorous marsupial mammals, in genus *Dasyurus*, of family Dasyuridae, often called tiger cats, marsupial cats, native cats, etc. See Appendix 1.

# Rr

**R factor** Short for **resistance factor**. A type of ★plasmid found in bacteria which contains genes giving the bacterium resistance to antibiotics. These plasmids can be readily transferred between bacteria, including bacteria from different genera.

**rabbit** A small leaping and burrowing mammal, *Oryctolagus cuniculus,* in the order Lagomorpha. A serious agricultural pest brought to Australia with the First Fleet but the most serious release occurred in 1859 at Geelong. Rabbits have been controlled to some extent by the disease ★myxomatosis and more recently by the rabbit ★calicivirus. See Appendix 1.

**rabbit calicivirus** See calicivirus.

**rabies** A lethal disease, not present in Australia, caused by a viral infection of the brain and nervous system. The virus is transmitted to humans by the bite of a rabid mammal. Can be treated with an unpleasant vaccine.

**race** 1 A subdivision of a species. See also subspecies, variety, breed, etc. 2 The divisions of *Homo sapiens*. There is no agreed racial classification for humans because of the many intermediates that exist. Most classifications include three main races, Caucasoid (e.g. European), Mongoloid (e.g. Chinese), and Negroid (e.g. Central African), plus a number of smaller groups such as Australoid (e.g. Australian Aborigine), Negritoid (e.g. Pygmy), Bushmanoid (e.g. Bushman of the Kalahari) and Papuan-Melanesian (e.g. New Guinea Highlander).

**racemose** Used to describe a type of ★inflorescence where the central axis, which bears flowers on shorter lateral stems, keeps growing and adding to the inflorescence. Hence the youngest flowers are at the top. There are a number of different types of racemose inflorescence: capitulum, corymb, panicle, spadix, spike, umbel and raceme. Compare cymose inflorescence. See Fig. I4.

**rachis** Also spelt **rhachis**. 1 The main axis of an ★inflorescence of flowers. See Fig. I4. 2 The main axis of a compound ★leaf or frond, to which the lateral leaflets or pinna are attached. See Fig. L1.

**radial symmetry** The arrangement of the parts of an organism so that any cut taken through the centre divides the organism into two similar halves. e.g. Sea anemone, sea urchin, flowers. Radial symmetry is seldom perfect. Compare bilateral symmetry. See also symmetry and Fig. S11.

**radiata pine** A valuable ★softwood tree, *Pinus radiata,* or Monterey pine, that is native to California, but is now grown widely in Australia for its timber.

**radiation** 1 The movement of energy in the form of parts of the electromagnetic spectrum, especially infra-red, light, ultra-violet, X-ray and gamma radiation. 2 The movement of radioactive particles such as alpha, beta and gamma radiation.

**radical** 1 Of leaves, arising together at the base of a plant stem, and producing a ★rosette plant. 2 A very reactive atom, or group of atoms, that have an unpaired electron. e.g. Free radicals such as $O^{\cdot}$ are important in the formation of photochemical smog.

**radicle** The root that appears from the embryo during the germination of a ⋆seed. It is normally the first organ to appear from the seed. Compare ⋆plumule.

**radioactive dating** Also called **radiometric dating**. The determination of the age of rocks, minerals and once-living material by measuring the levels of certain radioactive elements.

Fossils in the rock, or just above or below the rock, can be dated. Specimens up to about 50 000 years old can have their age determined by ⋆radiocarbon dating (also called carbon-14 dating). Older fossils have to be dated, if possible, from nearby volcanic rocks. e.g. The famous fossil hominid Lucy was estimated as being about 3 million years old, since the volcanic ash that lies just under the sedimentary rocks that contained her remains was dated by the potassium-argon methods as being about 3.3 million years old. Radioactive dating relies on the fact that radioactive decay occurs at a constant rate. Thus by measuring the rate of radioactive decay, the amount of the original radioactive element that is left, and how much of the stable daughter element has been produced, the age of the rock can be determined. e.g. In the potassium-argon method, radioactive potassium-40 decays to form stable argon-40 with a rate measured by its ⋆half-life of $11.8 \times 10^9$ years.

**radioactive decay** See radioactivity below.

**radioactive isotope** An ⋆isotope of an element that is ⋆radioactive, which may be naturally occurring, or artificially produced in a nuclear reactor. Isotopes of an element all have the same number of electrons and protons, but differ in the number of neutrons. e.g. Living things are mostly made of carbon-12, but there is always a small amount of radioactive isotope carbon-14 present as well. See also radioactive tracer and radiology below.

**radioactive tracer** A ⋆radioactive isotope used to trace or track where certain substances travel within an organism, during chemical reactions, etc. e.g. During early research into the chemical reactions that occur during photosynthesis, plants were grown in an atmosphere containing carbon dioxide that was made from radioactive carbon-14 instead of normal carbon-12. The reactions were stopped at different stages to see which compounds now contained the carbon-14.

**radioactivity** The property of certain nuclei (e.g. uranium-238; potassium-40; carbon-14) where they spontaneously give off small particles (alpha, beta, gamma) from their nuclei. The rate at which they decay is constant, and is measured by the ⋆half-life of the element.

**radiocarbon dating** A form of ⋆radioactive dating that uses an isotope of carbon called carbon-14.

High in the atmosphere carbon-14 is produced by cosmic rays bombarding the nitrogen-14 in the air. The carbon-14 enters plants as carbon dioxide and becomes part of their tissues. If the rate of formation of carbon-14 has remained constant, then when the plant (or animal which ate the plant) was fossilised, it should have had the same percentage of carbon-14 as do plants and animals today. Since that time, the carbon-14 has been decaying away with half-life of about 5700 years. Assuming that none of the carbon-14 has been removed (e.g. by leaching), we need only to determine the percentage left, and we can then calculate the specimen's age. As you can see, there are a number of assumptions made (and others not listed). When this radiocarbon 'clock' was recently used to date a tree of known age (i.e. by counting the tree-rings), it was found to be in error, and small corrections have had to be made.

**radioimmunoassay** An experimental technique used to measure the amounts of ⋆polypeptide hormones in the blood stream, and other body fluids.

**radioisotope** See radioactive isotope above.

**radiolarian** A group of marine ⋆protozoans, class Sarcodina, which form part of the ⋆plankton, and whose amoeba-like body protrudes through holes in a silica skeleton. Large deposits of these skeletons can collect on the ocean floor to form a radiolarian ooze. See Appendix 1.

**radiology** The science of using X-rays and the radiations from radioactive isotopes (e.g. Gamma rays from artificially made cobalt-60). Such materials can be used in research, materials testing, medicine (treating cancers), etc.

**radiometric dating** See radioactive dating above.

**radius** One of the two long bones in the lower arm or forelimb of vertebrates that connects the hand or forefoot to the ⋆humerus. Compare ulna. See Fig. S3.

**radula** The tooth-covered ribbon-like tongue found in most molluscs (except bivalves). See Fig. R1. This muscular tongue is used like a rasp or file to remove food e.g. algae from a hard surface. The radula is continually replaced as it is worn out.

**rain** Drops of water falling from the sky. Rain forms when water vapour condenses in the atmosphere to form tiny droplets. These combine to form water drops as they fall.

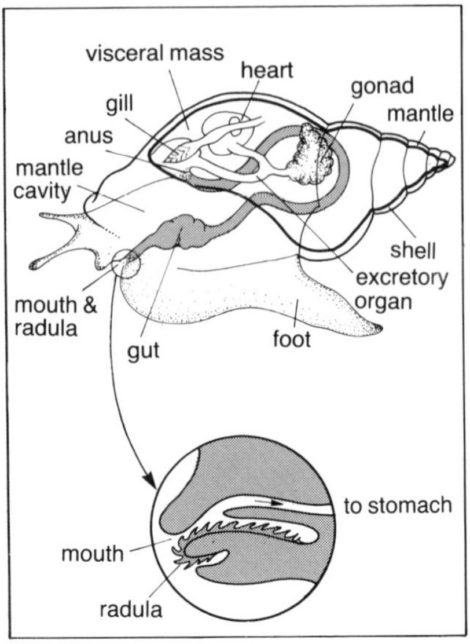


**Fig. R1 Radula of a mollusc**

**rain shadow** A dry region on the side of a mountain range normally sheltered from moisture-carrying winds.

**rainfall** The amount (depth) of rain, snow, etc. that falls in a particular place for a given period of time. e.g. The average rainfall in rainforests near Cairns is 500 mm in March.

**rainforest** A dense forest with a closed ⋆canopy (little light reaches the ground) often with ⋆lianas, ⋆epiphytes, and little ground cover.

Australian rainforests are divided into tropical, sub-tropical, and temperate rainforests. Tropical rainforests rarely include *Eucalyptus* species, and have an enormous variety of trees (often more than 50 species per hectare). The sub-tropical and temperate rainforests have relatively fewer species (which can include *Eucalyptus*), grow in areas of lower temperature, mostly on rich volcanic soils. Soils of tropical rainforests are sometimes poor due to the leaching effects of high rainfall. See Fig Z2.

***Ramapithecus*** A genus of fossil apes that were once thought to be ancestors of humans. Compare *Homo*.

**randomised trials** An experimental design in which different treatments are distributed in random order. e.g. In horticulture, treatment for each block or plot is chosen at random. In medicine, a drug is given to patients at random, usually with neither patient or doctor knowing. Other patients are given a ⋆placebo.

**range** 1 The area normally inhabited by a group of organisms. e.g. Home range. 2 A chain of mountains. e.g. Mountain range. 3 In statistics, the difference between the lowest and the highest values in a set of readings.

**raptor** A bird-of-prey. e.g. Hawk, eagle.

**rash** An area of skin coloured pink or red, sometimes with spots or lumps, caused by irritation, burning, infection, etc.

**rat** A group of long-tailed rodent placentals, especially *Rattus norvegicus*, or brown rat that inhabits cities and has transmitted serious disease through the fleas that it carries (e.g. ⋆plague). There is also a number of native bush and water rats in Australia. See Appendix 1.

**rat-kangaroo** Any of the small kangaroo-like marsupials, Family Potoroidae, smaller than a cat, and including the ⋆bettongs, ⋆potoroos, and Musky Rat-Kangaroo. They seem to be more omnivorous in their feeding habits than the ⋆macropods. See Appendix 1.

**ratite** An order of flightless birds lacking a keel on the sternum (breastbone). e.g. Emu, cassowary, kiwi, ostrich, rhea. See Appendix 1.

**ray** 1 In plants, rays are strands of living cells that connect the ⋆pith to outer layers of the stem. Rays carry water and nutrients from the vascular tissue to the inner parts of the stem. See also medullary rays. 2 In animals, a group of flattened fish, also called skates. See cartilaginous fish.

**reabsorption** To absorb again something that has already been absorbed. e.g. In the tubules of the ⋆kidney, water and other useful materials are reabsorbed into the blood capillaries.

**reactant** A substance that takes part in a chemical reaction.

**reaction** A chemical change brought about by the actions of two or more chemicals upon each other, or by heating a chemical, etc.

**reaction time** The time between the reception of a ⋆stimulus, and the beginning of an appropriate ⋆response. e.g. The time between a car driver seeing danger ahead, and applying the brakes.

**reagent** A chemical which can be used to test for the presence of some other chemical, for making certain new chemicals, etc.

**realm** One of the six major ⋆bio-geographical regions of the world, each with its characteristic variety of animals. See Australian, Oriental, Ethiopian, Palaearctic, Neotropical and Nearctic. See Fig. B5.

**recapitulation** A now-discredited theory that proposed that, as embryos, organisms pass through stages that the organism went through during its ⋆evolution. Thus, supposed gill slits in the embryos of humans represent structures from the fish stage in human ancestry. Proposed by Ernst Haeckel (1834–1919), it was made famous by statements like 'ontogeny repeats phylogeny', or 'in its development from zygote to adult, an organism climbs its own family tree'. In later life Haeckel admitted that he had 'touched up' the drawings of embryos he used to support the theory.

**Recent** An epoch in the Quaternary period, following the Pleistocene. It runs from the end of the last ice age to the present. See Appendix 2.

**receptacle** In flowering plants, the enlarged top of the flower stalk (peduncle) where the floral parts (sepals, petals, stamens and carpels) are attached. There is much variation in the relative positions of the ovary and other parts. See hypogyny, perigyny, epigyny. See Fig. F2.

**receptor** That part of the nervous system responsible for detecting changes in an animal's environment. Receptors are cells or parts of cells which respond to various stimuli by producing nerve *impulses, which are then sent to the central nervous system. The greater the number of impulses per second, the more intense the stimulus. Receptors include the *chemoreceptors (taste, smell), *photoreceptors (light detecting rods, cones), and *mechanoreceptors (stretch, balance).

**receptor potential** The tiny potential difference (voltage) created across the membrane of a stimulated receptor cell of a sense organ. The strength of the voltage increases with the strength of the stimulus. If the voltage induced is large enough (the *threshold), one or more impulses are produced in the next neurone.

**recessive** In genetics, the form of a gene (*allele) that is 'turned off' by the dominant form of the gene, thus not influencing the feature normally controlled by that gene. The recessive form of a gene only produces its effect when two are present (double recessive or *homozygous recessive), or if it is on the *sex chromosome (see sex linked). e.g. For the gene controlling height in Mendel's peas, tt. Compare dominant.

**reciprocal cross** A pair of matings where the genes present in the male and female parents are reversed in the second cross. e.g. In pea plants, if the sperm in the pollen was originally TT (where T produces tall plants) and the ova tt (where t produces short plants), then in the second cross the sperm would be tt and the ova TT. The reciprocal cross is used to see if the sex of the organism has any influence on the results of the cross.

**reclamation** 1 A term applied to the using of water and land resources for agricultural and other purposes, through activities such as irrigation, and drainage of swamps and tidal marshes. 2 In environmental studies, the restoration to more natural conditions of strip mine areas, construction sites, and areas damaged by other human activities.

**recognition site** A section of DNA where a specific base sequence occurs, allowing *restriction enzymes to cut the molecule. e.g. The enzyme *Eco* R1 cuts wherever it finds the sequence GAATTC, while *Sma* 1 cuts at CCGGG. See genetic engineering, gene splicing and sticky ends.

**recombinant DNA** A length of DNA molecule which contains genes that have been inserted using the techniques of *genetic engineering (or better, recombinant-DNA technology). e.g. The gene producing the hormone *insulin in humans has been inserted into the DNA of bacteria.

**recombination** In genetics, the formation of combinations of genes that are not present in the parents. This can take place during meiosis when *crossing over of chromatids occurs, thus exchanging genes between similar (homologous) chromosomes. See Fig. H3. It also occurs when the chromosomes line up along the equator during prophase 1 of meiosis, when

★independent assortment (also called random assortment) results in a general mixing of chromosomes (although one of each type ends up in each gamete). See also recombinant-DNA.

**rectum** The end section of the ★large intestine (or colon) in which faeces are stored. See Fig. D2. The rectum opens to the outside through the ★anus. In mammals, rings of muscle (sphincters) at both ends control movement of the faeces.

**rectus abdominis** A muscle of the abdomen that connects part of the pelvic girdle to the fifth to seventh ribs. It helps bend the vertebral column.

**recurved** Curved backwards or downwards.

**recycling** The reuse of empty bottles, metal, paper, etc. to make new products.

**red blood cell** Also called ★**erythrocytes** (Fig. B6). Red blood cells carry oxygen (attached to the haemoglobin) from the lungs to the cells of the body. They carry some carbon dioxide back again, but most is returned dissolved in the blood ★plasma. The red colouring is due to the pigment ★haemoglobin in the cells.

In humans there are about 5 000 000 red blood cells per cubic millimetre of blood. Each cell is flexible, being able to squeeze single file through narrow capillaries. They are shaped like flattened discs (5 μm across), with both faces curved inwards. Red blood cells lose their nucleus after they are produced in the bone marrow, and thus can only survive for a few months. They are replaced at a rate of more than a million per second. Compare leucocytes.

**red gum** Any of the *Eucalyptus* gums that have a hard red wood, such as *E. tereticornis*, or forest red gum from eastern Australia. See also river red gum.

**red-back spider** A small, but highly poisonous (venomous) spider with black body, and often an orange or red stripe on the back of the abdomen. The species, *Latrodectus hasseltii*, may have been introduced into Australia and is very similar to the katipo spider of New Zealand, and the black widow spider of America.

**redox reaction** Short for **reduction-oxidation reaction**. See oxidation-reduction reaction.

**reducing sugar** A sugar that is capable of reducing an oxidising agent. e.g. All monosaccharides, maltose (but not sucrose). ★Benedict's solution and ★Fehling's solution are both designed to detect reducing sugars.

**reduction** A chemical reaction in which one or more of the following occurs: **1** Oxygen is lost from a chemical substance. **2** Hydrogen combines with a chemical substance. **3** A chemical substance gains electrons. Compare with oxidation.

**reduction division** A reduction in the number of chromosomes from diploid (two of each type) to haploid (one of each type), as occurs in ★meiosis.

**redwood** *Sequoia sempervirens*, a conifer growing in western USA, specimens of which are reported to be the tallest trees in the world.

**reef** A relatively narrow ridge of rocks, sand or coral close to the surface of water or land. e.g. Coral reefs such as the Great Barrier Reef; on land, quartz reefs sometimes contain gold.

**reflex** A very simple form of behaviour present in almost all animals with a nervous system. The organism reacts automatically to a certain stimulus, which is not controlled by the brain. e.g. The blinking of the eye when an object is waved in front. The knee-jerk produced when one is struck just below the knee-cap. The nerve impulses from the sense organs travel to the spinal cord, and then directly to a muscle along a ⋆reflex arc. See Fig. R2.

**reflex arc** The nervous pathway along which the impulses for a ⋆reflex action pass. See Fig. R2.

Starting at the sense organ, the ⋆impulses pass along the ⋆sensory neurone (⋆afferent) to the spinal cord. Inside the cord they may connect directly to the ⋆motor neurone, or through a short intermediate neurone to the motor neurone. Thus, without going to the brain, the impulse passes out along the motor neurone (⋆efferent) to the muscle or other motor organ. See also reflex above.

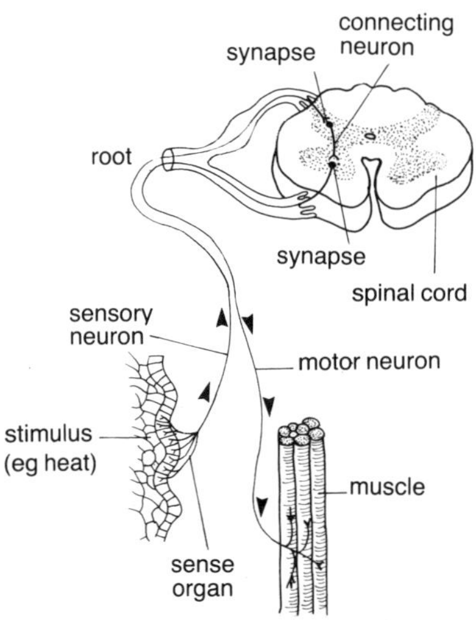


**Fig. R2 Reflex arc**

**reflexed** Bent backwards or downwards at a sharp angle.

**refractory period** The time taken by a ⋆neurone to recover after a nerve impulse has passed. During the first part of this time, or absolute refractory period, no impulse can pass no matter how strong the stimulus. The relative refractory period follows, during which time a particularly strong stimulus could generate an impulse. During the refractory period, the cell is pumping the sodium ions back outside the cell, and the potassium ions inside. See also impulse, action potential and Fig. I3.

**regeneration** The regrowth of parts of an organism lost through injury, damage, or by being shed to escape predators.

Regeneration is very common with plants, as we see after the lawn is mowed (see internodes), trees and shrubs are ⋆pruned, or cuttings are planted. Many invertebrates can regenerate parts of their bodies. Planarian flatworms can regenerate a complete body from only a small piece of tissue. Crustaceans (e.g. crabs), spiders and echinoderms (e.g. seastars) can regenerate lost arms. Vertebrates are not able to regenerate as much of their bodies. A lizard can regenerate its tail if it is shed (⋆autotomy) whilst escaping predators. Regeneration in mammals is restricted to wound healing, and the regrowth of damaged peripheral nerves. Some organs (e.g. liver, pancreas) will grow larger to compensate for a loss of part of them, which is not strictly regeneration.

**regosol** Very young soils developing from loose unconsolidated deposits of weathered material such as sand, dust and broken stone. Regosol covers about 45% of the country, especially the arid interior.

**regular** Used to describe flowers that have a radially symmetrical ⋆perianth.

**regulation** The control and co-ordination of ⋆development in the early stages of embryo growth. In some embryos, even severe damage can be repaired and normal growth can then continue. e.g. Sea-urchin embryos, if cut in half early enough, can complete normal growth. Other embryos can tolerate only a little damage. e.g. Frogs.

**regulator gene** A gene that helps control the rate at which another gene is transcribed through the production of a ⋆repressor molecule. The regulator gene is not always next to the genes it controls. See operon.

**regurgitation** The voluntary or involuntary return of partly digested food from the stomach to the mouth. e.g. In ⋆ruminants, the only partly chewed grass is returned to the mouth ('chewing the cud').

**reinforcement** In ⋆operant conditioning, the establishing of certain behaviour by rewarding acceptable responses.

**rejection** The actions of the ⋆immune system in attacking ⋆grafts or ⋆transplants as foreign tissues. See also immunological suppression.

**relative humidity** The percentage of moisture that is in the atmosphere at a particular temperature, compared to the amount that could be present if the air was ⋆saturated.

**relaxin** A hormone produced in the female body during the latter stages of pregnancy which makes the birth of young easier by breaking down cartilage and collagen fibre, thus causing relaxation of the pelvic ligaments, as well as the cervix of the uterus.

**releaser** Any stimulus which causes an instinctive pattern of behaviour in an animal. e.g. The appearance of a bird at a nest releases the begging behaviour of the nestlings.

**relict** Also spelt **relic**. A group of organisms, often scattered in small groups, that are the survivors from a formerly more widely distributed group. e.g. The ⋆Antarctic Beech (or negro-head beech) along the coast of eastern Australia; ⋆Wollemi pine.

**REM** Short for **rapid eye movements** during sleep. During periods of dreaming, the frequent and rapid movements of the eyes behind the closed eyelids. Persons deprived of REM sleep have to make up the 'debt' during future sleep, or they eventually suffer behaviour problems.

**remote sensing** The use of an instrument to record information about an object without actually touching that object. e.g. Using orbiting artificial satellites for weather prediction and oceanographic and agricultural studies. Infrared- and ultraviolet-sensitive systems allow satellites to photograph views of the Earth that can show the distribution of plankton in the world's oceans or the spread of plant diseases before they are visible on the ground.

**renal** Of or relating to the kidney. e.g. Renal artery carries blood to the ⋆kidneys, while the renal vein carries it away after filtering.

**renal portal system** A system of veins in fish and amphibians which drains blood from the rear of the body, sending it to the kidneys.

**renal tubule** The part of the ⋆nephron of a ⋆kidney which follows the ⋆glomerulus, where the useful parts of the filtered blood are reabsorbed into the capillaries. See Fig. K1. In desert-dwelling animals, the water absorbing part of the renal tubule, called the ⋆loop of Henlé, is much longer than in animals living in wetter regions.

**reniform** Resembling a kidney in shape.

**rennin** An enzyme secreted in the ⋆gastric (stomach) juices of young mammals which causes the milk protein casein to become lumpy (coagulate).

**repeated sequence** Any long sequence of base pairs found at a number of different places throughout the chromosomes.

**repetitive DNA** Widely scattered sequences of DNA, about 300 base pairs long, often found adjacent to genes that produce enzymes or other proteins. Their function is unknown.

**replication** During cell division, the forming of an exact copy (replica) of the DNA of the chromosomes (Fig. D3).

The DNA double helix unwinds, and the ⋆complimentary or matching bases align themselves with the existing single strands of DNA. They are then joined together by enzymes. Thus each strand directs the construction of another strand of DNA that is complimentary with itself, and we end up with two double helixes identical to the one we started with. Each double strand of DNA thus consists of one new half and one old half of DNA. See also semiconservative replication. Compare transcription.

**repressor** A protein produced by a ⋆regulator gene which binds to the ⋆operator site on a DNA molecule, between the ⋆promoter and genes to be transcribed. By binding to this site, the repressor stops the enzyme RNA polymerase from reaching the genes to be transcribed and no mRNA can be produced. See also operon, corepressor.

**reproduction** The production of offspring, or procreation. Divided into ⋆asexual and ⋆sexual reproduction. Asexual reproduction produces offspring genetically identical to their parents. e.g. ⋆Budding, ⋆binary fission, ⋆grafting, cuttings, etc. Sexual reproduction involves the joining of male and female gametes (sperm and ova). Many organisms alternate between sexual and asexual forms of reproduction. See alternation of generations, reproductive systems.

**reproductive isolation** The prevention of gene flow (mixing of genes) between all members of a population due to barriers that isolate one population from another. See isolating mechanisms.

**reproductive system** The system of organs involved in the production of male and female sex cells, and in the female, that which also protects and feeds the embryo during growth. e.g. **1 Human reproductive system. a Male reproductive system** is relatively more simple (see Fig. R3). ⋆Sperm are produced in the ⋆testicles which are carried in a sac called the ⋆scrotum. The sperm passes along the ⋆vas deferens (sperm duct) to the ⋆penis, with various chemicals added at the ⋆prostate gland, ⋆Cowper's gland, and ⋆seminal vesicle, thus forming ⋆semen. The semen is ejaculated into the female vagina during sexual intercourse (see erection). See also testosterone. **b Female reproductive system** is more complex (see Fig. R4). ⋆Ova (eggs) are produced each month, usually alternately from each ⋆ovary. If sexual intercourse occurs when the ovum is released, the sperm and ovum may meet in the ⋆oviduct, where ⋆fertilisation occurs. The female reproductive system is prepared each month to receive such a fertilised ovum (see menstrual cycle). The fertilised ovum (⋆zygote) grows to form a ⋆blastula, which embeds itself into the prepared walls of the ⋆uterus. Here a ⋆placenta develops to allow food and oxygen to pass from mother to embryo, and wastes to return. If fertilisation does not occur, ⋆menstruation renews the walls of the uterus. See also oestrogen, progesterone, mammary gland, labour, trimester. **2** Many mammals reproduce in much the same way as humans, but other animals as well as plants and fungi use quite different systems. See alternation of generations.

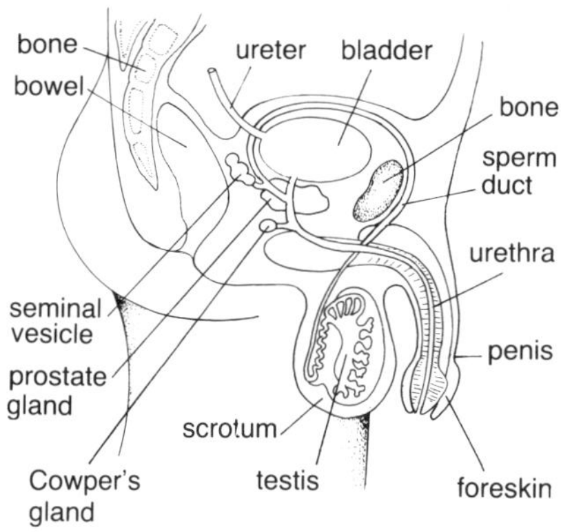


**Fig. R3 Male reproductive system**

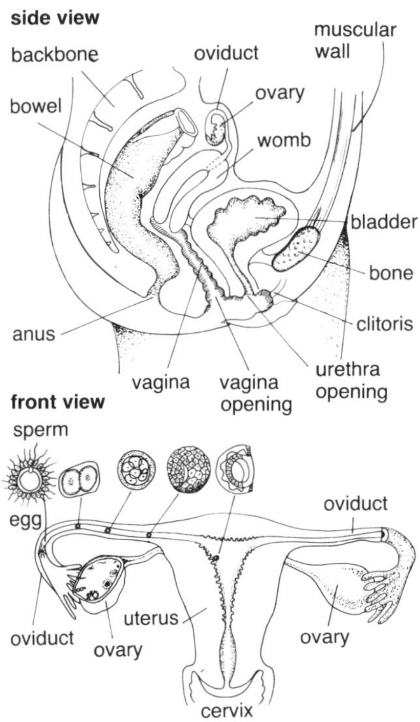


**Fig. R4 Female reproductive system**

**Reptilia** A class of vertebrates, distinguished by their dry scaly skin, being poikilothermic ('cold blooded'), and laying leathery shelled eggs. e.g. Turtles and tortoise, snakes and lizards, crocodiles and alligators, tuatara, and the extinct reptiles such as the dinosaurs. See Appendix 1.

**resident** An animal which remains in an area throughout the year. Compare migration.

**residual air** The volume of air remaining after we have forced as much as possible from our lungs. In human adults it is about 300 mL.

**resin** A complex mixture of compounds that is exuded by certain plants, especially when wounded, forming a hard coat and protecting them from insects and fungi.

Resins are insoluble in water, but will dissolve in alcohol and ether. On exposure to air, the easily vaporised components are lost, hardening the resin and thus protecting the damaged area. Resins are common in conifers (see resin canals below). If the resin from pine trees is distilled, it produces oil of turpentine, and the remaining material (rosin) is used in lacquers. The famous aromatic substance frankincense is a mixture of resin and their oils obtained from Asian and African trees of the genus *Boswellia*.

**resin canals** The ducts in bark and wood, especially in conifers, which carry the resin.

**resistance** The ability of an organism to resist or withstand the actions or effects of a certain activity. e.g. Certain insects have developed resistance to (are not affected by) ⋆insecticides. Various bacteria are no longer affected by (are resistant to) certain ⋆antibiotics.

**resolution** See resolving power.

**resolving power** Used to describe the ability of microscopes to allow two points that are close together to be distinguished from one another. e.g. The very best resolving power of a light microscope would be about 250 nm. Compare electron microscope.

**resource** The living and non-living materials available to support the life of an organism, especially humans. **Renewable resources** can be replaced, usually by natural means. e.g. Food, wood, wool, cotton, oxygen, water. **Non-renewable resources** are not able to replace themselves, and once used, are gone forever. e.g. Coal, oil, ores.

**respiration** One or all of the processes used to obtain energy from foods.

1 **External respiration a** The act of breathing, or drawing in and expelling air from a cavity such as the lungs. Inhaling and exhaling air. **b** The taking in of oxygen from the environment for cell respiration and giving off of waste carbon dioxide.

2 **Internal respiration** Also called **cell** or **tissue respiration**. See Fig. R5. The chemical reactions by which cells obtain energy. This energy is obtained from many foods, but especially sugars such as glucose. A complex series of enzyme-controlled reactions (usually in the cell's ⋆mitochondria) which combine glucose with oxygen obtained by external respiration. The energy is stored in the chemical bonds of ATP molecules, while waste carbon dioxide is lost through the external respiration system, and water through the ⋆excretory system. The overall reaction is: $C_6H_{12}O_6 + 6O_2 \xrightarrow{\text{enzymes}} 6CO_2 + 6H_2O$ + Energy (ATP)

The above reaction is called ⋆aerobic respiration as there is plenty of oxygen available. When there is insufficient oxygen (e.g. during a sprint race), then ⋆anaerobic respiration occurs, supplying a reduced amount of energy compared to the amount of glucose used. Anaerobic respiration also produces some byproducts, ⋆lactic acid in mammals (causing runners' cramps), and ⋆alcohol (ethanol) in plants. For details of aerobic respiration, see glycolysis, Krebs' cycle, electron transport chain, mitochondria.

**respirator** 1 A mechanical breathing device to assist patients whose breathing muscles are unable to work for some reason. 2 A device worn over the face to prevent the inhaling of dangerous fumes. e.g. Gas mask.

**respiratory chain** See electron transport chain.

**respiratory organ** The organ or structure across which oxygen and carbon dioxide pass during external respiration. e.g. Lungs (mammals, birds, reptiles), gills (fish, tadpoles, aquatic insects), trachea (insects), moist skin (amphibians, invertebrates).

**respiratory passage** The tubes leading to the lungs of land vertebrates. e.g. ⋆Trachea, ⋆bronchi, ⋆bronchioles.

**respiratory pigment** The coloured substance in the blood which increases the ability of the blood to carry oxygen and (some) carbon dioxide. e.g. ⋆Haemoglobin (Fig. H1) in red blood cells, and haemoglobin, haemocyanin, etc. in the blood plasma of various invertebrates. In the lungs, oxygen concentration is high, and oxygen is absorbed by haemoglobin (Hb), forming ⋆oxyhaemoglobin $HbO_2$. Oxygen is later given up at the cells of the tissues, where oxygen concentration is low. Thus the reaction is:

$$Hb + O_2 \underset{\text{tissues}}{\overset{\text{lungs}}{\rightleftharpoons}} HbO_2$$

**respiratory quotient** Called **RQ** for short. The ratio of the volume of carbon dioxide produced by an organism during respiration, to the volume of oxygen consumed, during the same interval of time.

**respiratory surface** A surface at which gases are exchanged between the environment and the body. This surface must remain moist to allow the exchange of gases e.g. ⋆Gills, ⋆alveoli of lungs, skin of amphibians.

**response** A change in a cell, tissue, or organism because of a ⋆stimulus.

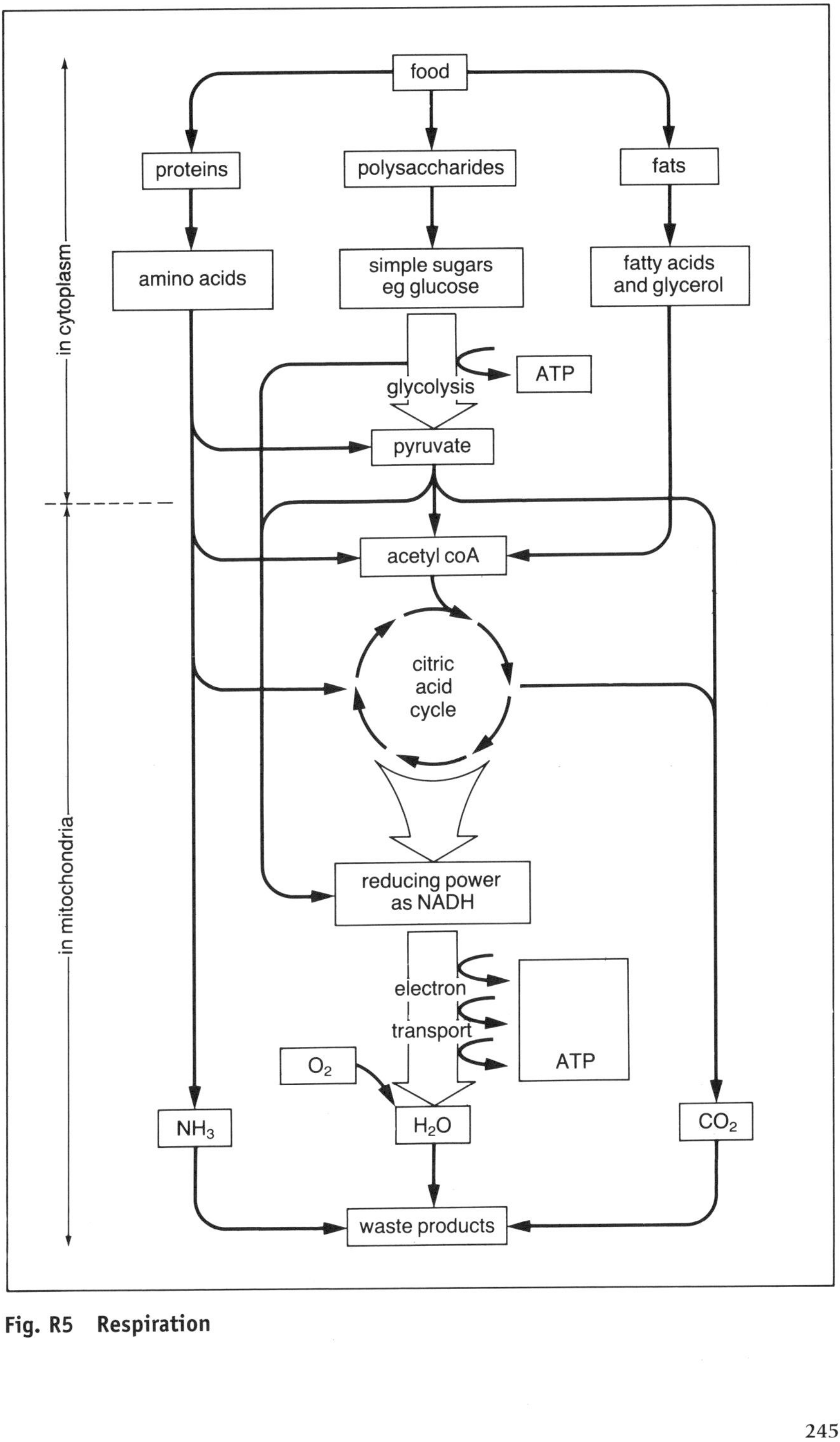


**Fig. R5 Respiration**

e.g. Animals moving, glands secreting, tropisms in plants.

**resting cell** A cell that is not undergoing cell division, and is thus in ★interphase. Actually, the cell is very active and the only thing it is resting from is mitosis or meiosis!

**resting potential** The potential difference (voltage) which exists across the cell membrane of a neurone when there is no impulse passing. The resting potential is often –70 mV, the inside of the neurone being more negative than the outside. The voltage arises because of the unequal distribution of ions across the membrane, with many sodium ions being on the inside, and many potassium ions on the outside. See impulse and Fig. I3.

**restriction enzyme** Also called **restriction endonuclease**. An enzyme that cuts a DNA molecule only at sites where specific base sequences occur (recognition sites). e.g. The enzyme *Eco* R1 cuts wherever it finds the sequence GAATTC, while *Sma* 1 cuts at CCGGG. Restriction enzymes occur naturally in bacteria where they help protect against attacks by a bacteriophage (they cut up the DNA of the virus). Restriction enzymes are very useful in ★genetic engineering, being used to help remove desired genes by cutting them from their chromosomes. They are also used for cutting open plasmids so that the isolated genes can be inserted. See also DNA ligases.

**resurrection plant** A plant the leaves of which wither and look dead during drought, but which quickly recovers after rains.

**resuscitation** Artificial respiration applied to someone who has stopped breathing.

**reticular formation** Also called the **reticular activating system**. A structure within the brain consisting of a network of nerve cells which are involved in consciousness. If impulses from one or more sense organs that have detected something of interest (e.g. danger) are received, then the reticular formation sends signals to other parts of the brain to alert them as well.

**reticulate** Resembling or having the form of a net. e.g. In leaves, the net-like pattern of the veins on a dicot leaf.

**reticulin fibres** Also called **reticular fibres**. Very fine fibres of the protein reticulin (similar to collagen) that form a net-like support that helps hold tissues together. e.g. Hold muscle fibres together.

**reticulum** 1 A net-like or honeycomb-like structure. e.g. ★Endoplasmic reticulum. 2 The second chamber in the four-chambered stomach of ★ruminants.

**retina** The light-sensitive layer lining the inside of the rear half of the vertebrate eye. (See Fig. E9.)

Closest to the outside of the eyeball is a layer of photo-sensitive cells called ★cones and ★rods. The cones detect colour (red, green, blue) and are concentrated near the back of the eye, especially directly opposite the pupil in an area called the ★fovea. Rods detect the intensity or brightness of the light, and are spread over most of the retina. The rods and cones are linked to the optic nerve by neurones that form a net on the inside of the photo-sensitive layer. Thus light has to pass through this tangle of nerve fibres which exit at the 'blind spot' (where there are no rods or cones present) to form the optic nerve connecting to the brain. Nocturnal animals (e.g. cats) often have a reflective layer behind the rods and cones to help capture more light (causing their eyes to 'shine' in beams of light). Cephalopods have a similar eye structure, except the light falls directly onto the photo-sensitive layer, the nerve fibres being located behind.

**retrovirus** A virus containing RNA (rather than DNA) within a complex membrane of lipids and protein, some of which can cause ⋆tumours in vertebrates while others (⋆HIV) cause ⋆AIDS in humans. The RNA has to be first transcribed into DNA (see ⋆reverse transcriptase) before it can be inserted into the DNA of the host cell. e.g. Rous sarcoma virus (RSV) which causes tumours in chickens. See also virus.

**reverse transcriptase** An enzyme present in some micro-organisms which is able to transcribe DNA from RNA (see ⋆retrovirus), the reverse of the normal direction of ⋆transcription. Widely used in genetic engineering to produce copies of a gene for which samples of its mRNA can be obtained. Reverse transcriptase is used to produce intron-free DNA from the mRNA, and this DNA is then ⋆cloned.

**revolute** In leaves, rolled backwards from the edge on the undersurface. See Fig. L1.

**Rh factor** Short for **rhesus factor**. A substance (⋆antigen) occurring on the surface of red blood cells (⋆erythrocytes) in about 85% of the Australian population. Such people are said to have Rh positive blood. Those without this substance are said to have Rh negative blood. The main gene producing Rh positive blood is dominant to that producing Rh negative blood.

If an Rh negative mother has an Rh positive child, some of the child's blood may seep across the placenta and enter her blood. This causes antibodies to be produced in the mother against Rh positive blood. This rarely causes problems with her first Rh positive child, but can cause problems if she has a second child with Rh positive blood. If she has a lot of antibodies against Rh positive blood, they can seep across the placenta into the blood of the foetus, and cause serious problems. For similar reasons, a person with Rh negative blood cannot receive a ⋆transfusion from someone with Rh positive blood.

**rhachis** See rachis above.

**rhesus factor** See Rh factor above.

**rheumatism** A general term for painful, disabling conditions affecting the joints, muscles, etc. Compare arthritis.

**rhizoid** A slender root-like anchoring structure in some lower plants. e.g. on the undersurface of the ⋆prothallus of a fern, and under moss and liverwort plants. Rhizoids may also absorb water and nutrients, and since they contain no ⋆vascular tissue, they are short.

**rhizome** A horizontal stem which grows along or under the ground producing leaves, shoots and roots. In some plants (e.g. couch grass) the rhizome acts to allow ⋆vegetative reproduction, and after the new plant is established, the old parts rot away. In other plants (e.g. iris), the rhizomes store food, allowing the plant to survive hard winters. Compare corm, rootstock, tuber.

**Rhizopoda** Any protozoan of the class Sarcodina (Rhizopoda), all of which possess ⋆pseudopodia to help with feeding, and sometimes with movement. e.g. *Amoeba*, Radiolaria, Foraminifera. See Appendix 1.

**rhodopsin** Also called **visual purple**. A light-sensitive pigment present in the ⋆rod cells of the vertebrate ⋆retina that detects the presence of light.

Rhodopsin is composed of retinal (a compound made from vitamin A) which is attached to the protein opsin. When light falls onto the retina, a series of reactions leads to the splitting of retinal and opsin, which generates an impulse in the nerve. An enzyme helps recombine the opsin and retinal. In bright sunlight, most of the rhodopsin is broken down. In the dark, most of the opsin and retinal can recombine to form rhodopsin, thus making the eye much more sensitive to light.

**rhomboid** Diamond shaped.

**Rhynchocephalia** An order of reptiles with one species, *Sphenodon punctatus*, or the ⋆tuatara of New Zealand. See Appendix 1.

**rhythm** A sequence of events or activities repeated at regular intervals. e.g. ⋆Biological rhythms.

**rib** One of a series of curved bones found in vertebrates, one end of which is connected to the vertebral column (backbone), the other end being free (amphibians) or attached to a sternum (breastbone). See Fig. S3.

**ribbon worm** Also called ⋆proboscis worm, a small phylum of mostly marine worms called ⋆Nemertea. See Appendix 1.

**riboflavin** Also called **vitamin B2**. A compound that is part of the vitamin B group that must be part of the diet of humans (green vegetables, yeast, milk, liver, etc.). Riboflavin is needed in the making of coenzymes FMN and FAD.

**ribonucleic acid** See RNA below.

**ribose** A five-carbon sugar (a pentose) which is a component of the 'backbone' of ⋆RNA, as well as being a part of various ⋆nucleotides (e.g. AMP, ATP) and ⋆coenzymes (e.g. NAD, FAD). See Fig. R6.

**ribose nucleic acid** More correctly ribonucleic acid. See RNA below.

**ribosomal RNA** Called ⋆**rRNA** for short.

**ribosome** A particle (organelle) in the cytoplasm of cells where proteins are made from amino acids. See Fig. R7.

In prokaryotic organisms (no nucleus) they float freely in the cytoplasm. In eukaryotic organisms (with a nucleus), ribosomes are usually attached to the ⋆endoplasmic reticulum. Ribosomes have also been found inside ⋆mitochondria and ⋆chloroplasts. Ribosomes consist of two parts of unequal size, and are about 50% ⋆rRNA and 50% protein. As protein is being synthesised, the ribosome moves along the mRNA ⋆transcribing it into chains of amino acids whose order is directed by the mRNA. The separate amino acids are joined together by an enzyme that is also one of the proteins found in the ribosome. See also translation.

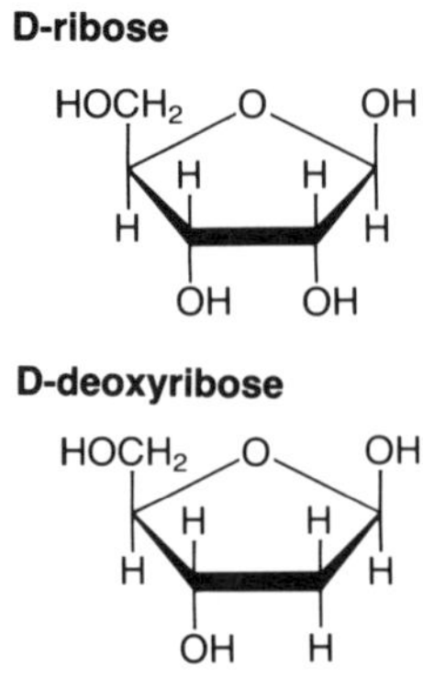


**Fig. R6 Ribose sugars**

**Fig. R7 Ribosome**

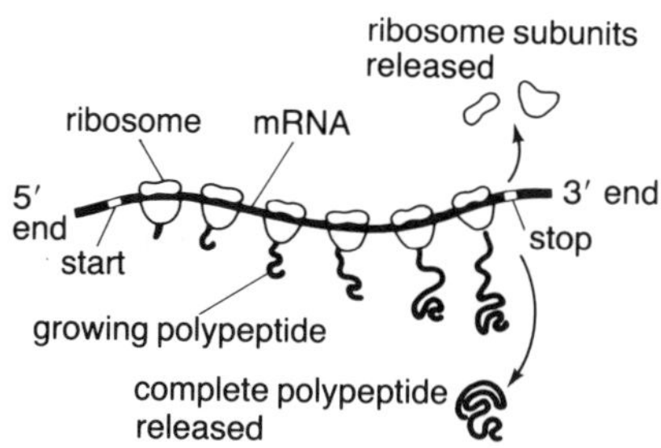


**ribozymes** An ⋆RNA molecule that also acts as an ⋆enzyme. e.g. RNA molecules can cut ⋆introns from the ⋆primary RNA transcribed from DNA and this intron is then able to act as an enzyme.

**rickets** A disease in infants caused by lack of vitamin D in the diet, or due to insufficient exposure to sunshine, which can create vitamin D from substances in the skin.

**rickettsias** A group of rod-shaped and very tiny bacteria, but which are a little like viruses in that they cannot multiply except inside the cells of their ⋆hosts (e.g. commonly ticks, lice and fleas). They do not appear to do much harm to their arthropod hosts, but if transmitted to humans or other mammals, they may cause severe and often fatal disease. e.g. Typhus, Rocky Mountain spotted fever, Q fever.

**ringbark** To cut away the bark all the way around the trunk or a branch of a tree thereby destroying the ⋆phloem function, so that the tree or branch will die.

**Ringer's solution** An aqueous solution of calcium, sodium, and potassium chlorides (and sometimes other salts). Used as a ⋆medium to allow living cells or tissues to survive *in vitro*. These are the main salts which normally bathe cells in the tissues, and need to have a concentration that matches the concentration of chemicals inside the cell (isotonic) to prevent ⋆osmosis.

**ringworm** Infection by various fungi in areas such as the skin, hair, nails, and scalp. It has nothing to do with worms.

**riparian** Relating to the bank or shore of a river or stream.

**ripen** 1 As far as humans are concerned, when a fruit or vegetable becomes mature and ready to be eaten, and thus ready to be harvested, picked or gathered. In, fruits, plant ⋆growth substances such as ethylene promote ripening. 2 As far as the plants are concerned, the fruits exist and ripen as an aid to seed ⋆dispersal.

**river oak** *Casuarina cunninghamiana*, a tree that commonly lines the banks of streams on the western slopes and tablelands of eastern Australia. *Casuarina* has needle-like photosynthetic stems (⋆cladodes) which look like pine needles, but have a number of segments along their length.

**river red gum** A large picturesque tree (often painted by artists), *Eucalyptus camaldulensis*, which lines the banks of inland streams and dry watercourses of Australia. It is the most widespread eucalypt.

**RNA** Short for **ribonucleic acid**, a complex single-stranded molecule important in the making of proteins (Figs T2 and T4).

There are three types or RNA: **a Messenger RNA** or mRNA is a linear molecule that moves from the nucleus to the cytoplasm carrying the instructions for making the protein. **b Ribosomal RNA** or rRNA is a much-folded molecule that helps make up the particles called ⋆ribosomes in the cytoplasm where the protein is made. **c Transfer RNA** or tRNA (also folded) brings in the amino acids needed to make the protein as the ribosome moves along the mRNA. Compare DNA.

Chemically, RNA has two main parts. Its 'backbone' is made of a long chain of alternating ⋆ribose sugar and phosphate units. Attached to each of the ribose sugars is one of the organic bases which makes up the genetic code: ⋆adenine (A), ⋆cytosine (C), ⋆uracil (U) and ⋆guanine (G). Thus RNA is a single-stranded polymer of a nucleotide composed of a phosphate, ribose sugar and organic base. The exact order of these organic bases is determined when the DNA in the nucleus is ⋆transcribed into mRNA (Fig. T2). After the mRNA is '⋆edited', it passes to the ribosomes in the cytoplasm where it is ⋆translated into protein (Fig. T3). Each three bases (triplet or ⋆codon) specify an amino acid, or a start/stop signal (Fig. G1). The tRNA and their attached amino acid each have a corresponding ⋆anticodon to match and thus allow them to be attached to the codons along the mRNA. Hence the codon CUU specifies that the amino acid leucine is to be attached whenever it is present along the mRNA. Since in RNA the base C always pairs with G, and A with U, the anticodon of the tRNA would be GAA (Fig. T4). At the other end of a tRNA molecule with anticodon GAA there will be attached a leucine molecule. See also protein synthesis, ribosomes, translation, transcription.

**RNA polymerase** Also called **RNA transcriptase**. An enzyme which joins together the ⋆nucleotides in an order specified by codons of the DNA to form a single-stranded chain of RNA. See operon.

**roan** A colouration pattern where the basic chestnut or bay colour is sprinkled with patches of white or grey. e.g Roan cattle have a mixture of red and white hair. See also co-dominance.

**rod cell** A light-sensitive cell present in the ⋆retina of most vertebrates. The cell body is attached to a rod-like part containing light-sensitive ⋆rhodopsin arranged in discs (lamella) at right angles to the incoming light. The other end of the cell body connects to nerve fibres which carry signals to the optic nerve. The rods are mainly responsible for detecting low intensity light and movement. There are about 125 million rod cells in a human eye. Compare cones.

**Rodentia** An order of placental mammals whose members have two sharp chisel teeth at the front of each jaw, and no canine teeth. e.g. ⋆Rats, ⋆mice, squirrels, beavers, etc. See Appendix 1.

**rogue** Also called a **sport**. An organism which is different from the standard type or variety, often because of a ⋆mutation.

**root** 1 In ⋆vascular plants, the part that grows usually downwards into the soil, anchors the plant, and absorbs water and dissolved minerals. See Fig. R8.

Some roots also store food (e.g. turnips, sugar beet), help stop plants falling over (e.g. prop-roots, buttress roots), while in waterlogged soils special roots protrude above the surface (pneumatophores) to allow oxygen to reach those parts underground. Roots often have symbiotic organisms attached that help the plant survive. e.g. ⋆Nitrogen-fixing bacteria in the ⋆root nodules of legumes; ⋆mycorrhizal fungi living in or around the roots of many trees. Roots generally differ from stems by not containing chlorophyll, or growing buds, and by having a different internal arrangement of vascular tissue. Roots have two basic forms: there may be one large taproot with a number of smaller lateral roots coming from it (often dicots), or there may be a fibrous pattern with many branching roots of more nearly the same size (often monocots). The root grows at the very tips, with a ⋆meristem producing new cells. Some new cells move forward to form the protective ⋆root cap. Those new cells left behind start to grow longer and force the root cap forward through the soil. These cells then differentiate to form ⋆root hairs where the water and dissolved minerals are absorbed from the soil (see active uptake), and vascular tissue (phloem and xylem) to help move food and water about the plant.

2 In mammals, the bottom part of an organ or structure that is buried in underlying tissues. e.g. Root of a tooth, hair, tongue. 3 In the nervous system of mammals, the channels through which nerves enter and leave the central

nervous system (Fig. R2). e.g. Each vertebra has a pair of dorsal (uppermost or rear) roots through which the sensory nerves enter, and a pair of ventral (bottom or forward) roots through which motor nerves leave. See reflex arc.

**Fig. R8 Roots**

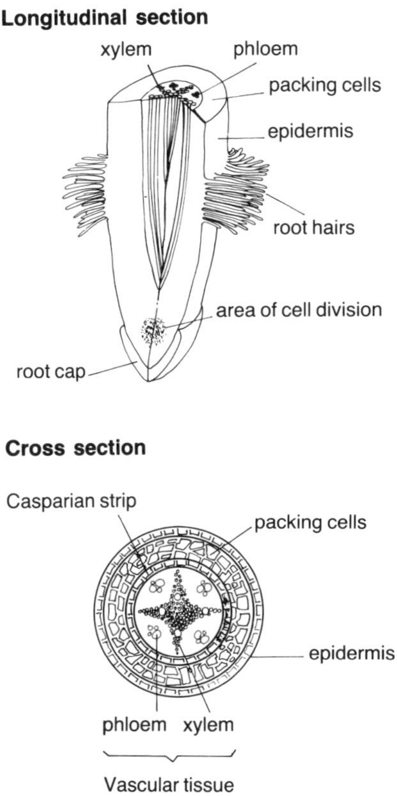


**root cap** The protective layer of cells covering and protecting the growing tip of the root where they push through the soil. The abrasion that results damages the cells, and they are continually being replaced by new cells from cell division in the ⋆meristem just behind the root cap. See Fig. R8.

**root hair** Microscopic, hair-like outgrowths of the surface cells of the roots in the zone of active growth just behind the root tip. Root hairs greatly increase the surface area of the roots. Being so small, they force themselves between the grains of soil to absorb the water and dissolved minerals that are found there. Root hairs disappear after a few days, to be replaced by new ones nearer the growing root-tip. See Fig. R8.

**root nodule** Small swellings in the roots of legumes such as peas, beans, clover and wattle, containing ⋆nitrogen-fixing bacteria. In legumes, a bacterium of genus *Rhizobium* uses a plasmid to induce the host plant to form the nodule. Inside, the bacteria are able to convert nitrogen from the air into soluble nitrates which they share with the host plant, receiving sugars etc. in return. See symbiosis.

**root pressure** The pressure which builds up within a plant due to ⋆osmosis at the roots. Since the cytoplasm has a higher concentration than the water outside, water enters by osmosis. This flow of water can be observed by cutting across a stem, or by observing the formation of water droplets along the edges of some leaves (see guttation).

**rootstock 1** A root used for ⋆grafting or other method of plant propagation. e.g. Wild-lemon rootstock is used with most citrus grafting as it is more resistant to soil parasites, etc. **2** A vertical underground stem or root from which plants can sprout in a new growing season. e.g. Dandelions and daisies.

**rosette** A cluster of leaves arising in a radial pattern from a short stem. e.g. Dandelion, daisy.

**rostrate** Beaked.

**rostrum** A beak, or beak-like structure found in many animals.

**Rotifera** A Phylum of invertebrates, microscopic aquatic animals with a ring of ⋆cilia around the mouth. They have a digestive system with separate mouth and anus. See Appendix 1.

**roughage** The parts of plants that we have eaten, but are unable to digest. Roughage contains a lot of cellulose, and is needed for proper functioning of the large intestine.

**roundworm** Any member of the phylum Nematoda. These cylindrical worms differ from annelids in having no body segments. See Appendix 1.

**royal jelly** A substance produced by worker bees which, when fed in large quantities to female larvae, results in their development as queens. Normally worker bees are inhibited from producing much royal jelly due to ⋆pheromones produced by the queen.

**RQ** See respiratory quotient.

**rRNA** Short for ⋆**ribosomal RNA**. A type of RNA which combines with certain proteins (⋆nucleoprotein) to form the ⋆ribosomes. The rRNA is transcribed from a section of the DNA in the nucleus called the ⋆organiser, along parts of the chromosome associated with the ⋆nucleolus. It then moves out into the cytoplasm where it is assembled into ribosomes.

**rubella** The virus responsible for causing German measles.

**rugose** Having many wrinkles or ridges.

**rumen** The first of the four chambers of the stomach of a ruminant, where food is stored and partly digested. Food can be sent back to the mouth for further chewing (chewing the cud).

**ruminant** 1 An animal which chews its cud. 2 A group of even-toed hoofed animals with a stomach of four chambers (rumen, reticulum, omasum, abomasum) which swallows its food after little or no chewing. This allows rapid eating, but a slower digestion rate when away from predators. During digestion, the food is returned to the mouth for thorough chewing, before it is re-swallowed. In the first two chambers of the stomach, the food is mixed with bacteria which help digest the cellulose. This mixture passes to the third chamber where water is absorbed, and finally into the fourth chamber where normal digestion occurs. e.g. Cattle, goats, antelope, deer.

**runner** Also called a **stolon**. A horizontal plant stem growing above the ground, and spreading by developing new plants at the ⋆nodes. e.g. Strawberry.

**rupture** A ⋆**hernia**.

**rust** Any of a number of parasitic fungi (or the diseases they cause) which produce brown or rust-coloured spots to develop on the stems or leaves of infected plants. e.g. Wheat rusts.

**rut** The period when certain male mammals are 'on heat', and sexually very active. e.g. The rut of deer during which fights for dominance of harems occur.

**Rutaceae** The citrus family of dicot flowering plants, including *Boronia* and wax flowers. See Appendix 1.

# Ss

**S-A node** Short for **sinoatrial node**. See pacemaker.

**Sabin vaccine** A vaccine against the disease polio which can be taken by mouth. The vaccine contains live, but weakened, polio ⋆viruses. Compare Salk vaccine.

**sabre tooth tiger** An extinct cat-like mammal with long curved upper canine teeth.

**sac** A pouch or bag-like structure.

**saccharide** An alternative name for sugar. e.g. ⋆glucose, ⋆sucrose. See Fig. C2.

**sacculus** Also called **saccule**. A chamber of the ⋆inner ear that lies on top of the ⋆cochlea. It contains sensory hairs that help record changes in the position of the head (i.e. balance). See Fig. E1 and vestibular system. Compare utricles.

**sacral** Relating to the lower back. e.g. Sacral vertebrae where the ⋆pelvic girdle is attached to the backbone.

**sacrum** A triangular bone made up of the sacral vertebrae that are fused (joined) together, and attach the hip girdle to the backbone.

**safranine** A red ⋆stain used to stain nuclei in plant cells etc.

**sagittal** A vertical plane through a ⋆bilaterally symmetrical animal that divides it into right and left parts. See symmetry and Fig. S11.

**sagittate** Used to describe leaves that are shaped like arrowheads. See Fig. L1.

**salamander** A lizard-like amphibian with a tail. Instead of scales, it has the normal moist skin of an amphibian. See Appendix 1.

**salination** The process where salt accumulates in the soil, eventually making it useless for agriculture. A significant problem in irrigation areas where rising water tables are bringing salt to the surface.

**saline** Containing salt. e.g. A salt solution.

**salinity** The amount of salt which something contains. e.g. High salinity waters contain relatively large amounts of salt.

**saliva** A watery fluid secreted into the mouth from the ⋆salivary glands, containing ⋆mucus to help the food slide down the food pipe (⋆oesophagus) more easily (i.e. a lubricant). In some land vertebrates and insects the saliva also contains ⋆amylase to start digestion of carbohydrates. e.g. ⋆Ptyalin in human. The saliva of some blood-sucking animals (e.g. Mosquito, leech) also contains ⋆anticoagulant to stop the blood clotting.

**salivary gland** A gland that secretes saliva, as found in some insects and land vertebrates. Some salivary glands also produce poisons (e.g. snakes), and others produce sticky substances (e.g. some nest-building birds). In humans there are three salivary glands on each side of the head: parotid, submandibular (or submaxillary), and sublingual.

**Salk vaccine** A vaccine against the disease polio that has to be injected into the bloodstream. The vaccine contains killed polio viruses. Compare Sabin vaccine.

***Salmonella*** A genus of ⋆pathogenic ⋆bacteria responsible for food poisoning, gastroenteritis, typhoid fever, etc.

**salt** 1 The compound sodium chloride (NaCl). 2 More generally, any ionic inorganic compound.

**salt gland** 1 A gland located above the eyeball of certain saltwater (marine) reptiles and birds that secretes excess salt which has been taken in with food. 2 Glands, especially in some saltmarsh plants and mangroves, that secrete salt.

**salt marsh** A coastal ⋆community of herbaceous plants which are regularly flooded at high tide. Salt marshes are often found on temperate coasts.

**salt pan** A shallow depression or basin lined with salt, often a dried up lake or marsh.

**saltation** 1 A jump or a leap. An abrupt change. 2 The way in which nerve ⋆impulses travel more rapidly in ⋆myelinated nerve fibres than in those without a myelin sheath. The ions can only move across the membrane at the ⋆nodes of Ranvier, so that the impulse appears to jump or leap along the ⋆neurone. 3 A type of ⋆evolution where, it is proposed, relatively short periods of rapid change followed longer periods with very little change. See punctuated equilibrium.

**saltbush** 1 A plant community of low shrubs found in the drier regions of Australia where rainfall is below 250 mm per year, and where the dominant plants are saltbush and ⋆bluebush. 2 Any of the many drought-resistant plants in the genus *Atriplex*, commonly found growing on ⋆saline and ⋆alkaline soils in the drier parts of Australia.

**samara** A winged ⋆achene. e.g. Maple, sycamore.

**sap** The sugary plant fluids that seep from the ⋆vascular tissue of a plant when cut. The sap is a watery mixture of sugars, minerals, and small amounts of more complex compounds moving about the plant on the ⋆xylem and ⋆phloem pipelines. Compare gum and kino.

**saponification** The hydrolysis of fats (or any ester) to yield fatty acids and glycerol.

**saprophyte** A plant or plant-like organism which feeds on dead organic material. Saprophytes are the main agents for breaking down and recycling dead plant and animal material. e.g. Bacteria and fungi. Compare scavenger.

**sapwood** The outer region of a tree trunk, containing the living cells, and especially the sap-carrying vascular tissue (xylem, phloem). Compare with the heartwood.

**Sarcodina** A class of ⋆protozoans (also called **Rhizopoda**), all of which possess ⋆pseudopodia to help with feeding, and sometimes with movement. e.g. *Amoeba*, Radiolaria, Foraminifera. See Appendix 1.

**sarcoma** A form of ⋆cancer that develops mainly in connective tissue. e.g. Cancer in bones.

**sarcomere** A contractile unit in the microscopic structure of striated ⋆muscle fibres. See Fig. M5.

**sarcoplasmic reticulum** A type of ⋆endoplasmic reticulum found in muscle cells which helps transmit the nerve impulse to the ⋆filaments.

**sartorius** A leg muscle that connects the pelvic girdle to the tibia, and used to move the leg.

**sassafras** Any of several Australian trees with fragrant smelling bark, such as the nutmeg-scented trees like the Southern Sassafras, *Atherosperma moschatum*.

**satellite DNA** A highly repetitive length of DNA, often located near the ⋆centromere.

**saturated** 1 Of fats and ⋆lipids, where there are no double bonds, and thus no further hydrogen can be added. See also lipids, polyunsaturated, unsaturated. 2 Of solutions, where as much solid (solute) has been dissolved as is possible for the volume and temperature concerned. 3 Of air, containing as much water vapour as it can for the temperature concerned.

**Sauria** The suborder of reptiles that includes the ⋆lizards. See also Squamata, Serpentes and Appendix 1.

**Sauropoda** The huge, semiaquatic but now extinct herbivorous reptiles that include *Brontosaurus* (18.5 m long) and *Diplodocus* (27.7 m long and 30 tonnes in weight). Single bones of even larger types have been recently found. Compare Theropoda.

**savanna** A type of ⋆grassland in tropical and subtropical regions with drought-resistant plants, and scattered trees and shrubs. e.g. Northern Australia away from the coast.

**scab** A hard coating (encrustation) which forms over a sore or cut during healing.

**scabies** An itch due to infestation by female mites that burrow holes in the skin and lay eggs in them.

**scabrous** Rough to the touch.

**scale** Thin flattened plate-like structures. e.g. Scales covering the skin of fish and reptiles, and the bodies and wings of butterflies and moths. Also many small leaf-like structures in plants such as the scales that make up most of a pine cone.

**scalpel** A small and very sharp blade used in dissections and medical operations.

**scape** The long leafless stem with a flower or group of flowers at the top. e.g. Tulip or dandelion.

**Scaphopoda** A small class of molluscs also called **tusk shells**. Their tube-shaped shell is open at both ends. At one end a pointed foot is used to burrow into the sand. The hole at the other end allows food and oxygen-carrying water currents to enter and leave. See Appendix 1.

**scapula** Also called the **shoulder blade**. A bone of the ⋆pectoral girdle with which the forelimb or arm is attached. It is also where most muscles of the forelimb are connected.

**scavenger** 1 An animal living on dead and decaying organic matter. e.g. Some beetles, gulls, crows. Compare saprophyte. 2 A ⋆leucocyte (white blood cell) which acts as a ⋆phagocyte, removing many unwanted materials from the blood.

**schistosomiasis** A chronic tropical and subtropical disease of humans and animals due to infestation by blood ⋆flukes which gain entry by burrowing through the skin. Also called bilharziasis.

**Schizomycetes** The phylum that includes the ⋆bacteria and ⋆rickettsias, unicellular ⋆prokaryotes. See Appendix 1.

**schizophrenia** A term for a group of serious mental disorders usually beginning in adolescence or young adulthood. Symptoms include disturbances in thought, perception, emotion, and interpersonal relationships. Literally, the term means 'split mind' but this does not mean a split personality, in the sense of someone acting like two different people.

**Schwann cell** A type of ⋆glial cell that wraps itself around the nerve fibres and produces the myelin sheath. See neurone and Fig. N1.

**sciatica** A pain in the back of the

thigh and leg along the path of the sciatic nerve. Caused by several diseases.

**science** Based on the latin word *scientia* which means knowledge. This knowledge is highly organised, and relates to humans and their environment (natural and artificial). The modern-day word science also includes the methods used to obtain that knowledge.

**scientific method** Originally concerned with the way in which scientists gathered facts and figures, developed ⋆hypotheses based on these observations, and tested the hypotheses by more observation and experiment. If such a hypothesis were successful it came to be regarded as a ⋆theory. Today it is recognised that scientists have no special order in which they work. However, they normally present their results and justify their hypotheses and theories in scientific journals, and at scientific meetings, using the steps in the scientific method. Compare law.

**scion** A bud or shoot that is removed from one plant, and then grafted onto the rooted part of another plant, called the ⋆stock.

**sclereid** A short **⋆sclerenchyma cell**, often found in woody parts of plants. e.g. Hard seed coats and shells of fruits; stem cortex. Compare fibres.

**sclerenchyma** The strengthening and supporting tissues of plants. The sclerenchyma cells form from ⋆parenchyma when the walls are thickened and often ⋆lignified. They frequently lose their cytoplasm, and the ⋆pits in the walls are quite obvious. There are two types: the longer ⋆fibres and shorter ⋆sclereids. Compare collenchyma.

**sclerophyll** Plants whose leaves are hard, stiff and leathery because they contain relatively large amounts of thick-walled sclerenchyma cells, thus helping the leaf to withstand drought by not shrivelling up (wilting). e.g. Many Australian plants including *Eucalyptus*, *Epacris*, *Hakea*, *Grevillea*, and *Banksia*. Both ⋆tall open (wet sclerophyll) forest and ⋆open (dry sclerophyll) forest is dominated by eucalypts, but the first is usually taller, and has a higher proportion of understorey species with soft thin leaves (⋆mesomorphic), while the second has understorey shrubs with small hard leaves (⋆xeromorphic).

**sclerosis** The hardening of tissue, especially by excessive growth of fibrous tissue. e.g. Nerve tissue and ⋆multiple sclerosis; linings of arteries and ⋆arteriosclerosis.

**sclerotic** Also called the sclera or the 'white' of the eye. The tough, fibrous outer layer that surrounds the eyes of vertebrates and cephalopods, helping to give them their rounded shape. The part at the front of the eye forms the ⋆cornea. See Fig. E9.

**scorpion** An ⋆arachnid of the order Scorpionida, with large claws, and a dangerous poisonous sting in the tail. See Appendix 1.

**scrapie** A disease in sheep thought to be transmitted by a ⋆prion. It leaves parts of the brain looking spongy and the animal loses coordination and dies. Compare Creutzfeldt–Jakob disease, kuru and mad cow disease.

**screen** A system for carefully examining and separating into different groups. In medicine the testing is carried out to see if the disorder is there even if there are no symptoms. e.g. Women are often screened for breast cancer or cervical cancer.

**scrotum** The external pouch that contains the ⋆testes of the male (only during the breeding season for some animals). They usually hang free of the body to keep them cool, as the higher body temperature inhibits development of the sperm. See Fig. R3.

**scrub** A general term describing various unrelated types of vegetation in Australia. e.g. *Acacia*-dominated communities; ⋆mallee; rainforest.

**scurvy** A deficiency disease caused by lack of vitamin C in the diet. Advanced symptoms include spongy and bleeding gums, and haemorrhaging under the skin. See ascorbic acid.

**scutellum** A shield-shaped structure. e.g. The single cotyledon of a grain seed.

**sea anemone** A group of cnidarians (coelenterates) of class Anthozoa, usually stationary, with an often colourful ring of tentacles around their mouth. See polyp and Fig. P8, and Appendix 1.

**sea cucumber** Any of the echinoderms in class Holothuroidea that have a long cucumber shape, and a tough leathery skin. Also called bêche-de-mer or trepang. See Appendix 1.

**sea floor spreading** A geological process occurring along the mid-ocean ridges where molten rock from within the mantle rises to the surface to form new ocean floor, which then spreads out (Fig. S1). See plate tectonics and Fig. P7.

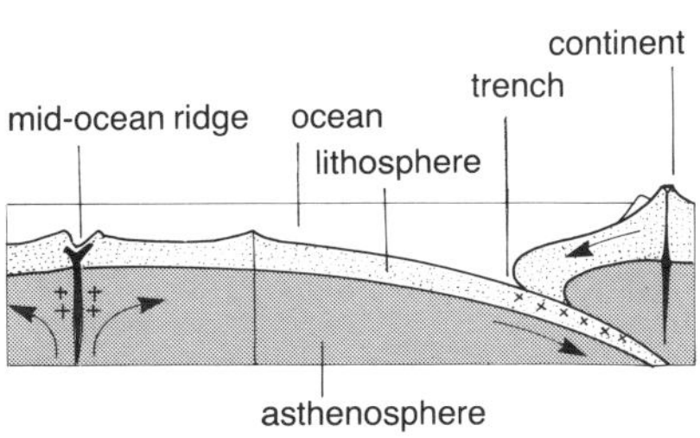


**Fig. S1 Sea floor spreading**

**sea grass** Any of about 12 genera of marine flowering plants generally with straplike leaves, and which are capable of existing in completely marine conditions.

**sea lettuce** The common name of the green seaweed, genus *Ulva*, in Phylum (Division) Chlorophyta which has broad fronds and hence the appearance of lettuce. See Appendix 1.

**sea lilies** A common name for crinoids, a class of ⋆echinoderms having a cup-shaped body with feathery arms and attached to the ocean bottom by a stalk. Compare feather stars.

**sea lion** A carnivorous marine placental mammal in order Pinnepedia which differs from ⋆seals in possessing ears. e.g. Australian sea-lion *Neophoca cinerea*.

**sea snake** Marine snakes in the family Hydrophiidae, with a flattened tail to assist swimming, and often highly venomous. See Appendix 1.

**sea star** The common name for ⋆echinoderms of the class Asteroidea, often incorrectly called star fish. Their arms contain many ⋆tubed-feet powered by a ⋆water vascular system. They use the tubed feet to attach themselves to prey, such as ⋆bivalves, and then force open the shell. See Appendix 1.

**sea urchin** Any echinoderm in class Echinoidea, being marine animals with a globular or disc-shaped body covered with small plates and having many spines projecting from the surface. See Appendix 1.

**seal** A carnivorous marine placental mammal in order Pinnepedia which lack the ears of sea-lions and fur-seals. They are not normally seen along Australian coasts. The animal normally called a seal is the Australian fur-seal *Arctocephalus pusillus*.

**season** A period which occurs each year when conditions of weather, temperature, etc. are usually similar. Caused by the Earth's movement around the Sun. e.g. Summer (hottest), autumn, winter (coldest), spring.

**seawasp** See **box jellyfish**.

**seaweed** Any large ⋆algae that grow in the ocean, or on coastal rock platforms. e.g. Kelp, sea-lettuce, Neptune's necklace, etc. See also Rhodophyta, Phaeophyta, Chlorophyta. Compare sea grass.

**sebaceous gland** Oil glands that release their oils into the ⋆follicle at the ⋆roots of hairs of mammals. Absent from palms of hands and soles of feet, but present on other hairless areas such as lips. Activity of the sebaceous gland increases during puberty, contributing towards ⋆acne.

**sebum** The fatty secretions of a ⋆sebaceous gland which oils the hair and forms a thin protective layer over the surface of the skin to cut down water loss under dry conditions, and water entry when wet.

**second growth** Forest growth that comes after fire, logging, storm, or other disturbance.

**secondaries** The smaller flight feathers attached to the ulna or forearm of a bird. Compare primaries.

**secondary growth** Also called **secondary thickening**. An increase in diameter of a plant resulting from the formation of secondary vascular tissues, due to cell division in the vascular ⋆cambium. Occurs in most gymnosperms and dicotyledons, and in some monocotyledons. Compare primary growth.

**secondary meristem** Regions in plants where cell division (⋆mitosis) regularly occurs, but which were once normal tissues. e.g. ⋆Cork cambium (⋆phellogen), healing of wounds. Compare primary meristem.

**secondary sexual characteristics** Characteristics of an animal that differ between males and females (excluding reproductive organs), and produced by male and female sex hormones. Examples include antlers in stags and bright plumage of male birds. In humans, secondary sexual characteristics develop at ⋆puberty, and for the male include deeper voice, potentially more muscular build, growth of pubic hair and relatively more hair on the body overall, widening of the shoulders, enlarged penis, and interest in the opposite sex. In females, they include development of breasts, widening of the hips, growth of pubic hair, an extra layer of fat beneath the surface of the skin, onset of menstruation, and interest in the opposite sex.

**secondary thickening** See secondary growth above.

**secretin** A hormone produced by the walls of the duodenum (due to stimulation by acid from the stomach) that increases the rate at which digestive juices from the pancreas, and bile from the liver, are produced.

**secretion** The giving off, discharging, extruding, etc. of a liquid substance by a living thing. The substance secreted is also called a secretion. The secretion can come from a special cell (e.g. neurotransmitter from the end of a neurone), from a gland (e.g. digestive juice from pancreas into the duodenum; insulin into the bloodstream), or from less easily defined regions (e.g. gum from the bark of a tree).

**section** 1 To cut through or divide into two or more parts. 2 A thin section cuts a slice so thin that light can pass through it and it can be viewed using a microscope. Sections can be either transverse (T.S., across a section of tissue) or longitudinal (L.S., along the tissue). See microtome.

**secund** Arranged along one side only. e.g. Leaves or flowers along a stem.

**sedative** A substance which calms, relaxes and puts one to sleep. e.g. Barbiturates, alcohol.

**sedentary** Concerning animals, fixed in one place for most of its life. e.g. sea anemone.

**sedimentary** Describing rocks formed from the cementing together of materials gradually deposited by wind or moving water. e.g. Sandstone, shale, conglomerate.

**seed** The often small grains produced by plants, especially the gymnosperms and angiosperms, from which new plants can grow. See also germination, dispersal and Fig. S2.

Seeds have a protective external coating (seed coat or ⋆testa), an embryo plant with a ⋆radicle (root) and ⋆plumule (shoot), and usually a store of food (in ⋆cotyledons or ⋆endosperm). In angiosperms (flowering plants), the seed develops from the ovule, and surrounds the embryo which has developed from the fertilised ovum. Surrounding both is the fruit which has usually developed from the ovary wall. Flowering plants which have seeds with one cotyledon (e.g. orchid, grass) are called ⋆monocotyledons, and those with two cotyledons (e.g. roses, plums, eucalypts) are called ⋆dicotyledons. In gymnosperms, the seeds develop between the scales of the female ⋆cone.

**seed coat** The ⋆testa or outer protective covering of a seed.

**seed dispersal** See dispersal.

**seed leaf** A ⋆cotyledon.

**seedling** A young plant. Seedlings often appear quite different from the adult. e.g. Young eucalypts bear juvenile leaves that are often a very different shape from the adult leaves.

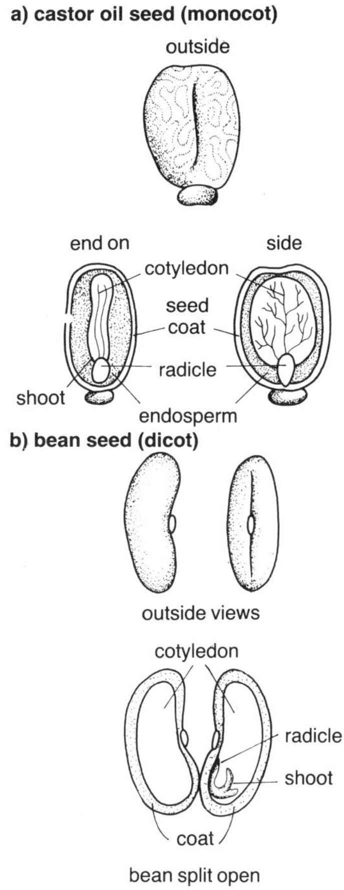


**Fig. S2 Seeds (a) Castor oil seed (monocot) (b) Bean seed (dicot)**

**segment** A part of an organism or structure which is divided or separated from parts next to it. Many organisms are divided into many segments, with internal structures often repeated (see metameric segmentation). In other cases, the segments are major functional divisions, such as head, thorax, and abdomen of insects. Organisms which readily display segmentation include millipedes, centipedes, annelids (e.g. earthworms), crustaceans, insect larvae, etc.

**segmentation** See segment above.

**segmented worm** Any round, segmented worm of the phylum Annelida which includes earthworms, bloodworms, leeches, and polychaete beach worms. Such worms have a separate mouth and anus. See Appendix 1.

**segregation** During meiosis, the separation of each member of the pairs of chromosomes, so that gametes result that have only one of each chromosome. See Mendel's first law.

**selection** Short for ⋆**natural selection** or ⋆**artificial selection**. Refers to any process that allows some organisms with particular features to survive and reproduce in preference to organisms which lack those features.

**selection pressure** The 'force' of ⋆natural selection acting on a ⋆population of organisms.

**selective agent** Any feature of the environment (e.g. living thing, climate) that actively selects which organisms survive e.g. Lions are selective agents for a population of antelopes, antibiotics are selective agents for strains of resistant bacteria.

**self-fertilisation** The union of male and female gametes produced by the same organism. e.g. Self ⋆pollination, ⋆autogamy in *Paramecium*. See also hermaphrodites, cross-fertilisation.

**self-pollination** See pollination.

**self-replicating** ⋆Plasmids and other non-nuclear DNA or RNA able to time the rate of their own duplication rather than be under the control of the DNA in the nucleus.

**self-sterility** The condition in ⋆hermaphrodite animals, and in flowers with both anthers and carpels, where male gametes cannot fertilise female gametes from the same organism. See also incompatibility.

**semen** The sticky, whitish fluid produced by male animals that contains the ⋆sperm cells (spermatozoa). In animals where there is internal fertilisation, the semen contains various secretions as well as the sperm. e.g. In men, fluids from the ⋆prostate gland, ⋆seminal vesicle and ⋆Cowper's gland.

**semicircular canal** The organ of the ⋆inner-ear of vertebrates that detects the rate of movement of the head. It consists of three semicircular tubes (canals) at right-angles to each other, each end of which connects to the ⋆utricle. A swelling at one end of each canal (called an ampulla) contains a receptor that is sensitive to movement. As the head moves, the fluid in the canals sets up currents which move the receptors in each ampulla. As the canals are at right angles to each other, information is provided about movement for each of the three-dimensions of space. See also sacculus and Fig. E1.

**semiconservative replication** The way in which DNA makes exact copies of itself (see replication) means that each new double helix consists or one new strand and one old strand. Thus the original molecule of DNA is semiconserved during replication. See Fig. D3.

**seminal fluid** The ⋆**semen**.

**seminal vesicle** 1 In various invertebrates, a sac-like structure where semen is stored. 2 In mammals, a pair of glands attached to the sperm duct (vas deferens) that produce a sticky glucose-rich fluid that is part of the semen. The glucose helps provide energy for the movements of the sperm cells. See Fig. R3.

**seminiferous tubule** A sperm-producing tube in the testes of vertebrates. In men, there are about 1000 tubules per testis, each about

50 cm long and 0.02 mm in diameter. Along the walls of the tubules, ⋆meiosis produces ⋆haploid sperm cells.

**semipermeable** Allows some substances to pass through, but not others. e.g. Cell and other membranes. See also dialysis.

**senescence** The complex process that eventually leads to death of an organism, or part of an organism. Within cells, old parts are removed by ⋆lysosomes. The death and fall of leaves and flowers is genetically controlled (see abscission). Also genetically controlled is the death of organisms after a period of hectic mating (e.g. Butterflies, *Antechinus*). However, for most organisms the precise cause of aging is not known.

**sense** The ability of an organism to be aware of the changes to the environment around using ⋆sense organs (receptors).

**sense organ** Also called a **receptor**. An organ which contains sensory cells, which when stimulated, generate impulses that can be interpreted by the brain. e.g. Eyes and sight; ears and hearing; tongue and taste; nose and smell; inner ear and balance; skin and touch, pressure, pain, heat, cold; muscles and position, etc.

**sensitive plant** A plant that moves when touched. e.g. *Mimosa pudica*, whose leaves collapse when touched. Trigger flowers, e.g. *Stylidium* sp. whose stamens move rapidly when touched.

**sensory** Of or relating to the senses.

**sensory nerve** See sensory neurone.

**sensory neurone** A neurone that carries impulses to the brain or spinal cord (⋆afferent). Such neurones reach the spinal cord through the dorsal (closest to the back) ⋆roots. Along this root is a ⋆ganglion where the cell bodies for the sensory neurones are located. See reflex arc and Fig. R2.

**sensory organ** See sense organ above.

**sepal** One of the leaf-like structures that protects the flower bud before it opens. Usually green, but sometimes of other colours, the sepals form the outer whorl of the ⋆perianth, and the sepals are known collectively as the ⋆calyx. See Fig. F2.

**separation layer** See abscission.

**septic** Infected by disease-causing microbes, especially the blood and tissues.

**septicaemia** Blood poisoning, with symptoms such as fever and extreme tiredness.

**septum** Plural is **septa**. An internal dividing wall or partition. e.g. The tissue between the two nostrils of the human nose.

**sequencing** Determining the order in which amino acids occur in a particular protein, or in which organic bases occur in DNA or RNA.

**seral stage** A particular stage in a plant ⋆succession. e.g. The first stage is called the pioneer community; the last stage the climax community.

**serine (ser)** One of the about 20 amino acids commonly found in proteins. It is also converted into a chemical that forms part of the ⋆myelin sheath of neurones. See Fig. A6.

**serotonin** A ⋆neurotransmitter found concentrated in the ⋆hypothalamus that is concerned with mood, and may be involved in inducing sleep. The drug ⋆LSD acts by increasing the concentration of serotonin in the brain.

**Serpentes** The suborder of reptiles that includes the ⋆snakes. See Squamata, Sauria and Appendix 1.

**serrate** A leaf edge (margin) with asymmetrical teeth which point forward. See Fig. L1.

**serratus anterior** A muscle of the chest region, connecting the upper eight or nine ribs to the scapula. It helps rotate the scapula or elevate the ribs.

**serrulated** A finely toothed edge (margin) to a leaf. See Fig. L1.

**serum** 1 The clear fluid of the blood, with blood cells and substances that cause blood clotting (e.g. fibrin) removed. Compare plasma. 2 A liquid containing *antibodies and *antitoxins that can be injected to counteract a disease already present. Normally prepared by injecting the disease-causing agent into an animal, and after the antibodies or antitoxins have been produced, the blood that contains them is extracted and purified. Compare vaccine.

**sessile** 1 Without a stalk, being used for organs such as flowers and leaves of plants, and the eyes of a crustacean. 2 *Sedentary. Animals that live their lives attached to a solid object. e.g. Sponges, corals, sea-squirts (tunicates), sea anemones, etc.

**seta** A usually stiff bristle, or bristle-like structure. 1 In plants, the stalk that holds the spore capsule of mosses and liverworts. 2 In invertebrate animals, the bristles that help annelids to move (called *chaeta), the 'hairs' of a caterpillar, tactile 'hairs' of a crustacean, etc.

**sex** The total characteristics, considering both anatomy and physiology, which distinguish male from female. Since it is not always easy to use anatomy to distinguish male from female, the most important difference is that males produce sperm (the moving or male gametes) while females produce ova (the stationary or female gametes). It is sometimes even hard to tell sperm from ova in some plants (then called *microspore and *megaspore respectively). Some organisms produce both male and female gametes, and are said to be *hermaphrodites.

**sex chromatin** Also called a **Barr body**. Only found in certain cells of females, and thought to be an inactive *X chromosome, as there is always one less Barr body than there is number of X chromosomes. Used to test if female athletes are actually female. Thus a 'superfemale' with XXX chromosomes has two Barr bodies, while a female with Turner's syndrome (only one X) has no Barr bodies. See also sex chromosomes.

**sex chromosomes** One of a pair of *chromosomes that determine the sex of an organism. One of the pair looks like the other chromosomes of the cell, and is called X. The other is much smaller and is called a Y. In many organisms, including humans, two X chromosomes (XX) produces a female while XY produces a male. Generally, the Y chromosome appears to contain few instructions, leading to *sex-linked features. In other organisms, different patterns of sex chromosomes exist. e.g. Certain grasshoppers have XX females, and XO males (the O indicating no Y chromosomes). In certain caddis flies, the males are XX and females XO. In butterflies, moths, birds, and some fish, the male is XX and the female XY.

**sex determination** The method by which the sex of an organism is determined.

In most cases, the sex of an organism is determined by the presence of the *sex chromosomes. However, sometimes male and female is determined by which of two alleles is present. In mosquitoes, the *homozygous recessive (ff) is female, while the *heterozygous animal (Ff) is male. In other organisms such as bees, the male is haploid and the female is diploid. The queen bee mostly lays fertilised eggs to produce females, but when a new hive needs to be established she lays unfertilised eggs to produce male ★drones. In some cases, organisms including certain fish, snails and marine worms can change sex. Certain fish found on the Great Barrier Reef live in schools with one male controlling many females. If the male should die, the head female changes sex and takes control of the school.

**sex hormone** One of a number of ★hormones produced by the gonads (testes, ovaries) and other organs which help control the development of ★secondary sexual characteristics, the maturing of reproductive organs, and mating behaviour. See androgens, testosterone, progesterone, oestrogen.

**sex linked** A hereditary character whose genes are carried on one of the (mostly X) sex chromosomes, and is usually only seen in one sex.

With humans, recessive genes that cause haemophilia, colour blindness, etc., occur on the X chromosome and the resulting diseases are nearly always seen in males. Males receive one X chromosome carrying the recessive allele. Since their Y chromosome carries few genes, there is no dominant normal allele to counteract the recessive allele's damaging effects. Females would need two of these rare recessive alleles to show the disease, a very unlikely occurrence indeed. Women who have the recessive allele on one of their chromosomes are said to be ★carriers.

**sexual dimorphism** The difference in size and appearance between male and female organisms. e.g. Satin bowerbird, many spiders.

**sexual intercourse** Also called **coitus** and **copulation**. The act of the male placing the penis into the vagina of the female. Usually followed by and associated with ★ejaculation of ★semen.

**sexual reproduction** Any form of reproduction where a sperm fertilises an ovum, and the resulting cell develops into a new organism. Usually the haploid gametes (sperm, ova) come from different organisms resulting in ★cross-fertilisation, with a resulting greater amount of genetic recombination, and more variety amongst organisms. Compare asexual reproduction.

**sexual selection** An example of natural selection proposed by Charles Darwin, where females choose the males, 'weeding out' those males without the desired characteristics. Darwin used this idea to explain, for example, the elaborate display and plumage of many birds. e.g. Lyrebird, peacock.

**sexually transmissible disease** Called **STD** for short. Diseases transmitted during sexual intercourse. e.g. Syphilis, Gonorrhoea, genital herpes, AIDS, Chlamydia, genital warts; hepatitis B.

**shade plant** 1 A plant that is able to thrive under conditions of low light intensity, such as in the shade of the closed canopy of a rainforest. 2 Also describes quick growing trees that are planted to shade and protect the growth of permanent trees until they are established. The quick growing trees are then removed.

**shaft** The long axis or rod-like part of a structure. e.g. The rod-like part of a bone. The rachis or midrib of a feather.

**shark** Predatory ★cartilaginous fish, order Selachii in class Chondrichthyes, including the white pointer, hammerhead, tiger, and grey nurse sharks. Their streamlined body contains a mouth that can be opened very widely. In some species, the often numerous sharp teeth are replaced from behind as they are broken or are worn out. See spiracle and Appendix 1.

**she-oak** Common name given to some species of the genus ⋆*Allocasuarina*.

**sheath** A case or covering, often to protect the object within. e.g. A tube-shaped fold or sac into which long protruding organs (e.g. proboscis, penis) can be retracted. The base of a leaf that is wrapped around a stem. e.g. grass.

**shed** To let objects fall off by natural processes. e.g. Trees shed bark, branches and leaves. Reptiles shed their skin.

**shell** The hard outer covering of organisms or parts of organisms. e.g. The hard covering of molluscs (mainly made of calcium carbonate) and some reptiles (bone). The hard ⋆exoskeleton of insects and crustaceans. The covering of an egg—hard for birds, leathery for reptiles. The hard outer covering of fruits and seeds, especially nuts.

**shin** In humans, the relatively sharp front part of the tibia, the long bone connecting the knee to the foot.

**shingles** A virus infection, *Herpes zoster*, of the nerve endings. The virus may be the same as that causing chickenpox.

**shipworm** Any of a number of wood-boring bivalve ⋆molluscs, especially those in genus *Teredo*. They are responsible for much damage to wooden structures such as boat hulls, and wharf pylons. See Appendix 1.

**shiver** To shake and tremble, usually due to cold. The muscles undergo involuntary contractions thus generating heat to warm the body.

**shoot** 1 In a seed plant, the stem with its branches and leaves. 2 A young branch with its leaves.

**short-day plant** A plant that must be exposed to light for periods shorter than a critical length (often less than 12 hours of sunlight) before it will flower. Perhaps better called a long night plant, since it actually needs long nights to flower, as even a short flash of light during darkness will inhibit flowering in some plants. Almost all plants that flower in late autumn or early spring. e.g. Chrysanthemum. Compare long-day plants and day-neutral plants. See also photoperiodism.

**short sighted** Eye problem where distant objects are out of focus, and cannot be clearly seen. Compare long sighted.

**shoulder girdle** Also called the ⋆pectoral girdle.

**shrew** A small mouse-like mammal in the order Insectivora. One member, *Sorex minutus*, is the smallest mammal known, weighing less than 25 grams. See Appendix 1.

**shrimp** One of a number of prawn-like crustaceans in the order ⋆Decapoda, including true shrimps, brine shrimps, fairy shrimps, etc. Compare prawns. See Appendix 1.

**shrub** A woody plant or bush, usually with several main stems rising near the ground. Not big enough to be called a tree. e.g. Most specimens of *Hakea, Dillwynia, Leptospermum*.

**SI units** Short for **Système International d'Unités**. The international metric system of units used in science and based on metres, kilograms and seconds.

**sibling** Called **sib** for short. Two or more offspring (male or female) having the same parent. Brothers and sisters.

**sickle cell anaemia** A human genetic disease which produces sickle-shaped red blood cells. These defective cells break easily and are removed from the blood, leading to anaemia. Caused by a ⋆point mutation where a glutamic acid at position 6 in the β-haemoglobin chain is replace by a valine.

**sieve cell** A type of ⋆sieve element that forms part of the ⋆phloem. More common in ferns and gymnosperms. See Fig. P4.

**sieve element** The cells in the ⋆phloem of the vascular system of gymnosperms and angiosperms, carrying (by ⋆translocation) sugars and other nutrients made in the leaves to the stem and roots (Fig. P4).

There are two types of sieve elements. ⋆Sieve cells are long and narrow, with very small pores (pits) arranged at random through the cell walls. ⋆Sieve tube elements are shorter and wider with a ⋆sieve plate on the ends containing much larger pores. Sieve cells and sieve tube elements contain cytoplasm, but no nuclei. The cells are controlled by the nucleus in adjacent cells—⋆albuminous cells for sieve cells in gymnosperm phloem, or ⋆companion cells for sieve tube elements in angiosperms.

**sieve plate** One of the flat cell walls at the end of a ⋆sieve tube element that contains numerous pores through which strands of cytoplasm connect adjacent cells of the ⋆phloem. See Fig. P4. Compare perforation plates.

**sieve tube** A long tube in the phloem made up of numerous ⋆sieve tube elements arranged end-to-end and connected by ⋆sieve plates. See Fig. P4.

**sieve tube element** A type of ⋆sieve element that forms part of the phloem. More common in angiosperms. See Fig. P4.

**sight** Vision, or the sense of seeing. See eye.

**sigma factor** An important part of the ⋆RNA polymerase molecule in ⋆prokaryotes.

**siliceous** Containing or made from material that is mainly silica (silicon dioxide). e.g. The skeletons of ⋆diatoms, and ⋆spicules in the flesh of many sponges.

**silk** 1 A fine, shiny fibre produced by the silk glands of various insects, especially butterflies and moth larvae, for the construction of their cocoons. e.g. Silkworms are the larvae of a moth. 2 A fine fibre produced by spiders and used to make their webs, egg sacs, shelters, etc.

**Silurian** The third period in the Paleozoic era, between the Ordovician and Devonian periods. See Appendix 2.

**silverfish** Any of the small wingless insects of genus *Lepisma*, order Thysanura, which cause damage to books, wallpaper, clothes, etc. See Appendix 1.

**simple** Of a leaf, not divided or branched.

**sinew** A ⋆**tendon**.

**sinoatrial node** The **S-A node** for short. See ⋆pacemaker.

**sinus** A space or cavity. e.g. **a** In leaves, a notch or gap between two lobes of a leaf. **b** In humans, the nasal sinuses are membrane-lined hollows in the bones around, and connected to, the nasal cavity. See Fig. D2.

**siphon** A tube through which certain animals pass fluids. e.g. The breathing siphon of a mosquito. Tubes of bivalve molluscs and cephalopods through which water is drawn in and expelled.

**Siphonaptera** The order of wingless insects that includes the ⋆fleas. They are external parasites of birds and mammals, having piercing mouthparts for extracting blood. See Appendix 1.

**Sirenia** The order of aquatic placental mammals also called sea-cows, including the dugong and manatee. These slow-moving herbivorous animals feed on seagrasses in warm coastal waters and tropical rivers. See Appendix 1.

**SIV** Short for **simian immunodeficiency virus** that produces symptoms similar to ⋆AIDS but in monkeys. Since SIV in monkeys is thought to acts like ⋆HIV in humans, it is often used to test theories or treatments for AIDS.

**skates** A type of ⋆ray.

**skeletal** Of or relating to the skeleton.

**skeletal muscle** A muscle that is attached to, or moves, part of the skeleton. In vertebrates it is made up of ⋆striated or striped muscle tissue.

**skeleton** A hard and rigid supporting or protective framework. In many organisms the skeleton is outside the body, and is called an ⋆exoskeleton. For vertebrates, the skeleton lies inside the body (endoskeleton), and acts as a rigid support for the soft tissues. The skeleton of humans can be divided into the axial skeleton, which includes the skull, vertebrae, and ribs, and the appendages such as the arms and legs. See individual names for the bones in Fig. S3, bone, joints, and pentadactyl limb.

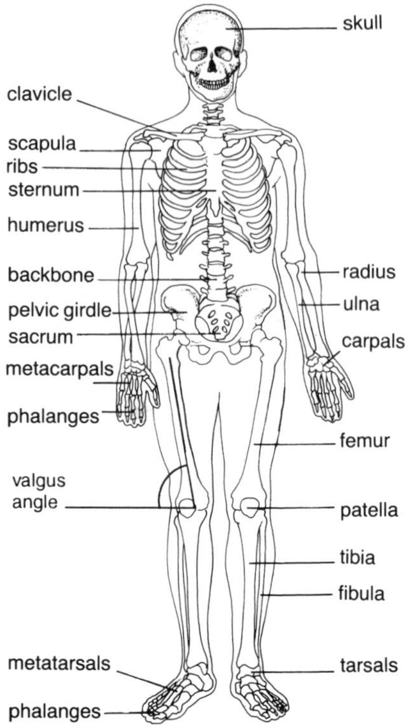


**Fig. S3 Skeleton**

**skin** Also called the **cutis**. The organ that covers almost the entire body surface of vertebrates. See Fig. S4.

In humans it consists of two layers: **a** The ⋆epidermis is an outer protective layer of surface (epithelial) cells. Lower levels of the epidermis produce new cells by mitosis, and contain the pigment ⋆melanin. **b** The ⋆dermis (or corium) is an underlying layer of connective tissue. Within this layer are sweat and sebaceous glands, hair follicles, and receptors for pain, touch, heat and cold. Blood ⋆capillaries supply oxygen and nutrients, and remove wastes. The blood also helps rid the body of excess heat by carrying warm blood to the skin from deeper tissues of the body. The skin of other vertebrates differs in a number of ways. The skin of fish and amphibians contains glands that secrete protective mucus. Scales cover fish and reptiles, feathers cover birds, while most mammals have much more hair than do humans.

**skink** A type of lizard, family Scincidae, with smooth scaly skin and a long tail. See Sauria and Appendix 1.

**Skinner box** A cage designed by American psychologist B.F. Skinner to test theories of learning called ⋆operant conditioning.

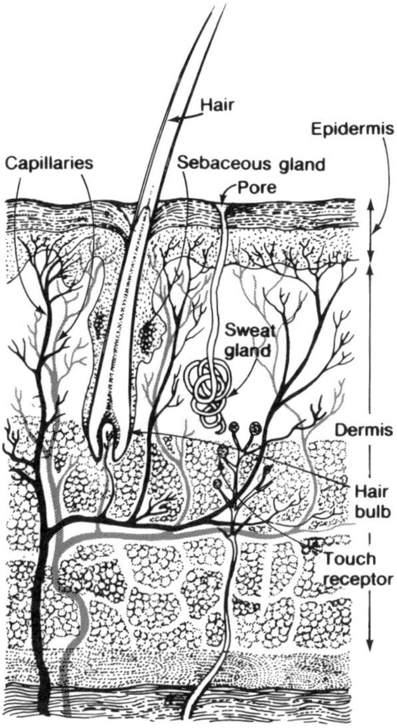


**Fig. S4 Human skin**

**skull** In vertebrates, the bony and/or cartilaginous covering of the head which encloses and helps protect the brain and sense organs. Consists of the ⋆cranium and facial bones. See Fig. S3.

**slater** Also called **woodlice**. Any of the small land-living crustaceans in order Isopoda, commonly found under stones and logs, usually greyish in colour, and mostly in genera *Oniscus* and *Porcellio*. See Appendix 1.

**sleep** The usually daily period when the body enters a state of unconsciousness to allow the body to be rested and rejuvenated. The metabolic rate decreases, and senses are dulled.

Each night humans go through 4 or 5 sleep cycles, each lasting about 90 minutes. Each cycle starts with a light sleep stage where muscles gradually relax, breathing is long, heartbeat is lowered, but during which time we are easily awakened. Deep sleep follows, the muscles being completely relaxed, but breathing faster. Finally, at the end of deep sleep, muscles begin to twitch, and there are rapid eye movements behind the eyelids. This last stage is called ⋆REM sleep. Then the body returns to light sleep, and the cycle starts again. In each cycle, sleep becomes less 'deep', until finally we awake.

**sleeping sickness** The chronic disease ⋆trypanosomiasis, caused by a protozoan of genus *Trypanosoma*, and in Africa transmitted by the tsetse fly. In South America it is called Chagas' disease, and is transmitted by a beetle eaten by natives for its supposed aphrodisiac qualities.

**slide** A ⋆microscope slide used for viewing thin sections.

**sliding filament mechanism** The mechanism of muscle contraction where ⋆actin filaments slide along ⋆myosin filaments, consuming energy supplied by ATP. See ⋆sarcomere and ⋆muscle fibre (Fig. M5).

**slime fungi** Sometimes called **slime moulds**. An unusual group of organ isms in the Phylum Myxomycota with both plant and animal features. The body (thallus) is a slimy mass of protoplasm (called a *plasmodium*) containing many nuclei, which moves slowly much like an *Amoeba*, and ingests food particles (by ⋆engulfing) like an animal. However, it reproduces by spores like a plant. See Appendix 1.

**slime mould** See **slime fungi**.

**slipped disc** Also called a **ruptured disc**. Occurs when the cushioning pad between the ⋆vertebrae pushes through a tear in the surrounding ⋆ligaments. See Fig. B2.

**slough** To shed or lose a mass of usually dead tissue. e.g. The bark sloughs from a tree, and the skin sloughs from a snake.

**slug** Any of the molluscs in class Gastropoda that are similar to snails, but lack a shell. See Appendix 1.

**small intestine** The relatively narrow but long section of intestine connecting the stomach to the large intestine (or colon). The three sections, starting from near the stomach are ⋆duodenum, ⋆jejunum, and ⋆ileum. See Fig. D2.

Digestive juices from the walls of the small intestine, and from the pancreas complete digestion in the early part of the small intestine. Bile from the gall bladder helps prepare fats for digestion by converting lumps into small globules. Along the whole length of the small intestine, the products of digestion are absorbed into the bloodstream. These processes of digestion and absorption are aided by the enormous internal area of the small intestine. Internal folding increases the area 3 times, while the ⋆villi (Fig. V3) make the area 10 times bigger again. The tiny ⋆microvilli increase the area another 20 times, making it 600 times that of a simple smooth-walled cylinder.

**smallpox** An acute, highly contagious disease that results in small pustules on the skin, which after they burst, leave small pits or scars. Caused by a virus. A ⋆vaccine has been developed, and it is hoped that this disease is the first to be completely eradicated by humans.

**smear** A technique used in preparing microscope ⋆slides, where a substance such as blood or bacterial culture is spread thinly over the slide.

**smell** The ⋆olfactory sense, also called organs of smell. See nose and Fig. D2.

**smooth muscle** Also called **plain** or **involuntary** muscle. Smooth muscle is made of long thin cells with no obvious internal structure (compare ⋆striated muscle), and occurs in sheets wrapped around hollow organs such as blood vessels, the alimentary canal, urinary and gall bladder. In vertebrates the contractions of smooth muscle are automatically controlled (involuntary) by the ⋆autonomic nervous system. Contraction occurs slowly, but the contracted state can be maintained for extended periods (tonus). Compare cardiac and striated muscle.

**smut** A disease of grain caused by a smut fungus, which attacks the 'heads' of grain. e.g. Wheat, corn. Compare rusts.

**snail** A mollusc in the class Gastropoda, and carrying a single spiral shell. e.g. Garden snail. See Appendix 1.

**snake** A long legless reptile in the order Squamata, suborder Serpentes. Snakes have dry scaly skins, a forked tongue, and their eyes are covered by a transparent eyelid. They lack an eardrum, and are able to open their mouths extremely wide to swallow their prey whole. Australian snakes include the harmless 'worm' snakes, the pythons which kill their prey by constriction, the rear-fanged snakes that are poisonous but are not very dangerous, and the very dangerous front-fanged land and sea snakes (e.g. Taipan, death-adder, tiger snake). See Appendix 1.

**snake lizard** Small burrowing ⋆legless lizards in family Pygopodidae. See Appendix 1.

**snout** The long pointed nose of animals, also containing the jaws. The muzzle.

**snow gum** Any of the several species of smooth-barked eucalypts found growing in the high altitudes of Australia, especially *Eucalyptus pauciflora*, usually with several stems from the base.

**social behaviour** Of or relating to the behaviour of organisms when in groups, especially those organisms that normally live in groups. e.g. Bees, baboons, elephants.

**social dominance** See peck order.

**social insects** Insects such as ants, termites, bees and wasps that live in highly organised groups, with the work often divided up amongst different ⋆castes (e.g. queen, drone, worker, soldier).

**society** An organised group of organisms, of the same species, which depend upon one another for survival.

**sociobiology** The study of how evolution can explain the origins of the behaviour of organisms, especially qualities such as ⋆altruism.

**sodium (Na)** A soft silvery metal. In living things it exists as the unreactive sodium ion ($Na^+$), where it is an essential element, especially in nerve cells. See sodium pump below.

**sodium pump** The ⋆active transport of sodium ions out of a neurone to maintain the ⋆resting potential across the cell membrane. The pump is needed because sodium ions continually diffuse into the cell when resting, and enter in enormous numbers when a nerve impulse passes.

**soft drug** Those drugs widely accepted and used in our society. e.g. Alcohol, nicotine in tobacco, caffeine in coffee and tea. Compare hard drugs.

**soft palate** The soft flap of skin at the rear or the roof of the mouth which closes off the nasal cavity when we swallow. See Fig. D2.

**softwood** Wood that is softer and easier to plane and shape due to its lack of ⋆fibres and ⋆vessels, such as that obtained from ⋆conifers. Compare hardwood.

**soil** The loose material of the surface of the Earth in which plants grow. It is made up of weathered (broken down) rock, decaying organic material (humus), water, dissolved minerals, and air. Also usually present is a large population of microbes, fungi, earthworms, etc. See under podsol, laterite, loam, earths, regosols, lithosols for details. See also soil profile.

**soil profile** The layered structure of ⋆soil as seen when a vertical section is cut or bored into the soil. Generally the upper layers become paler with time as minerals are washed to lower levels. Fig. S5 shows the various layers that may be present.

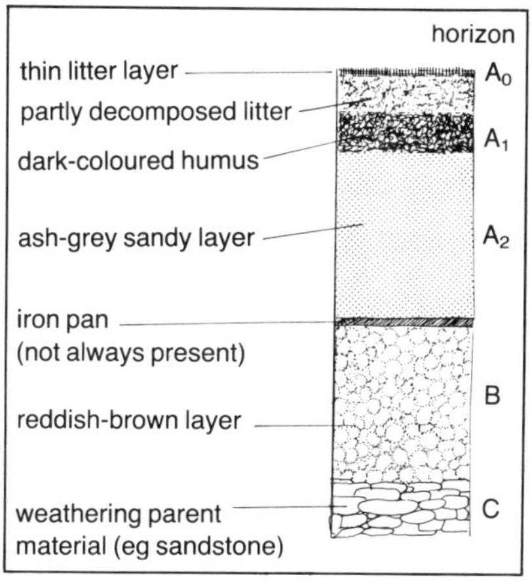


**Fig. S5 Soil profile**

**sol** A ⋆colloid in which small particles are spread out in a liquid. Compare gel.

**Solanaceae** The tomato family of dicot flowering plants. It also include potatoes and Petunia. See Appendix 1.

**soleus** A muscle beneath the calf-muscle at the rear of the leg.

**solitary** Existing or living singly or alone.

**soluble** Able to be dissolved to make a solution.

**solute** A substance that dissolves in a ⋆solvent to form a ⋆solution. e.g. The solute salt will dissolve in the solvent water.

**solution** A uniform (homogeneous) mixture of a solid, liquid or gas spread evenly throughout another solid, liquid or gas. The most common solutions are formed when solids (e.g. salt) or liquids (e.g. alcohol) are dissolved in water (a liquid). However, we can dissolve solid in solid (e.g. alloys), gas in liquid (e.g. oxygen in water), etc.

**solvent** A substance (usually a liquid) that allows another substance (solid, liquid or gas) to dissolve in it. e.g. Water is a solvent that dissolves the solute sugar.

**soma** 1 All the cells and tissues of a body except the reproductive cells. Compare germ plasm. 2 Cell body of a ⋆neurone.

**somatic** 1 Describes all the cells and tissues of a body except the reproductive cells. Compare germ plasm. 2 Describes the organs and tissues of the body other than those of the digestive system. e.g. ⋆Somatic nervous system.

**somatic hybridisation** The causing of two diploid body (somatic) cells from different species to fuse together in a tissue culture. e.g. Mouse and human cells are often joined this way to help determine which human genes exist on which chromosomes. Compare with fertilisation where haploid cells fuse.

**somatic nervous system** The part of the nervous system which allows us to control the muscles at will (voluntarily). It also carries the impulses from the sense organs back to the spinal cord and brain. This system consists of 12 pairs of cranial nerves, and 31 pairs of spinal nerves. Compare autonomic nervous system.

**somatotrophin** See ⋆growth hormone.

**sore** A usually relatively small wound, infection, bruise, etc.

**soredium** Plural is **soredia**. In ⋆lichens a single algal cell or group of them surrounded by fungal threads which is then able to grow into a lichen plant.

**sorus** Plural is **sori**. A means of vegetative reproduction. **1** A cluster of ⋆sporangia in ferns, usually found on the undersurface of the frond. The pattern of distribution of sori is useful for fern identification. **2** A mass of spores produced by certain smuts and rusts. **3** A cluster of ⋆soredia in lichens.

**spacer DNA** DNA that is located between genes, which, if it is transcribed, has to be 'edited' out of the resulting mRNA. See intron.

**spadix** A spike of flowers on a swollen axis. e.g. Arum lily. See Fig. I4.

**spathe** A large, often coloured leaf-like structure (⋆bract), which encloses the flowers of a spadix. e.g. Arum lily.

**spathulate** Of a leaf, spoon shaped, being larger and rounder at the upper end. See Fig. L1.

**spawn** 1 A mass of eggs laid by snails, fish, frogs, etc. 2 The act of laying such a mass of eggs.

**specialised** Having a special adaptation that allows an organism to survive in a particular, and often extreme, habitat. Such organisms tend to be very restricted in where they can survive, and any change to their habitat may result in their extinction. e.g. Marsupial 'moles' are adapted to survive only in sandy desert country. Mistletoe birds specialise in eating only fleshy mistletoe seeds.

**speciation** The origin of new ⋆species. The processes involved include ⋆mutation, ⋆natural selection, and ⋆isolation. See evolution.

**species** The smallest commonly used scientific unit of ★classification. A genus is made up of one or more species. 1 The ideal definition is a group of organisms that interbreed with each other to produce offspring that are themselves able to reproduce (i.e. are fertile). Thus members of different species do not normally interbreed, and if they do, their offspring are sterile (not able to produce offspring). 2 As not all organisms have been tested to see if they can successfully interbreed, a more practical definition is a population of organisms which closely resemble each other both structurally and physiologically and share a common gene pool. When using this method, all species have a type specimen (holotype) used for comparison. There are sometimes slight differences in appearances in organisms known to be able to breed together, and thus the species can be divided into ★subspecies, ★clines, ★races, ★varieties, etc. See binomial nomenclature.

**specific heat** The amount of heat (measured in ★Joules) that must be supplied to 1 gram of an object to raise its temperature by 1 degree Celsius. Water has one of the highest specific heats known for a liquid (4.17 Joules per gram per degree Celsius). This means that a large amount of heat must be supplied to living things before their temperature will rise very much, and a lot of heat must be removed before their temperature falls very much. This helps to protect sensitive plants and animals from very high or very low temperatures. Compare latent heat.

**specific immunity** See immunity.

**specimen** An organism, or part of an organism, taken for scientific study. e.g. A thin section placed on a microscope slide. A plant or animal considered typical of its kind, and collected for examination.

**spectroscope** A device for analysing the spectrum of light or other radiation being produced by a heated chemical. Such a spectrum can be used to identify the atoms and compounds present in a sample of unknown chemical.

**speech** The ability to communicate using sounds to represent objects, thoughts, emotions, etc.

**sperm** Short for **spermatozoa**. The male gamete, germ cell or reproductive cell. In animals, the sperm cell is typically made up of a head containing the haploid set of chromosomes, a middle section containing mitochondria, and a tail region to produce motion. See Fig. S6, semen, and acrosome.

**sperm bank** The storage of ★sperm at ultracold temperatures for later use in ★artificial insemination or ★*in vitro* fertilisation. When stored at the temperature of liquid nitrogen (−196°C) the sperm lasts indefinitely.

**sperm count** A laboratory procedure for estimating the number of sperm in a particular specimen of semen. There are typically 40 000 or more per cubic centimetre, but at least 15 000 per cubic centimetre of semen are needed for fertility.

**spermatheca** Also called **seminal receptacle**. A pouch or sac in certain female animals, or ★hermaphrodites, for the storage of sperm after mating.

**spermatogenesis** The formation of ★haploid ★sperm cells. Along the walls of the ★seminiferous tubules, meiosis produces haploid sperm cells, which mature as they pass along the tubules.

**spermatozoid** Also called **antherozoid**. Male ★gametes with flagella found in plants such as mosses, liverworts, and ferns.

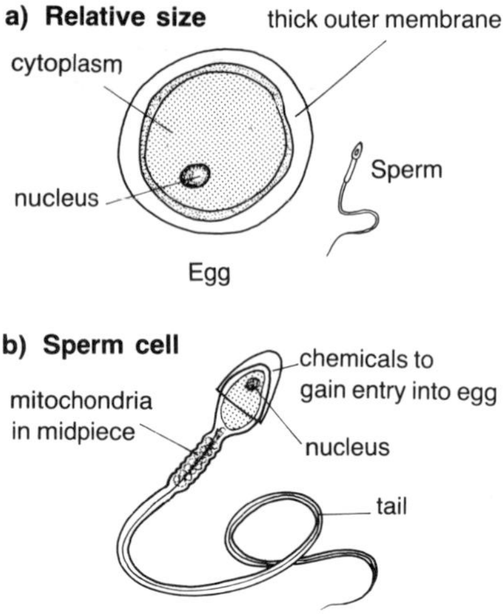


**Fig. S6 Sperm and egg**

**spermatozoon** Plural is **spermatozoa**. See sperm.

**sphagnum moss** Peat or bog moss common of genus *Sphagnum*, which builds up deep layers of peat. Widely used by gardeners in potting and packing plants because it can absorb so much water.

**Sphenopsida** A small class of vascular plants called horsetails, which do not naturally occur in Australia. See Appendix 1.

**sphincter** A ringlike band of smooth muscle tissue which surrounds a tube, and can close the tube when it contracts. e.g. Pyloric sphincter between stomach and intestine; anal sphincter.

**spicule** A hard sharp, pointed body (silicate or calcium carbonate) that forms the endoskeleton of certain sponges, corals and protozoans.

**spider** An arachnid of the arthropod order Araneae. Spiders have eight jointed legs (insects only six), and two body parts (insects have three). Most spiders use ⋆silk produced at their ⋆spinnerets, although only some use silk to make webs. Spiders immobilise their prey using venom, which also acts to digest the animal, which the spider then sucks dry. Apart from spiders that use webs to catch their prey, others actively chase and hunt, others are jumping spiders. Others hide under trapdoor or in funnel webs in the ground. See Appendix 1.

**spike** A type of flower head (inflorescence) where the flowers are borne on a long central axis. Such flowers have little or no stalk (sessile). e.g. Wheat, plantain, *Acanthus*. See Fig. I4. Compare spadix and catkin.

**spina bifida** A congenital malformation of the spine in which some of the vertebrae do not fuse, resulting in the spinal cord etc. protruding.

**spinal canal** The canal or tube within the backbone which contains the spinal cord.

**spinal column** The vertebral column or backbone.

**spinal cord** That part of the ⋆central nervous system of vertebrates that lies within the spinal column (backbone), helping connect the ⋆brain to the rest of the body.

The central grey matter is where synapses are made with sensory and motor nerves that connect the cord to the tissues of the body (see ⋆roots). Many of these synapses connect the motor and sensory nerves directly, establishing automatic reflex pathways. (See reflex arc and Fig. R2.) Other connections are with nerves in the surrounding white matter, which carry signals to and from the brain.

**spinal nerves** Nerves which connect the spinal cord to the body. Compare cranial nerves.

**spinal reflex** See reflex arc.

**spindle** In ⋆mitosis and ⋆meiosis, the delicate set of fibres (⋆microtubules) which connect the two poles, and which contract and thus drag the chromosomes that were lined up along the equator, to opposite poles of the cell. See Figs M1 and M3.

**spine** 1 The spinal column or backbone. 2 In plants, a sharp, pointed woody or hard body that has internal structures resembling a leaf. e.g. as in Cacti and holly. Compare prickle and thorn. 3 In animals such as sea urchins, insects, and fish, the often sharp, pointed structures that protrude from their bodies.

**spinifex** The common name given to two different grasses found in Australia. **a** Genus *Spinifex*, a grass that forms runners, and is used to bind sand on seashore sand dunes. **b** Genus *Triodia*, also called porcupine-grass, a tussock-shaped grass of inland Australia. The stomata are in deep grooves to cut down loss of moisture by ⋆transpiration.

**spinnerets** Organs on the end of a ⋆spider's abdomen where threads of silk are formed from secretions produced in the silk glands.

**spinulose** Bearing small spines.

**spiracle** 1 One of a pair of openings on the head of a cartilaginous fish (e.g. shark) through which water enters. It is the forward most of the gill slits. 2 In air-breathing arthropods, the external opening (with controlling valves) that allows the entry and exit of gases from the tube-like ⋆tracheae.

**spiral thickening** Also called **helical thickening**. Thickening in a spiral pattern of the cell walls of ⋆xylem vessels and ⋆tracheids to help give mechanical support to the plant.

**spirillum** Plural is **spirilla**. Any spiral or corkscrew-shaped bacterium. Compare bacillus, coccus, spirochaete, vibrio. See Fig. B1.

**spirochaete** Any unicellular ⋆bacterium of the order Spirochaetales, which are flexible, spiral bacteria that move by twisting their body. Some are free living, while others are parasitic and cause disease. e.g. Syphilis. Compare bacillus, coccus, spirillum, vibrio. See Fig. B1.

**spleen** A large organ of the ⋆lymphatic system located near the stomach, through which blood flows rather than lymph. The spleen filters the blood of invading organisms, and stores ⋆lymphocytes as they mature. It also stores red blood cells until needed, and helps remove and destroy those that are worn out.

**splicing** 1 The enzyme-controlled joining together of pieces of ⋆mRNA (exons) after the non-coding ⋆introns have been removed. 2 In genetic engineering, the enzyme-controlled joining together of pieces of DNA, especially when a gene is being inserted into a plasmid. See ligase.

**sponge** Any member of the phylum ⋆Porifera. See Appendix 1.

**spongy mesophyll** Internal leaf ⋆parenchyma with many air spaces between the cells. See Fig. L1. The spongy mesophyll allows gases (water vapour, carbon dioxide, oxygen) to move freely between the cells, and to the stomata, which allow the gases to enter and leave the leaf. Compare palisade cells.

**spontaneous generation** Also called **abiogenesis**. The theory that living things can arise from non-living, inorganic matter. Widely believed in ancient times, but by the late 1700s the only organisms whose reproduction could not be explained in alternative ways were the microbes. The theory was finally disproved by Louis Pasteur (1822–1895) when he showed that microbes which appeared in soup (broth) had come from the air, where they are always present. Compare biogenesis.

**sporangium** Plural is **sporangia**. Also called a **spore case**. A structure in which the asexual ⋆spores are formed. Some sporangia only produce one or a few spores (fungi), while others contain many thousands.

**spore** A small, usually microscopic, asexual reproductive body, consisting of between one and several cells, and which can develop into an adult without being ⋆fertilised. Their internal structure is quite different to that of a seed. In some organisms, spores are a resting stage in the life cycle that helps the organism survive drought or other difficult conditions. e.g. Bacteria, protozoa. Other spores are used to help spread the organism far and wide, usually being blown by the wind. e.g. Ferns, mosses, liverworts, fungi. Spores are usually part of a life cycle that involves ⋆alternation of generations. See also sporophyte and life cycle.

**spore case** A ⋆**sporangium**.

**sporocyst** Sac-like structures in the life cycle of many organisms (e.g. protozoans), inside of which they survive hard times.

**sporogonium** The sporophyte (spore producing) generation in the life cycle of mosses and liverworts. Usually it grows out of the gametophyte plant, and comprises a stalk with a spore-containing capsule at the top.

**sporophyll** A leaf on which sporangia (spore cases) occur. In ferns this is generally the normal leaf or frond, but plants such as the club mosses (*Lycopodium* and *Selaginella*) have special sporophylls.

**sporophyte** Also called **spore plant**. The diploid stage in a life cycle which involves ⋆alternation of generations, and which produces the asexual ⋆haploid spores. In mosses and liverworts, the sporophyte plant is often inconspicuous, and depends on the gametophyte for nutrients. In ferns, the sporophyte is the most obvious plant, the gametophyte being inconspicuous. See life cycle and alternation of generations.

**Sporozoa** A class of ⋆parasitic protozoans. They have complex life cycles which involve the formation of spores, often with two hosts. e.g. *Plasmodium* which causes malaria. See Appendix 1.

**sport** Also called a **rogue**. An organism that is different from the standard type or variety, often because of a ⋆mutation.

**sprain** Tearing or laceration of ligaments that hold the bones of a ⋆joint together.

**spreading** Leaves or branches which tend to stand or diverge out from the stem at almost right angles.

**Squamata** The order of reptiles that includes the ⋆lizards and the ⋆snakes. See Sauria, Serpentes and Appendix 1.

**squamous** Flat, scale-like. e.g. squamous epithelium is composed of flattened cells.

**squamule** A small scale.

**squarrose** Leaves or scales that are widely spread.

**squash** The flattening of a specimen before examining it with a microscope. e.g. Root tip squash. See also macerate.

**squid** Any of a number of ⋆cephalopod molluscs with a long body bearing ten arms on one end, and containing an internal shell. The giant squid, *Architeuthis princeps*, is the largest invertebrate at more than 15 m long. See Appendix 1.

**stage** 1 The platform on which ⋆slides are mounted when being viewed with a ⋆microscope. 2 One of the major steps into which ⋆mitosis and ⋆meiosis have been divided.

**staining** In microscopy, the adding of coloured dyes (stains) to specimens to make the details easier to see.

Non-vital staining is the colouring of dead tissue, while vital staining is the colouring of living tissue without harming or killing the cells. Acid stains tend to colour the cytoplasm of a cell, while basic stains tend to colour the nuclei. Neutral stains are a mixture of acid and basic stains, thus colouring both cytoplasm and nucleus. e.g. Leishman's stain is a mixture of eosin (acid) and methylene blue (basic) in alcohol. Certain stains need an additional substance called a mordant before they will colour certain tissues. e.g. Haematoxylin needs iron alum. It is also possible to use two or more stains one after the other (counter-staining) so as to colour different parts of the specimen. e.g. Safranine can be counterstained with light green in plants.

**stalk** Any slender connecting or supporting part. e.g. The main stem or stalk of a plant. The stalk of a stamen in a flower (see filament). The stalks that support the eye of snails and crabs.

**stamen** The male part of a ⋆flower, the stamen consists of the stalk (or filament) supporting the pollen-producing ⋆anther. The stamens of a flower together are called the ⋆androecium. See Fig. F2. Compare carpel.

**staminate** A flower possessing male parts (⋆stamens) but no female parts. e.g. Male flowers of pumpkin. Compare pistillate flowers.

**staminode** A sterile ⋆stamen, usually modified in form.

**stand** A relatively small area of vegetation, especially forest. e.g. A stand of trees.

**standing crop** The total amount of living material in a given population at a particular time, expressed either as its ⋆biomass, or in terms of the equivalent amount of energy. It represents the energy potentially available for use by other living things, or as fuel. See Appendix 1.

**stapes** Also called the **stirrup**. The innermost of the ear ⋆ossicles that transmit the vibrations detected by the eardrum to the inner ear. See Fig. E1.

***Staphylococcus*** A genus of spherical bacteria (⋆coccus) which form masses that resemble a bunch of grapes. Many are pathogens. e.g. Boils, septicemia, 'golden staph'. See Fig. B1.

**starch** The main energy-storing ⋆polysaccharide of plants, it is made of long chains of glucose molecules. It is found in special storage organelles called ⋆leucoplasts, and in ⋆chloroplasts where it is produced. Starch is particularly common in seed cotyledons and endosperm (e.g. grains such as wheat, rice, etc.), as well as plant storage structures such as tubers (potato), roots (cassava), rhizomes (arrowroot), stem piths (sago palm), etc. See carbohydrates and Fig. C2.

**starfish** A name once used for ⋆**sea-stars**.

**STD** Short for ⋆**sexually transmissible disease**.

**stele** The central cylinder or core of vascular (conducting) tissue and associated ⋆ground tissues found in the stems and roots of plants. See Fig. S7.

The stele is made up of ⋆xylem, ⋆phloem, ⋆pericycle, and, in some steles, ⋆pith and ⋆medullary rays. It is surrounded by the ⋆endodermis which is the innermost layer of the ⋆cortex. The arrangement of tissues within a stele varies between plant types, and even within the same plant (e.g. stem and roots).

**stellate** Star shaped. Parts radiating outwards from the centre.

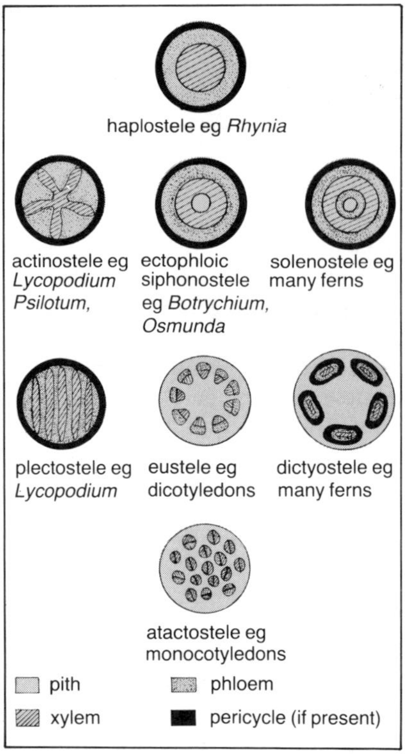


**Fig. S7 Stele—plant systems and roots**

**stem** That part of a plant, normally above ground, that forms the axis of the plant, and differs from ⋆roots in carrying the leaves, buds and reproductive structures, and in the arrangement of the vascular tissue inside. ⋆Corms and ⋆rhizomes are unusual underground stems. See thorn, cladode, runners.

**steppe** A large area of flat treeless ⋆grassland as found in some parts of southeastern Europe and western Asia.

**stereoisomers** ⋆Isomers are chemicals that are so similar that they have the same ⋆functional groups, but they are arranged differently in space, being mirror images of each other. See Fig. S8.

Stereoisomers are classed as D-forms or L-forms, depending on the way they influence polarised light. When chemists prepare compounds, they usually get about 50% each for the L and D forms, unless they create special conditions. However, molecules in living things are almost all the L-form. e.g. glucose.

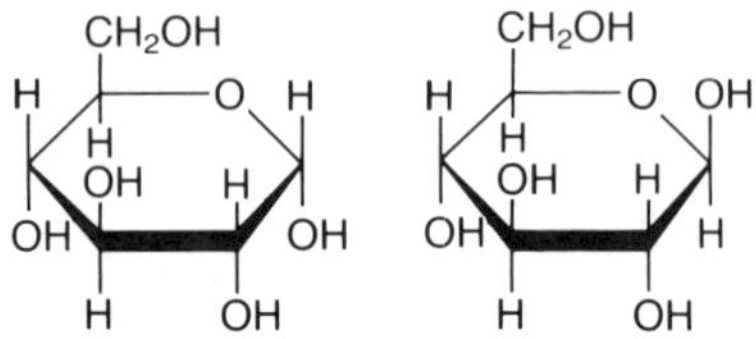


**Fig. S8 Stereoisomers (e.g. glucose)**

**stereoscopic vision** Ability to see objects in three dimensions. This is only possible when the two eyes have overlapping fields of view. Animals that live above ground level (e.g. fly, in trees) usually have good stereoscopic vision.

**sterile** 1 Not fertile. Unable to produce functioning sperm or ova. 2 Free from microbes, especially those which could cause disease (pathogens). 3 A flower that lacks stamens or carpels.

**sterilisation** To make ⋆sterile. e.g. With humans, the male sperm ducts (vas deferens) are cut (vasectomy), or the female fallopian tubes are cut (tubal ligation).

**sternum** Also called the **breast-bone**. The bone (or cartilage) of terrestrial vertebrates in the middle of the chest to which, at one end (rear), most of the ribs are attached, and at the other (front), the ⋆pectoral girdle is attached. In humans, the ⋆clavicle (collarbone) is attached to the top, while most of the ribs are attached further down. See Fig. S3.

**steroids** A large group of complex lipids, all containing the same basic 17-carbon group in the form of a series of rings. Steroids have a wide range of functions. They include ★cholesterol (Fig. C7); ★vitamin-D; ★bile acids that aid in the digestion of fats; sex hormones such as ★androgens (e.g. ★testosterone, Fig. T1), ★progesterone (Fig. P9) and ★oestrogen (Fig. O2); and various hormones produced by the adrenal cortex that are made from cholesterol.

**stethoscope** A device used to convey sound of the body (e.g. heartbeat, breathing) to the ear of a person (e.g. doctor).

**stick insect** Any member of the herbivorous insect order Phasmatodea which is camouflaged due to looking like a stick or twig for camouflage. See Appendix 1.

**sticky ends** The ends of a double strand of DNA that has been cut, by enzymes called ★restriction endonucleases, in such a way that new sections of DNA can be inserted. See genetic engineering and Fig. G2.

**stigma** Plural is **stigmata**. **1** The tip of the ★style of a flower and the surface on which the ★pollen lands and germinates. The stigma surface may be sticky, hairy or have other features which help pollen attachment. See also self-pollination. Compare stamen. **2** The eyespot of certain algae and protozoans.

**stimulant** Any food or drug which quickens the body functions. e.g. Caffeine in tea and coffee. Drugs such as amphetamines and cocaine.

**stimulus** Something that excites an organism, or part of an organism, to action (the response). See also conditioning.

**stinging cell** See nematocyst.

**stipe** A short stalk or supporting structure. e.g. The stalk supporting the fruiting body (cap) of a mushroom or toadstool, or connecting the frond of a large alga (e.g. kelp) to the holdfast.

**stipule** One of a pair of leaflike structures at the base of the petiole (stalk) of the leaves of some plants.

**stirrup** See stapes above.

**stock** **1** The population of organisms from which new varieties have been developed. **2** The rootstock or rooted part of a plant to which a ★scion or shoot is grafted.

**stolon** **1** The horizontal stem of a plant that grows along the surface of the ground, and sometimes produces roots at the ★nodes. e.g. A strawberry runner. **2** In hydroid cnidarian (coelenterates, e.g. hydra), a horizontal growth where ★budding takes place.

**stoma** Plural is **stomata**. One of the numerous minute openings in the surface (epidermis) of a plant leaf, through which gases enter and leave. Stomata are normally located on the underside of a leaf. Each stoma consists of a central pore surrounded by two ★guard cells. The guard cells open and close the pore to control the entry and exit of gases. See Fig. S9.

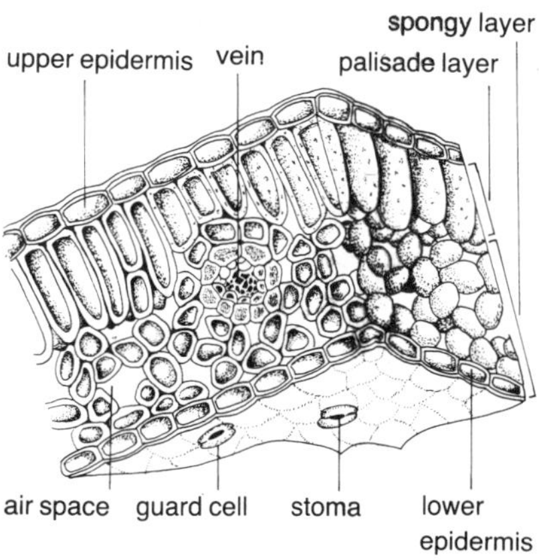


**Fig. S9 Stoma and leaf cross section**

**stomach** A muscular saclike organ of the ★alimentary canal where food is mixed, and partly digested. See Fig. D2.

In vertebrates, the stomach is located between the ★oesophagus (food pipe) and the ★small intestine. Muscles in the walls help mix (churn) the food, while cells in the stomach walls produce hydrochloric acid and ★digestive juices (e.g. ★pepsin) to help digestion, especially of proteins. The walls also produce a ★mucus to protect themselves from being attacked by their own digestive juices. At the end of the stomach is a ring of muscles called the ★pyloric sphincter that controls release of the partly digested food (called ★chyme) into the duodenum. In birds the stomach is divided into two, the rear portion forming a ★gizzard. In ★ruminants the stomach is divided into four chambers.

**stomata** See stoma above.

**stomatal pore** The small hole between the guard cells of a ★stoma, through which gases enter and leave the leaf of a plant. This pore opens when the guard cells fill with water, i.e. are ★turgid. See Fig. S9.

**stomate** More correctly, **stoma**. See above.

**strain** A group of very similar organisms with a common ancestor, within a species. e.g. A strain of pure-bred hairless ('nude') mice bred for experimental work. A strain of bacteria resistant to the antibiotic penicillin.

**stratification** The state of being in layers. e.g. Vegetation in a rainforest.

**stratosphere** The layer of the Earth's atmosphere above the ★troposphere but below the ★ionosphere, extending from about 10 km to 50 km, and where the temperature remains approximately constant.

**stratum** Plural is **strata**. A single layer. e.g. A stratum or bed of rock. A layer of vegetation in a rainforest.

***Streptococcus*** A genus of ball-shaped bacteria (★coccus) arranged in thread-like chains. They cause a number of diseases, including throat infections, scarlet and rheumatic fever. See Fig. B1 and necrotizing fasciitus.

**streptomycin** An ★antibiotic that functions by attaching itself to bacterial ribosomes, and causing codons to be misread during protein synthesis.

**striated** Marked with stripes, cross bands, grooves, etc. e.g. Striated muscle.

**striated muscle** Also called **striped**, **skeletal**, or **voluntary** muscle. Muscle composed of fibres with a characteristic banding pattern, and which is responsible for the voluntary movement of the bones of the skeleton, and thus movement. See muscle fibre and Fig. M5. Compare cardiac and smooth muscle.

**stridulation** The production of sound, in many insects, by the rubbing of two hard parts of the body together. e.g. ★Crickets rub rasps, found along their forewings, together.

**stringybark** Any of the eucalypts with bark which peels off in long fibrous strips. e.g. The red stringy-bark *Eucalyptus macrorhyncha*, common on the foothills and tablelands (to 1000 m) of eastern Australia.

**striped muscle** See striated muscle, above.

**strobilus** 1 A ★cone as found in gymnosperms. 2 A cone-like grouping of ★sporophylls ('leaves' bearing spore producing sporangia) found on horsetails and clubmosses.

**stroke** A disease of the ★cerebral blood vessels which results in a poor supply of blood to the brain. e.g. A clot cuts off the flow of blood resulting in death of the brain cells. A cerebral haemorrhage occurring after an artery ruptures. In serious cases it leads to partial or complete paralysis.

**stroma** 1 In plants, the part of the chloroplast that lies between the ★grana, and is the site of the dark reactions of ⋆photosynthesis. 2 A mass of thread-like hyphae of a fungus from which the fruiting body (e.g. cap of a mushroom) develops. 3 The network of connective tissues that forms a framework within an organ to hold it together.

**stromatolites** Layered structures formed in certain limestone due to the activity of ⋆blue-green algae in warm ocean waters. Interest has arisen in stromatolites because similar structures have been found in Pre-Cambrian rocks in Western Australia, and since they have been dated at more than 3000 million years old, they may be evidence for some of the earliest living things.

**structural adaptation** Any ⋆evolutionary change in the structure of an organism that makes it better able to survive in its environment. e.g. Plant leaves with a thick waxy cuticle or mammals with a layer of fur or hair.

**structural gene** A gene (length of DNA) that codes for an enzyme or other type of protein. See operon.

**stump** 1 The lower part of a tree after the trunk and branches have been removed. 2 The part of a body limb that remains after part has been amputated.

**style** The stalk of a flower ⋆carpel that connects the stigma to the ovary. See Fig. F2.

**stylet** A sharp piercing mouthpart. e.g. The sucking mouthpart of a mosquito.

**subcutaneous** Immediately below the skin. e.g. Subcutaneous tissue is the layer of connective tissue beneath the ⋆dermis of the skin, usually containing fat cells, large blood vessels and nerves.

**suberin** A fatty or waxy substance present in the cork tissue of bark, and the ⋆Casparian strip of roots, which functions as a waterproofing agent.

**sublittoral** 1 The zone in a lake or pond going down from the lower edge of growth of rooted plants. 2 The marine habitat, from low tide mark (edge of ⋆littoral zone) down. See Fig. O1.

**subspecies** A group of similar organisms (usually a geographic race) of the same species. This subspecies is only slightly different from other members of the species and can interbreed with them. e.g. The koala in Queensland (*Phascolarctos cinereus adustus*) is smaller and lighter in colour than koalas in Victoria (*P. cinereus victor*). See also cline, race, breed.

**substrate** 1 The chemical upon which an ⋆enzyme acts (Fig. A4). The enzyme has an ⋆active site into which the substrate fits, exposing and weakening the chemical bonds where the reaction is to occur. 2 See substratum.

**substratum** The material upon which an organism lives or to which it is attached. e.g. The rock to which barnacles are attached on a rock platform.

**subulate** Of a leaf, narrow and tapering to a fine point. See Fig. L1.

**succession** In ecology, the series of changes in the composition of the plant and animal life in an area.

**a** A primary succession is one that occurs in an area not previously occupied. e.g. New volcanic island. Usually the first visible organisms present are lichens and algae, followed by mosses once a little soil develops. Since these successions take so long, it is easier to study them in a location where they are gradually extending the range of the community, and all or most stages can be observed spread out in a linear sequence. e.g. Sand-dune (Fig. S10) or rock succession.
**b** A secondary succession occurs where the original community has been destroyed, and the climax phase is reached more quickly. e.g. Regeneration after bushfire, logging.

**Fig. S10 Sand dune succession**

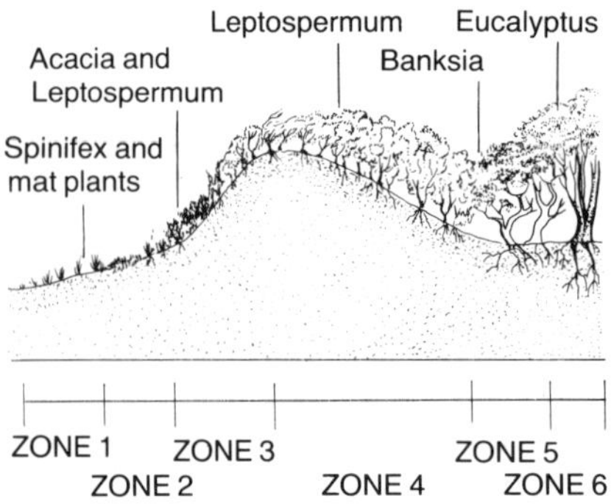


**succulent** Used to describe plants that store water in fleshy tissues. Such plants occur in desert regions of Africa and the Americas, and in areas of high salinity such as saltmarsh. e.g. Cactus, pigface.

**sucker** 1 A plant shoot that develops from the roots or the lower part of a stem. Often occurs in trees such as eucalypts after fire, or after they are cut down. 2 A cup-shaped structure that relies on suction to attach animals to another object. e.g. Suckers on the arms of sea-stars, on the ends of leeches.

**sucrose** Also called **cane sugar**. A double sugar (★disaccharide), $C_{12}H_{22}O_{11}$, made of glucose and fructose joined together. Obtained from sugar cane (and sugar beet overseas). See carbohydrate and Fig. C2.

**sugar** Any ★carbohydrate with a sweet taste, especially ★glucose and ★sucrose.

**sugar glider** A small pollen and nectar-eating possum, *Petaurus breviceps*, Family Petauridae, is a marsupial capable of gliding flight for short distances. See Appendix 1.

**suicide gene** See **apoptosis**.

**sulfonamides** Also called **sulfa drugs**. A group of compounds used to interfere with the growth of certain bacteria by stopping them from producing folic acid. Bacteria that don't need folic acid, or can use it from their environment, are not affected by sulfa drugs.

**sulfur** (S) A yellow element that is a non-metal. It makes up an important part of many proteins, sulfur 'bridges' helping to hold them in their three-dimensional shapes.

**summation** In the ★nervous system, suppose that two ★neurones connect by ★synapses to the same third neurone. Then summation is the additive effect of two impulses arriving at the third neurone at the same time. **a** In 'excitatory' summation, if either arrived on their own, they can not produce a new impulse in the next neurone. If they arrive together, their combined effect can stimulate an impulse. **b** In 'inhibitory' summation, if an impulse of excitatory neurone arrives alone, then an impulse is produced in the next neurone. If, however, impulses on both an excitatory and an inhibitory neurone arrive together, no new impulse is generated. Summation is one of the main ways in which nervous ★integration occurs. There is often many more than just two neurones arriving. Compare facilitation.

**sunburn** Inflammation of the skin caused by ultraviolet light from the sun. See also melanin.

**supercoiled DNA** A method of packing a coil of double-stranded DNA where it is twisted about its own axis, such as when a rubber band is twisted about itself.

**superior** Used of ★flowers where the ovary is above the attachment point for sepals, petals, and stamens. See Fig. F2.

**suppository** A medical substance, usually solid or in capsule form, that is inserted into the rectum or vagina.

**suppressor T-cells** See T-cells.

**surface area/volume ratio** The ratio of the surface area of an organism to its internal volume. Organisms living in colder climates tend to have a smaller ratio than similar organisms living in the tropics. This proportionally less surface area helps retain heat in colder climates. In warmer climates the relatively greater surface area helps radiate excess heat. e.g. Koalas of the tropical Queensland are smaller, and hence have a bigger ratio than koalas in colder southern parts of Australia.

**surface tension** The property of a liquid that makes it behave as though it has an elastic skin covering its surface. Thus very light insects (and other animals) can walk on the surface of water as long as they do not break through this surface (e.g. water striders). Surface tension is also partly responsible for the capillary force that helps force water up the narrow ⋆xylem tubes in the vascular tissue of plants. See cohesion and transpiration.

**surgery** The treating of disease, injury, or deformities by physical means, usually by cutting open the body to reach or remove the parts concerned.

**survival of the fittest** See natural selection.

**suspension** A mixture in which very small particles (but bigger than molecules) are suspended in a liquid. e.g. A mixture of clay and water.

**suture** 1 A line where two structures or organs join. e.g. Junction between bones of the skull. Line of closure between the two shells of a bivalve. Line along which the parts of a seed pod split open. 2 To sew up a wound, or the threadlike material used to achieve this.

**SV40 virus** A small animal virus widely used in genetic research. The DNA is double stranded, and contains six genes.

**Svedberg unit** Symbol **S**. A unit used to indicate the size of macromolecules and cell particles as measured in an ultracentrifuge. e.g. The 70S ribosome (found in prokaryotes) is made up from 30S and 50S subunits.

**swallow** To pass food from the mouth cavity to the stomach or crop. As food is swallowed the ⋆epiglottis closes over the windpipe to prevent food entering the lungs.

**swamp** 1 An area of wet, spongy and marshy ground saturated with water. 2 An area of still, often stagnant water.

**sweat** To perspire or produce liquids from sweat glands. See perspiration.

**sweat gland** A gland in the skin of mammals that produces sweat, mainly to help control body temperature through evaporation, but also to excrete some wastes. This ⋆exocrine gland is a tube that ends in a coiled section in the ⋆dermis where the sweat is secreted from surrounding cells, under the control of the ⋆sympathetic nervous system. The numbers of sweat glands varies from one part of the body to another, and from one type of mammal to another (e.g. Humans have many, a dog has only a few located mainly on its feet).

**swim bladder** Also called an **air bladder**. A bladder or sac found in fish. When filled with air the fish becomes less dense and rises. When air is pumped out, the fish becomes more dense and sinks.

**swimmeret** A small paddle-like swimming appendage on the abdomen of ⋆crustaceans such as lobsters and prawns.

**symbiosis** The living together in close association of organisms from different species. ⋆Mutualism, ⋆commensalism, and ⋆parasitism are all types of symbiosis.

**symmetry** A likeness or correspondence in size, shape, or structure of an organism, or part of an organism about a plane, line or point. See Fig. S11, bilateral and radial symmetry, ventral, dorsal, anterior and posterior.

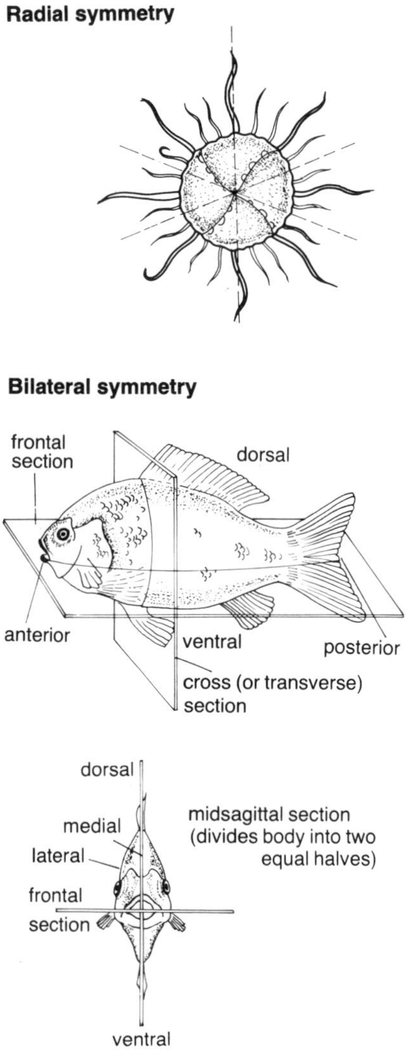


**Fig. S11 Symmetry**

**sympathetic nervous system** A division or the vertebrate ⋆autonomic nervous system which generally excites body functions and thus prepares the body for emergencies. The sympathetic nervous system decreases the amount of saliva produced, increases heart rate, and constricts blood vessels, thus balancing the opposite and relaxing effects of the ⋆parasympathetic nervous system. The nerve endings produce ⋆adrenaline or ⋆noradrenaline ⋆neurotransmitters.

**sympatric** Concerning similar species that occupy the same or overlapping areas, and would be expected to interbreed, but do not because of differences in behaviour, flowering time, etc. Compare allopatric.

**sympetalous** Having petals fused together.

**symphysis** A type of joint which allows only slight movement. e.g. Between the vertebrae of the spine. See Fig. V2.

**symptom** A sign or indication of something unseen. e.g. Symptoms of a disease, such as pain, stiffness, giddiness.

**synapse** The area of contact between the ⋆axon of one ⋆neurone, and the cell body or ⋆dendrites of another. The incoming impulse causes the secretion of a ⋆neurotransmitter from ⋆vesicles in the axon, which passes across the gap in the synapse and stimulates (or inhibits) the production of a new impulse in the next neurone. See Fig. S12 and synaptic vesicles.

**synapsis** See pairing.

**synaptic vesicles** Organelles produced in the cell body of a neurone, and transferred to the synapse area of the axons. The synaptic vesicles contain ⋆neurotransmitters which are released into the gap between the two neurones when the vesicle fuses temporarily with the cell membrane. See also exocytosis.

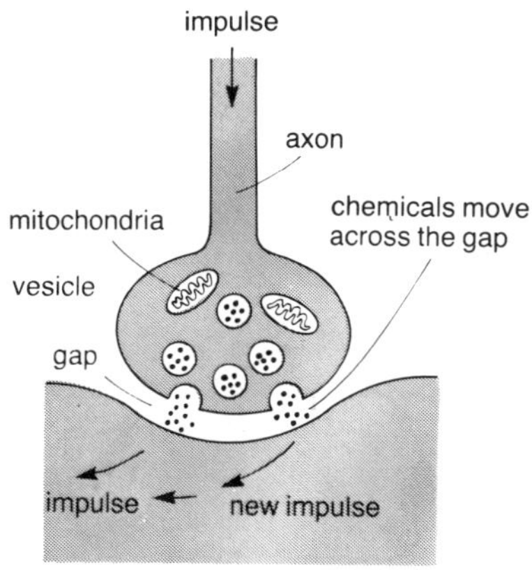


**Fig. S12 Synapse**

**syncarpous** A flower in which the ⋆carpels are fused together. Compare apocarpous.

**syndrome** A set of symptoms which occur together and collectively help in identifying the disease. e.g. Down's syndrome.

**synergism** The joint action of two or more substances so that, when acting together, their effect is greater than the sum of the parts. e.g. Exhibited by some drugs, hormones, plant growth substances.

**synovial membrane** The membrane which lines the bag (synovial sac) that encloses the space between the bones of a freely moving ⋆joint (e.g. elbow, knee). The sac is filled with fluid (synovial fluid) that helps lubricate the smooth cartilage surfaces where the bones make contact. See Fig. J1.

**synthesis** The formation of a chemical compound from simpler compounds. Not restricted to living things. e.g. The synthesis of protein from amino acids. See also biosynthesis and anabolism. Compare catabolism.

**syphilis** A chronic, infectious ⋆sexually transmissible disease (STD) caused by a spirochaete bacterium, *Treponema pallidum*.

**system** A group of ⋆organs that function together as a unit. e.g. Leaf of a plant. Digestive system of animals.

**systematics** The study of the classification and relationships of organisms; taxonomy.

**systemic** Distributed throughout the body of a organism. e.g. Systemic ⋆circulation carrying blood from the heart to the body in general. Compare pulmonary circulation.

**systole** The contraction phase of the heartbeat forcing blood into the arteries. Compare diastole. See also blood pressure.

# Tt

**2,4-D** Short for **2,4-dichlorophenoxyacetic acid**. A synthetic ⋆auxin widely used as a weed killer (herbicide). At appropriate concentrations it can be sprayed on lawns to kill broad-leaved (dicot) weeds as it has little effect on grasses. See also 2,4,5-T.

**2,4,5-T** Short for **2,4,5-trichlorophenoxyacetic acid**! A synthetic ⋆auxin that is widely used as a weed killer (herbicide) and defoliant. Its use has been banned because a very poisonous contaminant, ⋆dioxin, is often present. See also 2,4-D.

**T-cell** Also called **T-lymphocyte** as they mature in the ⋆thymus. They are ⋆lymphocytes which can act both directly against invading microbes, as well as help control other parts of the immune system.

There are four types which work together to provide cellular ⋆immunity. **a** Killer T-cells (also called cytotoxic T-cells) destroy the invading cell directly, by binding to it, and causing a ⋆lesion that destroys it. **b** Long-lived T-memory cells (a type of killer T-cell) form so that the immune system can react more rapidly by recognising a particular antigen on future exposures. **c** The T-helper cells regulate the immune system—when they recognise an antigen, they stimulate other components of the immune system, including ⋆B-cells and killer T-cells. **d** T-suppressor cells help regulate the immune system by 'turning off' the actions of these other cells when the infection is past. See also CD-4 and CD-8 cells.

**T-lymphocyte** See T-cell.

**tactile** Of or relating to the sense of touch.

**tadpole** The ⋆larva of a ⋆frog or ⋆toad, which differs from the adult in having gills and tail but no legs.

**taiga** A ⋆biome in the northern hemisphere, especially Eurasia, with large areas of conifer forests, long cold winters, and short summers. Grades into ⋆tundra further north.

**tail** The long, relatively narrow rear (posterior) part of an animal. e.g. The skin-covered extension of the backbone in mammals, reptiles, etc. The long feathers at the rear of a bird. See also coccyx.

**tailing** In eukaroytic cells, the adding of 10–20 ⋆adenosine molecules to the 3′ end of a messenger RNA after it has been transcribed from DNA. This 'tail' remains after the ⋆introns have been removed during 'editing'. Compare capping.

**tall open forest** Also called **wet sclerophyll forest**. A forest dominated by tall eucalypts. The understorey has relatively more soft leaved (mesomorphic) species than does ⋆open forest. See also sclerophyll.

**tallowwood** A large tree of the coastal forests of northern NSW and Queensland, *Eucalyptus microcorys*, prized for its strong timber, and the flavour of the honey that bees make from its nectar.

**talon** A sharp claw, especially of a bird of prey.

**tannin** Bitter tasting compounds present in the leaves, bark, wood, and fruits of plants. They are derived from ⋆phenol, and help protect plants from plant-eating insects, etc. They are used by humans in tanning and dying.

**tapeworm** A parasitic flatworm in class Cestoda, lacking any alimentary canal as it lives inside the alimentary canal of vertebrates, including humans. It absorbs nutrients through the surface of its body. e.g. ⋆hydatid worm. See Appendix 1.

**taproot** The main root of many dicot plants which grows directly downwards with smaller side (lateral) roots. Swollen taproots often act as food stores. e.g. Carrot.

**tarantula** In Australia, a large huntsman spider of arachnid genus *Isopoda*, common in houses after rain, and noted for running sideways across walls and ceilings.

**tarsal bones** The bones in the region of the ankle of the hind foot of a tetrapod vertebrate. In humans there are seven tarsals. See pentadactyl limb and Fig. P2.

**Tarsipedidae** The family of marsupial mammals also called honey possums. The single species in this Family feeds on nectar and pollen using a brush-tipped tongue. See Appendix 1.

**tarsus** 1 The ankle region of the hind foot of tetrapod vertebrates where the tarsal bones are located. See Fig. P2. 2 A segment in the leg of an insect.

**Tasmanian devil** The carnivorous marsupial *Sarcophilus harrisii*, with black coat and white markings, now only found in Tasmania. It is mainly nocturnal, feeding on carrion. See Appendix 1.

**Tasmanian tiger** See Tasmanian wolf.

**Tasmanian wolf** Also called **thylacine**. The extinct wolf-like carnivorous marsupial *Thylacinus cynocephalus*, with tiger-like stripes across its back. See Appendix 1.

**taste** The perception of flavour using signals from the taste buds of the tongue. Taste buds can only detect sweet, salty, bitter and sour. Other tastes are due to the combined senses of taste and smell, odours passing up the rear of the mouth to the nasal cavity.

**taste-bud** The receptor for taste in vertebrates. Taste buds are normally found on the tongue of land vertebrates, but with fish they can be located anywhere on the body. In humans, the taste buds are located along the walls of small pores in the surface of the tongue (see papillae). There are four types of taste bud, salty, sweet, sour and bitter. See also taste.

**taxis** The movement of a mobile organism in response to a directional stimulus, which can be away from (negative taxis) or towards (positive taxis) the stimulus. See chemotaxis, geotaxis, phototaxis. Compare tropism.

**taxon** A unit or level in the classification system of living things. e.g. Phylum, Class, Order, Family, Genus, Species.

**taxonomy** The science of the rules, principles and practice of classifying living things.

**TB** Short for ⋆**tuberculosis**.

**TCA cycle** Short for **tricarboxylic acid cycle**. Also called **citric acid cycle** and ⋆**Krebs cycle**. Part of the reactions of respiration.

**tea-tree** Any species of the genus *Leptospermum*, and several similar species in the genus *Melaleuca*. So called because of their use, during the early European settlement of Australia, as a substitute for tea.

**tectonic** Relating to the forces within the Earth's crust causing the crustal plates to move. See plate tectonics and Fig. P7.

**teeth** Plural of ⋆**tooth**.

**telomere** The end of a chromosome where the DNA forms a loop-like structure.

**telophase** The final stage of ⋆mitosis and ⋆meiosis where two new nuclei are formed, and the cells and their cytoplasm divide into two.

**temperate zone** The parts of the Earth's surface lying between the tropics and the north and south polar regions. Compare tropical.

**temperature** A measure of the hotness or coldness of a material, when referred to some standard. Temperature determines the direction in which heat will flow, i.e. from hotter to colder.

**temperature inversion** An unusual increase in temperature in the lower atmosphere. Normally the temperature gradually decreases with height, but under some circumstances, a layer of warmer air can overlay colder air beneath. This causes problems as pollution-laden air is trapped under the inversion layer, causing a gradual build-up in the concentrations of potentially toxic gases.

**template** A pattern from which a copy can be made. e.g. During ⋆replication of DNA, each strand acts as a template so that a copy can be made. See also complementary.

**tendon** A fibrous cord or band of connective tissue connecting a muscle to a bone. Made mainly of ⋆collagen fibres. e.g. ⋆Achilles tendon.

**tendril** A thread-like part of climbing plants that coils around solid objects, thus giving support to the growing plant. e.g. Sweet-pea.

**tension** A force that stretches or strains. e.g. ⋆Surface tension.

**tentacle** Any of the slender and flexible processes or 'arms' which invertebrates use to catch prey, pass food to the mouth, defend themselves, etc. e.g. Cnidarians (coelenterates) such as jelly-fish and sea-anemones, molluscs such as octopus and squid. See Fig. P8.

**terete** Cylindrical, or almost so.

**term** The end of a normal period of ⋆pregnancy when birth occurs.

**terminal** Forming the end or extremity of something. e.g. Terminal bud is at the end of a stem. Some diseases are called terminal as they end in death.

**termination codon** The sequence of three bases in the ⋆genetic code that causes the synthesis of protein along the mRNA to cease. There are three such codons: UAA, UAG and UGA.

**termite** Also called a **white ant**. Any of the pale-bodied, soft, wood-eating insects of the order Isoptera. These social insects have a complex caste system, and build complex 'mounds' with a highly controlled internal environment. See Appendix 1.

**terrestrial** Living or growing on land. Compare aquatic.

**territory** An area or space in a habitat occupied by an organism, pair or group of organisms, and defended by them against members of the same species, and occasionally other species. Territories are common at breeding time, and can be large or small. e.g. The territory of a wedge-tailed eagle is often more than 100 hectares, compared to less than 1 square metre for a sea-gull at its nest in a rookery.

**Tertiary period** The first period of the Cainozoic era following the Cretaceous, which ended the Mesozoic era. See Appendix 2.

**tertiary structure** The three-dimensional structure of a protein molecule. See also primary and secondary structure. See Fig. P10.

**tessellate** In the form of small squares or blocks, forming a mosaic pattern. e.g. Some tree barks.

**test** 1 A hardened shell or outer covering. e.g. Sea urchin. 2 The experimental checking of predictions made by a theory.

**test cross** A cross (mating) between an organism of uncertain genetic composition, and a homozygous recessive to find out the unknown genotype. If the cross is with a parent, then it is a type of ⋆backcross.

**test tube baby** See ***in vitro*** **fertilisation**.

**testa** The outer covering or coat of a ⋆seed. Compare aril.

**testes** Singular is **testis**. The male reproductive organs (male ⋆gonad) which produce the male gametes. See Fig. R3.

In vertebrates, the testes are paired, and apart from producing ⋆sperm, they also produce male sex hormones. e.g. ⋆Testosterone in men. They are carried in a bag called the ⋆scrotum, and connect to the ⋆penis by the ⋆sperm duct (or ⋆vas deferens). ⋆Sperm cells (spermatozoa) are produced in ⋆seminiferous tubules, scattered between which are the ⋆interstitial cells which produce the male sex hormones. Sperm pass from the tubules to the ⋆epididymis where they are stored and mature. The epididymis has walls of smooth muscle to push the sperm into the vas deferens during ⋆ejaculation.

**testicle** A testis. See above.

**testosterone** The main natural ⋆androgen or male sex hormone, produced in the testes. See Fig. T1. Testosterone is one of the hormones responsible for the male ⋆secondary sexual characteristics.

**tetanus** Also called **lockjaw**. A serious infection caused by the ⋆toxins of an anaerobic soil ⋆bacterium called *Clostridium tetani*, resulting in violent muscle spasms and possibly death. The toxins are a natural waste product of the bacterium. It only enters through deep wounds as it must remain out of contact with air. See also triple vaccine.

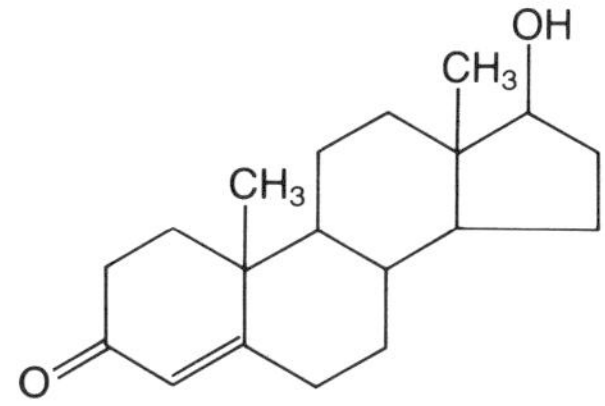


**Fig. T1 Testosterone**

**Tethys Ocean** The ancient ocean that is believed to have surrounded the supercontinent of Pangaea, and later separated the continents of Gondwana and Laurasia which formed from the splitting of Pangaea. See plate tectonics and Fig. P7.

**tetracyclines** A group of ⋆antibiotics derived from the bacterium *Streptomyces*. They interfere with protein synthesis. See also ⋆Cholera.

**tetrad** A group of four. e.g. The four ⋆chromatids present when a homologous pair of chromosomes lines up on the equator during metaphase I of meiosis. The four pollen grains formed as the result of meiosis in the anther.

**tetrapod** A four-limbed vertebrate, most of which are land living. e.g. Amphibians, reptiles, birds (counting their wings) and mammals.

**thalamus** Located in the rear portion of the vertebrate forebrain, the thalamus is a relay centre, sending sensory information to other parts of the brain. See Fig. B7. Compare hypothalamus.

**thalassaemia** Any of a group of genetic disorders of ⋆haemoglobin production causing mild to severe ⋆anaemia. Mild forms are due to a series of ⋆recessive genes, while severe forms are caused when a person is ⋆homozygous recessive for the gene. Tends to be more common in people of Mediterranean, black, or southeast Asian ancestry.

**Thallophyta** A Division in plant classification, no longer widely used, which included all the plants where the body is not differentiated into wood, stem and leaves. It contained the algae and fungi, and in some classifications, the bacteria and lichens as well.

**thallus** A plant body that does not have true leaves, stems or roots. Although they may have parts which look like a leaf, they do not have the same internal structure. Algae, moss, liverworts, and the gametophyte stage of ferns all have the thallus structure.

**thaw** 1 The melting of ice and snow. 2 To be freed from the effects of extreme cold. i.e. to thaw out.

**theory** A proposed explanation for a group of connected observations that has been successfully tested. In biology, most often applied to phenomena that are unable to be observed directly as they are too small (e.g. atomic theory) or take too long (e.g. theory of evolution). Compare hypothesis and law.

**Therapoda** A suborder of bipedal dinosaurs with small grasping forelimbs, which includes the genus **Tyrannosaurus*. Compare Sauropoda.

**therapy** Treatment for a disease that either aids healing or helps remove the suffering.

**thermocline** A layer of water in deep lakes or the ocean where the water becomes rapidly cooler.

**thiamine** Also called **vitamin $B_1$**. A water-soluble vitamin which is part of the vitamin B group. It is converted into a ⋆coenzyme involved in carbohydrate ⋆metabolism. Humans and many other vertebrates cannot make thiamine, and it must be in their diets. Lack of thiamine in people results in the disease ⋆beri-beri.

**thigmotropism** Also called **haptotropism**. A tropism (a movement produced by plant growth) stimulated by physical contact. e.g. A ⋆tendril in contact with a twig will curl around it. The growth of roots away from objects in their path.

**thinking** To contemplate, meditate, conceive in the mind, turn over in the mind, to form an idea, consider, reflect.

**thoracic cavity** In mammals, the body cavity containing heart and lungs, separated from the ⋆abdominal cavity by the muscular ⋆diaphragm.

**thoracic duct** The main lymphatic duct or vessel which collects lymph from most of the body, and empties it into the bloodstream at the posterior vena carva close to the heart.

**thorax** 1 In terrestrial vertebrates the cavity containing the heart and lungs (called the thoracic cavity). Only in mammals is it separated from the ⋆abdomen by a ⋆diaphragm. 2 In insects, the often narrow section between the head and abdomen. See Fig. I5. To the three segments in the thorax are attached three pairs of legs, and the wings (if present).

**thorn** A sharply pointed woody structure with an internal structure similar to that of a stem. e.g. Hawthorn. Compare spine, prickle.

**thread cell** A special stinging cell found in the outside walls of ⋆tentacles on cnidarians (coelenterates) such as hydra, sea-anemone, coral, jelly-fish. Inside the thread cells are the ⋆nematocysts.

**threadworm** Any of the roundworms in phylum ⋆Nematoda. See Appendix 1.

**threonine** (thr) One of the about 20 common ★amino acids found in protein. Also converted to the amino acid isoleucine. See Fig. A6.

**threshold** An intensity of stimulus below which there is no response in the tissue being stimulated. Compare generator potential.

**thrip** Any minute insect in the order Thysanoptera, often the cause of damage to plants. They feed on the sap of the plant using their piercing mouthparts. See Appendix 1.

**thrombin** An enzyme only present in blood near an injury, which converts ★fibrinogen into ★fibrin, thus aiding clotting. See prothrombin and blood clotting.

**thrombosis** The formation of a clot in a blood vessel, blocking the flow of blood. A major danger is that such a clot should move to the heart, or form near the heart, thus cutting off the flow of blood, and causing a heart attack. See coronary.

**thrombus** A blood clot formed within a blood vessel or the heart. See also ★occlusion.

**thylacine** See Tasmanian wolf.

**Thylacinidae** A family of marsupial mammals including the ★Tasmanian wolf. See Appendix 1.

**thylakoid** One of the flattened sac-like structures making up the ★grana of ★chloroplasts. Chlorophyll molecules and enzymes of the ★electron transport chain occur in the thylakoid. Fig. C5.

**thymine** One of the four nitrogen-containing organic bases that form the four 'letters' of the genetic code along DNA. Chemically, thymine is a ★pyrimidine. It is replaced by ★uracil in RNA. See Fig. B3.

**thymus gland** A ★lymphatic and ★endocrine organ of vertebrates, it is usually located in the neck region, but in mammals it is near the heart. In humans the thymus helps T-★lymphocytes mature, and secretes a number of hormones whose function is little known. The thymus is most active at puberty, but this activity appears to decrease after that time.

**thyroid gland** A large endocrine gland, with two lobes, that lies in the neck next to the wind pipe (trachea) and food pipe (oesophagus). In mammals it is controlled by the pituitary (★thyroid stimulating hormone), and secretes iodine-containing hormones, including ★thyroxine and ★calcitonin. In amphibian tadpoles it secretes a hormone that starts ★metamorphosis. Lack of this hormone in human infants results in mental and physical stunting (cretinism), while lack of at least a little iodine in the diet results is hyperplasia, a type of ★goitre.

**thyroid stimulating hormone** Called **TSH** for short. A hormone secreted by the front (anterior) lobe of the ★pituitary which stimulates the initial growth of the thyroid and secretion of its hormones.

**thyroxine** An iodine-containing hormone secreted by the ★thyroid gland which increases oxygen consumption and energy production in a number of tissues.

**Thysanoptera** An order of insects that includes the small sap-feeding thrips. See Appendix 1.

**Thysanura** An order of insects that includes the silverfish. See Appendix 1.

**tibia** Also called the **shinbone**. The main bone of the lower part of a land vertebrate's rear leg. It is next to the smaller ★fibula, and at the knee connects with the femur at the top end, and with the tarsals of the foot at the other end. See Fig. S3.

**tibialis posterior** A calf muscle connecting the metatarsals with tibia and fibula. Helps move the foot.

**tick** A bloodsucking eight-legged arachnid of the order Acarina. See Appendix 1.

**tidal air** The volume of air normally inhaled and exhaled with each breath. Compare residual air.

**tiger cat** Better called the spotted-tailed ⋆quoll. See Appendix 1.

**timberline** The upper limit of normal tree growth, both with altitude on mountains, and with latitude as one proceeds towards the poles. There is often a band of stunted trees for a little distance beyond the timberline. Compare treeline.

**tinea** Also called **athlete's foot**. Any of a number of fungus infections of the skin.

**tissue** A group of similar cells organised into a functional and structural unit. Combinations of tissues make up ⋆organs in multicellular organisms. In animals, tissues include ⋆epithelial, ⋆connective, ⋆lymphatic, ⋆muscular, ⋆fat and nervous. In plants, tissues include ⋆parenchyma, ⋆collenchyma, ⋆sclerenchyma, ⋆ground and ⋆vascular.

**tissue culture** The cultivation or growth of cells or small masses of ⋆tissue outside a normal body. Tissue cultures are often grown in ⋆Ringer's solution supplied with oxygen and nutrients, and kept free of microbes.

**titre** A relative measure of the amount of antibody in a solution. e.g. Serum.

**TMV** Short for ⋆**tobacco mosaic virus**.

**toad** A tailless amphibian, Family Bufonidae, similar to a frog. There are no native toads in Australia, but a serious pest, *Bufo marinus*, or cane toad, has been introduced and is widespread in Queensland, Northern Territory and northern NSW.

**toadstool** Common name given to the fruiting body of fungi that belong to the family Agaricaceae, which also includes mushrooms. There is no precise meaning to the words toadstool and mushroom. Generally, if gilled fungi are edible, they are called mushrooms. However, some toadstools are poisonous, including some that look like mushrooms.

**tobacco** Any plant of genus *Nicotiana*, especially *N. tabacum*, whose leaves are used for smoking, snuff and chewing. Many of the active ingredients in tobacco, and its smoke, are carcinogenic, while the nicotine present is addictive.

**tobacco mosaic virus** Called **TMV** for short. A rod-shaped RNA virus which infects tobacco plants, and is widely used in genetic research.

**tolerance** **1** The ability of a body to become used to a drug, and thus need greater amounts of that drug before it has an effect. e.g. Drug addicts. See also dependence. **2** The failure of the body's ⋆immune system to respond normally to an ⋆antigen. Often due to very early exposure to the antigen before the immune system developed. **3** In ecology, the ability of an organism to adjust to changes in the environment.

**tomentose** Covered with closely matted short hairs.

**tomentum** The hairy covering of some plant surfaces.

**tone** **1** The normal state of the body or one of its parts. **2** See tonus below.

**tongue** A muscular organ of vertebrates that is usually attached to the floor of the mouth. See Fig. D2.

It plays an important role in chewing and swallowing food, and its upper surface carries the ⋆taste buds. In humans it also helps produce the sounds of speech. In other vertebrates it helps catch food, as with the sticky tongues of ant-eating echidnas, numbats, and certain lizards and frogs.

**tonoplast** The membrane that surrounds the ⋆vacuoles of a cell.

**tonsil** Small masses of lymphatic tissue, found in many vertebrates, situated at the rear of the mouth (walls of the ⋆pharynx), at the back of the tongue, and at the rear of the nose. In humans, there is a pair of tonsils at the junction of the mouth and pharynx, and at the back of the nose (see adenoids). Tonsils contain many ⋆lymphocytes and ⋆macrophages to help fight infection.

**tonus** In muscles, the continuous state of slight tension or resistance to stretch, helping to maintain the posture of the animal.

**tool** Any non-living object used by humans or animals which assists in survival. e.g. In animals, the use of twigs by chimpanzees and certain Galapagos finches to extract grubs from crevices in wood.

**tooth** Plural **teeth**. Any of the hard structures in the mouths of vertebrates (except birds) which are mainly used for biting and chewing food, but in some organisms, for catching prey, defence and grooming.

In fish and amphibians the teeth are found all over the ⋆palate, but in reptiles and mammals they are arranged around the edge of the jaw. There are two main parts to a typical mammalian tooth. The crown is the exposed part, and is covered with a layer called ⋆enamel, the hardest substance in the body. Beneath the bony ⋆dentine is the ⋆pulp cavity containing blood vessels and nerves. The part of the tooth embedded in the jaw is called the ⋆roots. In mammals there are four main types of teeth, each with its own function: ⋆incisors, ⋆canines, ⋆premolars, and ⋆molars.

**tooth decay** The action of bacteria in dissolving away the enamel and dentine of a tooth. See also toothache.

**toothache** The pain caused when ⋆tooth decay reaches the pulp cavity of a tooth and exposes the nerves.

**topography** The shape of the surface of the ground in an area.

**torpor** A state of sluggishness and inactivity in an animal. Compare hibernation, aestivation, dormancy.

**torso** The ⋆trunk of a body, lacking head or limbs.

**tortoise** Any land-living or freshwater reptile of the order Chelonia, with its body enclosed in a shell, and which can be told apart from ⋆turtles by having toed (webbed) feet rather than flippers. See Appendix 1.

**touch** The sense of feeling that determines if contact has been made with an object. Touch receptors are those that detect pressure, especially those located in the fingers. e.g. Merkel's disks, Ruffini's end organs, Meissner's corpuscles, and nerves attached to the roots of body hairs.

**toxaemia** The presence in the blood of bacterial ⋆toxins without the bacteria being there themselves.

**toxin** A poisonous substance usually originating from microbes, animals or plants. Toxins from microbes include those which cause botulism, a form of food poisoning. This toxin (from the bacterium *Clostridium botulinum*) is perhaps the most poisonous substance known, 1 gram being able to kill 1 million guinea pigs. Other toxins are produced by spiders, snakes, some octopus (e.g. Blue ringed), cnidarians (coelenterates e.g. sea 'wasp'), ⋆dinoflagellates (red-tides), certain toadstools, etc. See also antitoxin.

**trace element** Also called a **micronutrient**. An ⋆inorganic substance required in minute amounts for the healthy growth of an organism. Trace elements are often needed as parts of enzymes, etc. Essential trace elements for plants include copper, zinc, molybdenum, manganese, and boron. Compare essential elements.

**tracer element** A ⋆radioactive element used to trace or follow the processes occurring in living things. e.g. Tracer elements can follow the movement of sap through the ⋆phloem of a plant. Tracer elements are incorporated into compounds (called labelling) to follow the path of atoms through complex chemical reactions such as photosynthesis.

**trachea** 1 The windpipe of vertebrates. See Fig. L4. The trachea extends from the larynx (voice box) to where it divides to form the ⋆bronchi just before entering the lungs. To stop the trachea from collapsing when animals inhale (breathe in), the walls are lined with rings of stiffening cartilage. They are also covered with ⋆cilia, fine hairs which help remove mucus and trapped dust and smoke particles from the lung. These cilia can be killed by constant exposure to cigarette smoke. 2 In insects, the tubes which start from the ⋆spiracles on the outside of the body and lead into the interior where oxygen is supplied, and waste carbon dioxide removed.

**tracheid** A dead water-conducting cell in ⋆xylem. See Fig. X1. A long thin cell with tapering ends, and thick walls, it overlaps with other tracheids allowing water to pass from one to the other through ⋆pits which are concentrated in this area. Tracheids are most common in ferns and gymnosperms. Compare vessels.

**Tracheophyta** A Phylum (Division) of plants including all those with true vascular tissue. See Angiospermae (flowering plants), Coniferopsida (conifers), Filicopsida (ferns), and related plants as listed in Appendix 1. Compare Bryopsida.

**trachoma** A contagious disease of the eyes, caused by a virus-like bacterium (*Chlamydia trachomatis*) which attacks the lining membranes of the lids and eyes, leading to ulcers and blindness.

**tract** A pathway or channel. e.g. The alimentary tract or canal. A tract or group of nerve fibres connecting two points.

**trait** A distinguishing character or feature. e.g. A heritable trait.

**tranquilliser** A non-medical term for a group of drugs which calm, relax and sedate without necessarily causing sleep. e.g. Valium. Compare stimulant.

**transcriptase** A RNA ⋆polymerase which joins together organic bases as DNA is being transcribed into RNA. Compare reverse transcriptase.

**transcription** The production of RNA from DNA in the nucleus. See Fig. T2.

A single strand of DNA acts as a ⋆template for the synthesis of the corresponding (or ⋆complementary) RNA sequence by the enzyme called ⋆RNA polymerase. The genes which code for ribosomal RNA (⋆rRNA) and ⋆transfer RNA (tRNA) are first tran-scribed into long sections of ⋆primary RNA which are then split into many rRNAs or tRNAs. The DNA that codes for messenger RNA (⋆mRNA) in ⋆eukaryotes also produces a primary RNA from which the ⋆introns have to be removed, a process called ⋆editing. See operon for details. Compare translation, replication.

**transduction** The transfer of DNA from one bacterium to another by a bacteria-infecting virus called a ⋆bacteriophage. Occurs naturally to a small extent, but is an important technique in ⋆genetic engineering. See also oncogenes.

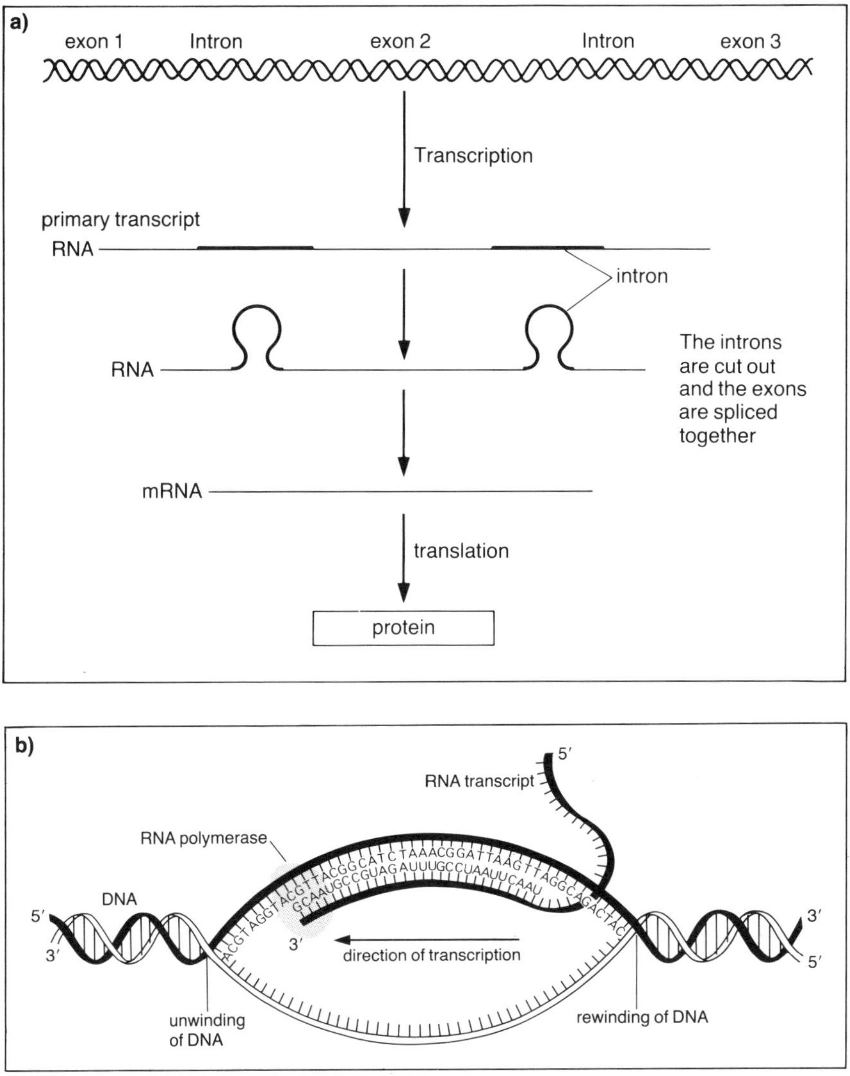


**Fig. T2 Transcription (a) Protein synthesis (b) Transcription**

**transect** In ecology, a line, a strip, or a profile for counting and mapping. e.g. The distribution of plants and animals can be measured at different distances along the line. Compare quadrat.

**transfer RNA** Called *tRNA for short.

**transferase** Any enzyme that is a catalyst to a reaction where a *functional group (e.g. phosphate) is transferred from one molecule to another.

**transformation** 1 A permanent genetic change in a cell after it has gained some new DNA. e.g. If DNA from a disease-causing (virulent) bacteria called *Pneumococcus* is added to the non-virulent rough-surfaced form of the bacteria, then the rough form grows a smooth coat. When first performed in the late 1920s, this experiment showed that DNA was the genetic material. **2** The changing of a normal cell to a cancer cell. This can be achieved by radioactive particles, cancer-causing chemicals (*carcinogens), and by viruses inserting *oncogenes into the cell's DNA.

**transfusion** The introduction of whole blood, *serum, or *saline solution into a blood vessel of the body. See also blood groups.

**transgenic** Describing a cell or organism which contains *DNA from a different organism that has become integrated into its own genetic material. e.g. Transgenic mice can contain human genes. Transgenic cotton contains genes from a bacterium that produces a natural insecticide. See genetic engineering and vectors.

**transition** A point mutation where one organic base of the genetic code is replaced by another. e.g. Adenine replaces guanine or vice-versa (both purines), or thymine might replace cytosine or vice versa (both pyrimidines). Compare transversion below.

**transitional form** An organism that shows intermediate characteristics between two groups of organisms by having features of both. e.g. *Archaeopteryx* is a fossil with features of both reptiles and birds. Transitional forms help show a possible evolutionary sequence of life from common ancestors.

**translation** In *protein synthesis, the 'reading' of the *messenger RNA (mRNA) to produce the correct order of amino acids in a protein. See Fig. T3.

Translation occurs as *ribosomes move along the mRNA, and can be divided into four stages: **a Activation** Amino acids are attached to their own *transfer RNA (tRNA) in the cytoplasm. **b Initiation** A ribosome is attached to the mRNA that was transcribed in the nucleus from DNA, at a special starting *codon. **c Elongation** Amino acids are joined together as the ribosome moves along the mRNA, thus elongating the peptide chain being formed (see tRNA). **d Termination** A special terminal or ending codon causes the release of the completed protein, and the ribosome to separate from the mRNA. Compare transcription, replication.

**translocation** **1** The transfer of soluble food materials (e.g. sugars, vitamins, etc.) from the leaves where they are made during photosynthesis, down through the *phloem to the stem and roots. Growth substances may also be carried. Compare transpiration. **2** A type of chromosome mutation where a length of chromosome becomes detached and then added to another (non-homologous) chromosome. Most often there is a mutual swapping of chromosome segments. In Down's syndrome, chromosome 21 may become fused to any of chromosomes 13 to 15 or 21 and 22. Compare deletion, duplication, inversion, gene mutation. See also chromosome mutation and non-disjunction.

**transmitter** See neurotransmitter.

**transmutation** A change of a species or type to another species or type.

**transpiration** The loss of water vapour by evaporation from the surface of a plant, especially the leaves.

Plants must carefully control loss of water through *stomata to prevent *wilting. The amount of $CO_2$ in the air spaces of the leaf is a major factor in controlling the rate of transpiration. **a** High amounts of $CO_2$ result from more *respiration than *photosynthesis, as occurs at night and during droughts, when there is little water available.

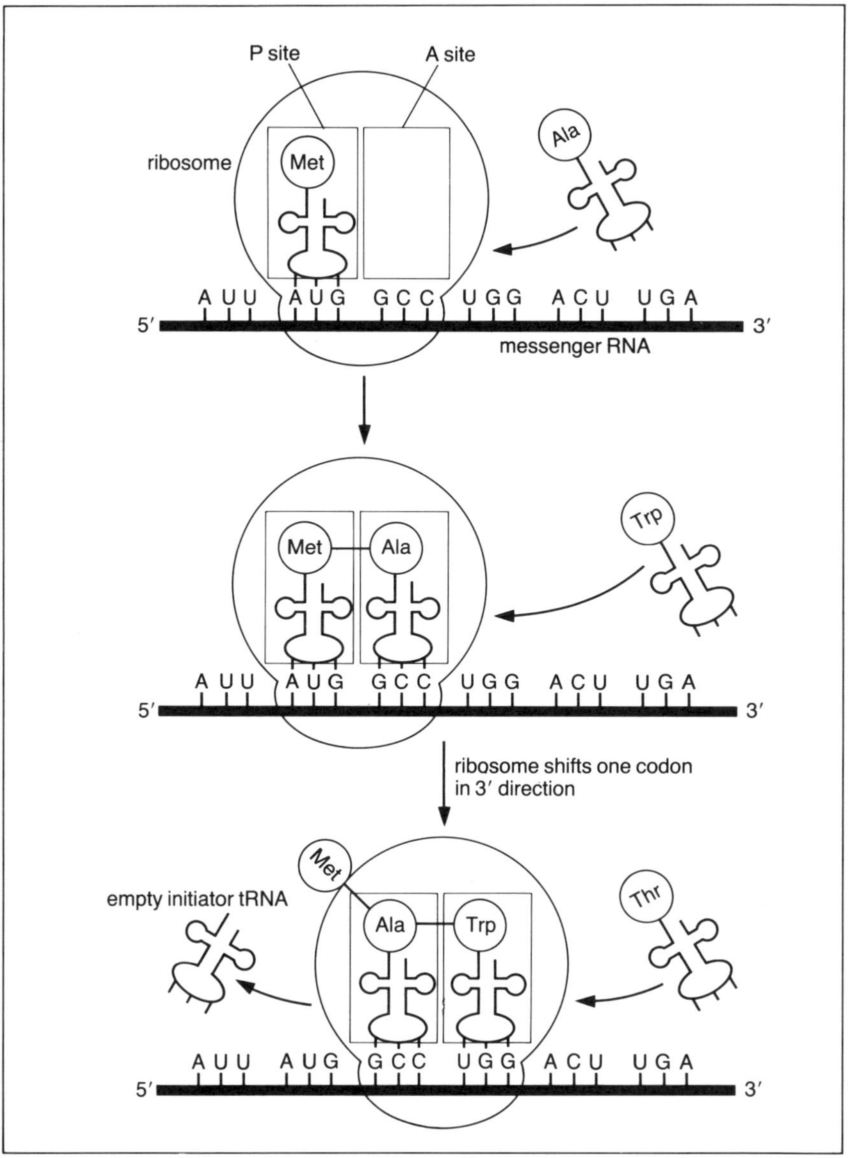


**Fig. T3 Translation**

In both cases the ⋆guard cells of the stomata lose water, becoming less ⋆turgid, and the pore closes. **b** During daylight, and when there is plenty of water, photosynthesis consumes most of the leaf's $CO_2$, and the guard cells absorb water. They become more turgid and open the pore, allowing more $CO_2$ to enter. A different mechanism works in plants that open their stomata at night and close them by day. See also transpiration stream.

**transpiration pull** The forces needed to pull water to the top of a plant (nearly 100 metres in the tallest trees). See transpiration stream.

**transpiration stream** The flow of water from the roots of a plant to the leaves via ⋆xylem vessels. A number of forces keep the water moving. **a** ⋆Osmosis at the roots produces ⋆root pressure—a pushing force. **b** The strongest force, however, is a pulling force due to ⋆cohesion forces within the water (water is a polar molecule forming ⋆hydrogen bonds). As water evaporates from the surface of cells inside the spongy layer of a leaf, the cohesive forces pull up more water to replace what has been lost. The effect is roughly similar to siphoning petrol from a tank. As long as the thread of water in the xylem is continuous (helped by its small diameter and pits in each cell's side walls), this process can continue. **c** Capillary action due to ⋆adhesion of water molecules to the walls of the xylem tubes also assists the transpiration stream.

**transplant** 1 Surgical technique for transferring an organ or tissue from one person to another (e.g. Kidney transplants are common), or from one part of the body to another (e.g. Skin ⋆graft). 2 Carefully digging a plant from the ground and replanting it in another location.

**transport** The movement of a substance from one place to another. e.g. ⋆Active transport.

**transposon** A length of DNA that is able to insert itself into other sections of DNA. e.g. **a** Regulatory genes in maize. **b** Genes for bacterial resistance to antibiotics can pass from DNA to ⋆plasmid, the plasmid can then be transferred to another bacterium, where the gene then inserts itself into the DNA of the new host cell.

**transversion** A point mutation where an organic base that is part of the genetic code, one of the purines (adenine or guanine), is replaced by one of the pyrimidine bases (thymine or cytosine), or vice-versa. Compare transition above.

**trapdoor** Any member of the ⋆spider family Ctenizidae, which live in silk-lined burrows with a 'lid' or trapdoor at the top.

**trauma** 1 Medically, an injury or wound. 2 Psychologically, a startling or unpleasant experience which has lasting effects on the emotions.

**tree** A woody plant, usually with a single main axis or trunk, usually fairly tall, which bears branches and leaves. ⋆Perennial. Compare bush, shrub, mallee.

**tree-kangaroo** Any of the tree-living kangaroos of north-eastern Queensland, and New Guinea, belonging to the genus *Dendrolagus*. To aid life in trees, they have a ⋆prehensile tail and large friction pads on their hind feet. See Appendix 1.

**treeline** The upper limit of tree growth, both with altitude on mountains (subalpine/alpine boundary), and with latitude as one proceeds towards the poles (taiga/tundra boundary). The treeline includes any band of stunted trees that may extend for a distance beyond the timberline. Compare timberline.

**Trematoda** A class of flatworms in phylum ⋆Platyhelminthes, called the flukes, which are all parasitic. e.g. ⋆Liver fluke. See Appendix 1.

**tremor** Involuntary quivering and trembling of groups of muscles, other than for obvious causes such as shivering from cold. Useful in diagnosis of disease.

**Triassic** The first period of the Mesozoic era, following the Permian period, and succeeded by the Jurassic period. See Appendix 2.

**tribe** 1 In plant classification, a group below the level of a Family, and containing related genera. 2 A social group consisting of a number of related families or clans.

**tricep** A muscle with three lobes where it is connected to bone. e.g. The muscle at the back of the upper arm which balances the ⋆bicep. See Fig. B4.

**trichinosis** A parasitic disease due to the eating of ⋆roundworm larvae (*Trichinella spiralis*), protected in cysts, and most often present in raw or undercooked pork.

**trichome** A fine plant ⋆hair formed by the outgrowth of a surface cell.

**Trichoptera** The order of insects that includes the ⋆caddis flies. See Appendix 1.

**tricuspid valve** A valve, made of three membrane-like flaps, which stops the flow of blood backwards from the ventricle to the atrium on the right side (as far as the owner is concerned) of the mammalian ⋆heart. See Fig. H2.

**trifoliate** A compound leaf having three leaflets. e.g. Clover.

**trigger plant** A general name given to plants where the stamens of the flowers move rapidly when touched, thus dusting visiting insects with pollen. e.g. *Stylidium*.

**triglyceride** A chemical name given to ⋆lipids (fats and vegetable oils) where three fatty acids are attached to a glycerol molecule. See Fig. L3.

**trilobite** Any member of a class of extinct ocean-bottom-dwelling arthropods abundant from the Cambrian to the Silurian. See Appendix 2.

**trimester** Three months, or one-third of the nine months of pregnancy. The nine months of pregnancy are often divided into the first, second and third trimesters.

**tripinnate** Pinnate three times. The secondary leaflets or ⋆pinnules of a pinnate leaf are further divided. e.g. Certain ferns. See Fig. L1.

**triple vaccine** A vaccine which combines agents against diphtheria, whooping cough and tetanus, and that can be administered in one injection.

**triplet** The genetic code where each amino acid is specified by three of the four organic bases. See codon.

**trisomy** A condition where one or more chromosomes exists three times in each cell instead of the normal twice (diploid). e.g. Down's syndrome is due to the presence of three copies of chromosome 21 (Trisomy 21).

**tRNA** Short for **transfer RNA**. A type of RNA which gathers ⋆amino acids from the cytoplasm, brings them to the ⋆ribosomes, and matches them to the RNA code as the messenger RNA is read (⋆translated). See Fig. T4.

Each amino acid has its own tRNA. At one end of the clover-leaf-shaped molecule the amino acid is attached. At the other end is a sequence of three organic bases called the ⋆anticodon. As the ribosome moves along the mRNA, the tRNA whose anticodon matches the ⋆codon on the mRNA is attached to a site on the ribosome. See translation for details.

**trophic level** A feeding level; a step in the energy flow of a ⋆food chain in an ⋆ecosystem. Organisms obtain their food in the same general manner at the same number of stages from the producers (usually plants). See food chains and food webs.

**tropical** Of or relating to organisms living in the humid tropics, roughly between the tropic of Cancer north of the equator, and the tropic of Capricorn south of the equator. Compare temperate.

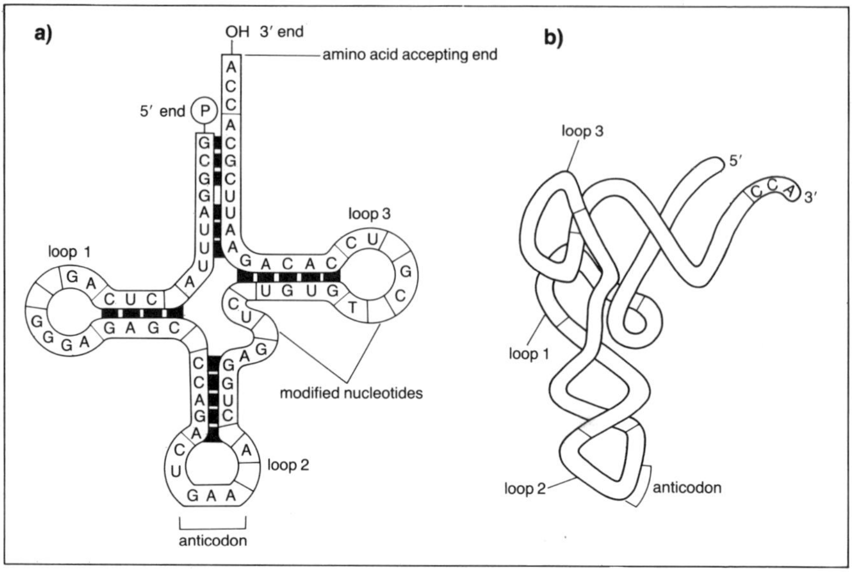


**Fig. T4 Transfer RNA (a) tRNA showing the bases (b) Three-dimensional shape of tRNA**

**tropism** A growth movement of a plant where it responds to an external stimulus in a way that depends on the direction of the stimulus. If it grows towards the stimulus it is called a positive tropism, and if it grows away from the stimulus it is called a negative tropism. e.g. ⋆Phototropism, ⋆geotropism, ⋆hydrotropism, ⋆chemotropism, ⋆haptotropism, etc. Compare taxis and photoperiodism.

**troposphere** The inner atmosphere nearest the ground, which varies in thickness from about 7 km near the poles to 28 km at the equator. In the vertical direction, the temperature drops steadily, about 2°C per 300 metres, and most of the Earth's weather occurs in this layer.

**truffle** Any of the fungi in order Pezizales, of the class Ascomycetes that produce their fruiting bodies underground. Especially the edible varieties which are considered a delicacy.

**truncate** Cut off sharply so that the end appears broken off, square, etc.

**trunk** 1 The main stem or axis of a woody tree. 2 The body without head, limbs, or other ⋆appendages, also called the ⋆torso. 3 A ⋆proboscis, such as that of an elephant.

**trypanosome** Any flagellate protozoan of the genus *Trypanosoma*, including blood parasites infecting humans and other vertebrates. Diseases caused by trypanosomes include African sleeping sickness and Chagas disease of South America. See also tsetse fly.

**trypanosomiasis** Any disease resulting from infection by trypanosomes.

**trypsin** A protein-digesting enzyme found in the duodenum. It is produced by the pancreas in a nearly inactive form called ⋆trypsinogen. An enzyme produced by the walls of the duodenum converts trypsinogen to trypsin as trypsin only works in alkaline conditions.

**trypsinogen** The ⋆precursor to trypsin. See above.

**tryptophan (trp)** A complex amino acid that is one of the 20 common ★amino acids occurring in proteins. It is converted into ★indole acetic acid (an ★auxin) in plants, and the hormone ★serotonin in mankind. See Fig. A6.

**tsetse fly** A blood-sucking fly in insect genus *Glossina*, and order Diptera. It is responsible for carrying the ★trypanosome flagellates that cause sleeping sickness in Africa.

**TSH** Short for *thyroid stimulating hormone*. See above.

**tuan** Also called the brush-tailed phascogale. A rat-sized carnivorous marsupial, ★*Phascogale tapoatafa*. See Appendix 1.

**tuatara** A lizard-like reptile, *Sphenodon punctatus*, found on a few islands off the coast of New Zealand, and renowned for its long life. It is often considered a living fossil, as its nearest relatives lived during the Cretaceous. See Appendix 1.

**tube feet** Also called **podia**. Hollow tubes that extend from the walls of echinoderms, and are connected to a water vascular system. They can be extended or retracted by varying the pressure of the water inside. The ends of the tubed feet of sea-stars (starfish) and sea-urchins have suckers to assist in movement and feeding. The feather-stars (crinoids) and brittle-stars normally have pointed tubed feet for feeding. The tubed feet around the mouth of sea-cucumbers are used for feeding.

**tuber** The swollen underground stem of a plant acting as a food-storing structure. In some plants, the plant reproduces vegetatively from the tuber. (e.g. potato—the 'eyes' are buds). Swollen roots (e.g. dahlia) are sometimes referred to as tuberous roots.

**tubercle** A small nodule or rounded structure. **1** The mass of small rounded cells produced by ★tuberculosis-causing bacteria. **2** A rounded structure on a bone. **3** A nodule on the root of a ★legume.

**tuberculate** Warty; covered with tubercles or knob-like structures.

**tuberculosis** Called **TB** for short. Also called **consumption**. An infectious disease caused by the ★tubercle-producing bacterium, *Mycobacterium tuberculosis*, which mainly affects the lungs. The ★lungs become inflamed and the ★alveoli fill with fluid.

**tubule** A small tube, or a minute tube-shaped structure.

**Tubulidentata** The order of placental mammals that includes the nocturnal termite-eating aardvark. It is native to Africa. See Appendix 1.

**tubulin** The protein that makes up the ★microtubules of the ★cytoskeleton.

**tumid** Swollen, inflated.

**tumour** Also called a **neoplasm**. A swelling in an organism that is produced by an abnormal and uncontrolled cell division. Tumours which do not invade or destroy other parts of the body are called ★benign, while those which do are called ★malignant. See also cancer.

**tundra** An extensive ★biome in the arctic regions, composed of level or slightly undulating treeless plains, and with the subsoil permanently frozen. Plants are mainly mosses and lichens. Compare taiga.

**tunicates** Also called **sea-squirts**. Any of the marine chordates which have free-swimming larvae with a ★notochord. They usually settle down to a sedentary life. e.g. Cunjevoi (*Pyura praeputialis*) common on ocean rock platforms. See Urochordata and Appendix 1.

**Turbellaria** A class of flatworms in phylum ★Platyhelminthes which includes the ★planarians. See Appendix 1.

**turbid** Used to describe water that is clouded or unclear due to suspended living or non-living particles. e.g. Microbes, mud.

**turbinate** Top-shaped, or an upside down cone.

**turgid** Swollen, distended. Used to describe cells into which water has diffused so that the walls are stretched and the cell is fairly rigid. With plant cells, the cellulose cell wall prevents the cell swelling too much, the extra water being held in the vacuoles. Turgid cells are one of the main ways in which many plants without ⋆lignified cell walls (e.g. moss) maintain their rigidity. However, animal cells (in experiments) can swell so much that they burst. Hence the concentration of liquids around animal cells must be carefully controlled (⋆homoeostasis). Compare wilting and flaccid.

**turgor** The state of being ⋆turgid. The outwardly directed pressure due to a swelling cell is called turgor pressure. This pressure is needed to help maintain the shape of soft plants (e.g. moss) and plant parts (e.g. flowers, young leaves and stems), for the expansion of growing cells, and the opening of stomata and of many flower parts.

**Turner's syndrome** A genetic anomaly in some females, who have only one ⋆sex chromosome (XO rather than XX). Turner's syndrome causes reduced stature, webbing of the neck, infertility, and some degree of mental retardation.

**turpentine** 1 Any oily liquid extracted from pine resin and used as a solvent. 2 A tree, *Syncarpia glomulifera*, which grows in rainforests or tall eucalypt forests along the east coast of Australia.

**turtle** Any marine reptile of the order Chelonia with its body enclosed in a shell. It has flippers rather than webbed toes like a ⋆tortoise. See Appendix 1.

**tussock** A tuft or clump of grass, twigs, hair, etc. e.g. Grass in the genus *Poa*, which grows as tussocks.

**twig** A small, often woody, branch of a tree.

**twin** One of two organisms from the same birth. **a Identical twins**, have the same genetic makeup (⋆genotype), and are the same sex. This occurs when the embryo splits in two soon after fertilisation, each part growing into separate but identical organisms. **b Fraternal or non-identical twins** are genetically as different as normal brothers and sisters. They occur when two eggs (ova) are produced, and then fertilised, at the same time by different sperm.

**twitch** Describing the type of contraction in ⋆muscle fibres. **a Slow twitch** muscle fibres can contract repeatedly, or maintain their contraction for long periods. **b Fast twitch** fibres give faster and stronger contractions, but tire more readily.

**Two-four-D** See beginning of letter T entries.

**Two-four-five-T** See beginning of letter T entries.

**tympanic cavity** Also called the **tympanum**. The drum-like cavity around the middle ear including the ⋆eardrum. See Fig. E1.

**tympanic membrane** The ⋆eardrum. See Fig. E1.

**type specimen** Also called a **holotype**. The specimen on which the original description of a species is based, and hence used for naming a ⋆species.

**typhoid fever** An acute feverish illness caused by bacteria, *Salmonella typhi*, usually transmitted by faeces-contaminated water, or food prepared by unknowing ⋆carriers of the disease.

**typhus** A disease caused by small bacteria, *Rickettsia prowazekii*, and transmitted to humans by lice and fleas, with rats and mice as intermediate hosts.

***Tyrannosaurus*** A genus of large carnivorous dinosaurs that walked on their hind legs. e.g. *T. rex*. See also Theropoda.

**tyrosine (tyr)** One of the 20 common ⋆amino acids found in proteins. Tyrosine is converted into many compounds, including noradrenaline, adrenaline, melanin, thyroid hormone, etc. See Fig. A6.

# Uu

**ulcer** An open sore with an often raw, running surface. It can occur on the skin (e.g. varicose ulcer), or on a mucous membrane such as the lining of the stomach (e.g. ⋆peptic ulcer).

**ulna** The larger of the two bones in the lower forelimb (forearm) of vertebrates. It connects the humerus at the elbow to the carpels of the hand. Compare radius. See Fig. S3.

**ultracentrifuge** A high-speed centrifuge (60 000 rpm) used to separate macromolecules (e.g. proteins), and ⋆colloids.

**ultrasound** Sound waves (called ultrasonics) with a frequency greater than 20 000 Hz. This very high frequency sound is used in medical ⋆diagnosis, especially during pregnancy when use of X-rays would be dangerous. Compare CT, PET and MRI.

**ultrastructure** The very fine and detailed structure of cells and tissues that cannot be seen when using a light microscope (e.g. the internal structure of mitochondria). Special techniques, especially electron microscopes, must be used.

**ultraviolet light** Called **UV** for short, and sometimes termed **black light**. Light with wavelength shorter than normal light that cannot be detected with human eyes. Occurs as part of sunlight, and is needed in small amounts to produce vitamin D in the skin. Too much UV light causes sunburn. Since UV light can cause some substances to fluoresce, it is used in ultraviolet fluorescence microscopy.

**umbel** A type of flower head (⋆inflorescence) where the many flowers form a cluster, all connected to the same point at the end of a stem. See Fig. I4.

**Umbelliferae** See Apiaceae and Appendix 1.

**umbilical cord** The cord-like structure that connects the ⋆embryo to the mother's ⋆placenta in placental mammals. See Fig. E4. The blood vessels inside carry food and oxygen from the mother to the embryo, and wastes back again. The umbilical cord is cut at birth, and the stump on the baby shrivels to form the naval.

**underbrush** Small shrubs and trees growing beneath large trees.

**understorey** A general term which includes all shrubs and trees growing beneath the ⋆canopy of large trees.

**undulate** Possessing a wavy surface or edge.

**ungulate** A now rarely used classification of hoofed, grazing placental mammals. Now grouped as the even-toed ⋆Artiodactyla, and the odd-toed ⋆Perissodactyla. See Appendix 1.

**unguligrade** A type of walking where an animal walks on the tips of its ⋆digits. e.g. A hoof, such as cows and horses. Compare digitigrade and plantigrade.

**unicellular** Consisting of only one cell.

**unisexual** A plant or animal species that has male or female reproductive organs, not both. e.g. Used especially of flowers. Compare hermaphrodite.

**univalve** Any member of the mollusc class Gastropoda, with only one shell. e.g. Snail, whelk, periwinkle. See Appendix 1.

**universal donor** A person with type O blood which, in an emergency, can be given in a ⋆transfusion to a person with any of type A, B, O or AB blood. However, there are many *blood groups as well as ABO (e.g. Rh, M, N) which also have to be matched when transfusions are given in non-emergency situations. See also universal receiver.

**universal receiver** A person with type AB blood who, in an emergency, can receive during a ⋆transfusion, blood of A, B, AB, or O types. However, there are many ⋆blood groups as well as ABO (e.g. Rh, M, N) which also have to be matched when transfusions are given in non-emergency situations. See also universal donor.

**unmyelinated** ⋆Neurones that lack a ⋆myelin sheath, although they are still surrounded by their ⋆Schwann cell. They conduct impulses more slowly than equivalent myelinated neurones. See Fig. N1.

**unsaturated** 1 Compounds that can react with more hydrogen because they have double and/or triple bonds within the molecule. e.g. Unsaturated fats. See saturated, polyunsaturated, and Fig. P9. 2 A solution where more solute can be dissolved in a solvent at a given temperature. Also applied to the air when more water vapour can enter the air for a given temperature. Compare saturated.

**uracil** One of the nitrogen containing organic bases which forms the genetic code along RNA molecules. Classed as a ⋆pyrimidine, it is replaced by thymine in DNA. See Fig. B3.

**urban** Of or relating to a city or town. e.g. Urban ecology.

**urchin** An echinoderm. See sea-urchin.

**urea** A nitrogen-containing, water-soluble organic compound, $H_2NCONH_2$. It is the main method by which mammals, and other animals such as amphibians, remove wastes formed by the breakdown of amino acids. Urea is formed in the liver, and filtered from the blood by the kidneys to form urine. See also excretory system.

**ureter** The duct which carries urine away from the kidney. In most mammals the urine drains into the ⋆urinary bladder, while in birds and some mammals it drains into the ⋆cloaca. See Fig. E8.

**urethra** The duct in mammals which carries urine from the urinary bladder to the outside. In males it connects to the sperm duct (⋆vas deferens) before passing down the penis to the exterior. See Figs E8 and R3.

**uric acid** A nitrogen-containing, almost water-insoluble organic compound, the main method by which birds, lizards, and snakes remove wastes formed by the breakdown of amino acids. In primates, the breakdown of purines from nucleic acid also produces uric acid, which under certain circumstances can build up in the body joints to produce ⋆gout.

**urinary bladder** A muscular sac in mammals for the storing of urine that reaches it from the kidney through the ⋆ureter. A ⋆sphincter (ring of muscle) at the beginning of the ⋆urethra controls the flow of urine to the exterior. See Fig. E8.

**urine** The waste-containing liquid produced in the kidneys and carried away by the ⋆ureters to the ⋆bladder or ⋆cloaca.

The urine consists of various mixtures of urea, uric acid, salts and other substances. Humans produce about 1 to 1.5 L per day depending on temperature, perspiration, and the amount of liquids drunk. Fish living in freshwater have to urinate continuously to rid their bodies of water that enters by osmosis, whereas salt water fish excrete highly concentrated urine.

**Urochordata** The subphylum of chordates which includes the sea-squirts and sea-tulips, with a ★notochord during part of their life-cycle. See Appendix 1.

**urogenital system** Also called the **urinogenital system**. The organs which comprise the urinary (★excretory) and reproductive systems.

**uterine** Of or relating to the ★uterus.

**uterus** Also called the **womb**. In female mammals, the muscular organ in which the embryo is nourished and develops. See Fig. R4.

The uterus is paired in most mammals (e.g. rats, mice), but is single in humans. The eggs (ova) travel along the ★oviducts (Fallopian tubes) that connect the ★ovaries to the top of the uterus. Sperm reach the uterus along the birth canal (★vagina) and through the cervix at the bottom. The baby is eventually born through the same passage. During the ★menstrual cycle the walls of the uterus become thick and spongy ready to receive the fertilised ovum. If pregnancy does occur, a ★placenta grows into the wall of the uterus, connecting mother and baby. At birth, the muscular walls contract to expel the baby. If pregnancy does not occur, then the walls break down (★menstruation) to be replaced and made ready for the next chance for receiving a fertilised ovum.

**utricle** Also called the **utriculus**. A chamber on top of the ★cochlea of the ★inner ear from which the three ★semicircular cells arise. Together with the ★saccule, it contains sensitive cells which help determine the position of the head relative to gravity, and its rate of movement.

# Vv

**vaccination** Introduction to the bloodstream (by direct injection or via the stomach after swallowing) of a *vaccine to cause the body to develop *immunity to a disease. See also immunisation.

**vaccine** Any material made from bacteria or viruses that is injected (or otherwise administered) so that the body will develop *immunity to that bacteria or virus.

The material in the vaccine stimulates the body to produce *antibodies against the organism. More importantly it causes certain *lymphocytes (white blood cells) to 'remember' the organism, so that if a real infection should occur, the immune system can respond quickly and remove the invader before it causes sickness in the body. Virus vaccines are made from either killed virus (e.g. influenza), or live but very weakened forms (*attenuated) of the virus (e.g. measles). Bacteria can also be killed (e.g. typhoid fever), or weakened (e.g. tuberculosis). Some bacterial vaccines act against the toxins that they produce (e.g. tetanus), or against a chemical extracted from their cell wall (e.g. meningitis). Compare serum.

**vacuole** A fluid-filled space within the cytoplasm surrounded by a membrane (the *tonoplast). See Fig. C4. Plant cells usually have one or more large vacuoles filled with salts, sugars, pigments, wastes, etc. Vacuoles in animal cells are much smaller, and are often called *vesicles. They do not occur in bacteria or blue-green algae. See also contractile vacuole.

**vagina** Also called the **birth canal**. A muscular tube that connects the womb (uterus) with the outside. See Fig. R4. It acts as a passageway for a baby to be born, and also receives the male penis during sexual intercourse. The *ejaculated sperm can then move up through the cervix into the uterus and to the ova in the oviducts.

**vagrant** *Nomadic or wandering from place to place. e.g. Applied to birds blown off course by storms and appearing in places where they are not normally seen. Also applied to introduced or *exotic plants or animals.

**vagus nerve** A nerve starting in the *medulla oblongata of the hindbrain and passing out into the body cavity. A mixed nerve, it contains sensory fibres from the heart, lungs, etc., and motor neurones of the *parasympathetic system which connect to the *pacemaker of the heart, to the bronchi, oesophagus, gall bladder, pancreas, small intestine, and secretory glands of the stomach.

**valence** Also called **valency**. The valency measures the combining power of an atom, or group of atoms, compared to hydrogen which is set equal to +1. With ions the valency is the same as its electronic charge. e.g. Since the formula of water is $H_2O$, then the valency of oxygen (O) must be –2 to balance the combining power of two hydrogens (H), each +1.

**valgus angle** Also called the **carrying angle**. Even though the leg bones (femur) start at the hip, they are close together near the knees. The angle the femur makes with the horizontal is the valgus angle. See Fig. S3.

**valine (val)** One of the 20 common

★amino acids found in protein. Valine is converted through many steps to coenzyme A. See Fig. A6.

**valvate** 1 Having valves or valve-like structures. 2 Concerning petals, sepals, etc. in bud, with the edges touching.

**valve** 1 In animals, a structure in the heart or veins that permits blood to flow in only one direction. 2 In plants, part of the lid of a eucalypt seed capsule (fruit) which lifts off to allow the seeds to escape.

**vane** The flat part of a feather.

**vapour** 1 A gas, especially gases that are visible as a mist, fog, steam, etc. 2 Any substance in the gaseous state.

**variable** Part of an experiment that is able to be changed. **a Controlled variables** are kept constant during the experiment. **b Experimental variables** or **manipulated variables** are changed in a predetermined way as part of the experiment. **c Uncontrolled variables** are ones that change, either because the researcher forgot to include them, or there was no known way of controlling them.

**variation** The differences between members of the same species of plants, animals, etc. **a Environmental variation** can be produced by many factors including diet, accidents, light intensity, population density, etc. Differences acquired during one's lifetime are not passed on to the next generation if the genes are not affected. **b Genetic variation** is inherited. Variation is produced mostly through mixing of genes during meiosis (★independent assortment), and again at ★fertilisation. Occasionally new variation is introduced by ★mutation. When a feature is controlled by many genes, then ★continuous variation is produced. e.g. The wide variety of normal heights in well-fed humans. When there is only one or two genes present in two or more forms (alleles), then only a few distinct varieties may be produced, and ★discontinuous variation results. e.g. Human ABO blood groups. See also normal curve and Fig. N3.

**varicose** Swollen or knotted. e.g. Varicose veins.

**variegated** Irregular in colour pattern, such as patches or streaks of different colours. Can be a natural feature. e.g. Variegated leaves of *Coleus* are due to differences in the concentration of chlorophyll present in different parts of the leaf. Can also be due to infection. e.g. Tobacco mosaic virus (TMV) discolours the leaves of the tobacco plant.

**variety** A subdivision of a species, where the organisms can be crossbred to produce fertile offspring, but differ in some minor feature of appearance. e.g. Varieties of rose plants. Compare breed, race.

**vas deferens** Also called the **sperm duct**. A muscular tube in males that carries sperm cells from the ★testes to the ★penis during copulation. The vas deferens starts at the ★epididymis and connects to the ★urethra near the prostate gland. See Fig. R3.

**vascular** Of or concerning tubes or vessels which carry fluids. e.g. Blood or sap.

**vascular bundle** A strand, bundle or ring of vascular tissue that runs along most of the length of a plant. Made mainly of ★xylem and ★phloem, the arrangement of vascular bundles can be important in classification. e.g. Dicots generally have ★vascular tissue arranged around the edge of the stem, whereas monocots generally have vascular tissue scattered throughout the stem. See also stele.

**vascular cylinder** See stele.

**vascular plants** The plant Phylum (Division) Tracheophyta, which includes those plants with ⋆vascular tissue. e.g. Ferns, gymnosperms, angiosperms, etc.

**vascular system** 1 In vertebrates, the ⋆circulatory system including heart, arteries and veins. 2 In echinoderms, the water vascular system that controls the ⋆tubed feet. 3 In plants, the system of ⋆xylem and ⋆phloem tubes that carries the ⋆sap around the plant. See vascular tissue below.

**vascular tissue** The sap-carrying tissues of ⋆vascular plants consisting mainly of water-carrying ⋆xylem and food-carrying ⋆phloem, along with strengthening cells (sclerenchyma) and packing cells (parenchyma). See Fig. V1. See also stele, meristem, secondary growth.

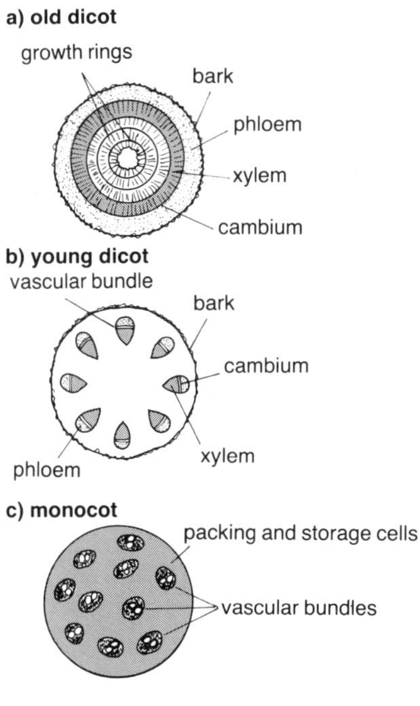


**Fig. V1 Vascular tissue**

**vasectomy** The sterilisation of a man by cutting the vas deferens.

**vasoconstriction** The reduction in diameter of a blood vessel. Compare vasodilation.

**vasodilation** The increase in diameter of a blood vessel, important during the ⋆inflammatory response. Compare vasoconstriction.

**vasopressin** Also called **antidiuretic hormone** or **ADH**. A ⋆peptide hormone secreted by the hypothalamus, and stored in the posterior (rear) lobe of the ⋆pituitary before being released into the bloodstream. Its effects include increasing bloodpressure, and increasing the reabsorption of water from urine in the kidneys, thus helping to conserve water. Compare oxytocin.

**VD** Short for **venereal disease**. Now called ⋆sexually transmissible disease.

**vector** 1 An organism that carries a disease-causing organism (pathogen) from one living thing to another. e.g. Certain mosquitoes are the vectors for the malaria parasite. A woodboring wasp is the vector for a fungus helping cause ⋆dieback in some Australian forests. 2 An agent for transferring ⋆genes from one organism to another. e.g. ⋆Plasmids and ⋆viruses can naturally transfer genes. Genes can also be injected directly into cells (⋆microinjection) using very fine pipettes, or DNA can be coated onto tiny particles of tungsten (microprojectiles) and fired into plant cells. See also transgenic and genetic engineering.

**vegetable** 1 Of or relating to plants. e.g. Vegetable matter. 2 A plant cultivated for food, or the edible part of such plants. Technically the edible parts may be fruits (e.g. tomato), tubers (e.g. potato), swollen taproots (e.g. carrot), buds (e.g. onion), etc.

**vegetation** All the plants in an area or region. Compare flora.

**vegetative growth** Any type of plant growth that does not involve sexual reproduction. e.g. Growth of a stem or leaf.

**vegetative propagation** Also called **vegetative reproduction**. **1** In plants, reproduction by any means other than sexual reproduction. e.g. Stems (rhizomes, tubers, corms, bulbs); root tubers; stem, leaf, and root cuttings; grafting and budding; etc. **2** In animals, non-sexual reproduction. e.g. Budding in cnidarians (coelenterates).

**vein** **1** In animals, a blood vessel that carries blood from the capillaries to the heart. Compared to arteries, veins have a larger diameter but thinner less muscular walls, and ⋆valves to prevent the reverse flow of blood. **2** A strand of vascular tissue visible in, or on the surface of a leaf, petal, etc.

**veldt** The ⋆savanna grasslands and ⋆steppes of southern Africa. i.e. open countryside in southern Africa covered with grass and thinly covered by shrubs or trees.

**vena cava** In land vertebrates, a large ⋆vein that returns blood to the right atrium of the ⋆heart. The posterior (or inferior) vena cava returns nearly all the blood from behind the forelimbs (below the arms in humans). The anterior (or superior) vena cava returns blood from the forelimbs and head. See Fig. H2.

**venation** The arrangement or pattern of veins in a leaf. Generally, dicotyledons have a net-like pattern, while monocotyledons have veins that are almost parallel to each other along the leaf.

**venereal disease** Called **VD** for short. Now called ⋆sexually transmissible diseases.

**venom** The poisonous fluid which some animals secrete, and usually inject into the bodies of their victims by biting, stinging, etc. e.g. Snakes, spiders, certain fish and octopus, scorpions, etc. Venom acts in a number of ways. Some act on the linings of blood vessels, allowing blood to escape into surrounding tissue. Others destroy red blood cells, while others cause paralysis.

**venomous** Having glands that produce ⋆venom. Capable of giving a poisonous bite.

**venous** Of or relating to veins or the blood within them.

**vent** Also called a **cloaca**. The cavity where the excretory ducts (ureter), alimentary canal, and reproductive organs all connect to the outside of the body. Present in most land vertebrates except mammals.

**ventilation** The process of supplying fresh air. e.g. Supplying fresh air to the lungs during breathing.

**ventral** In most animals, the side nearest the ground or substrate. e.g. Worms, lizards, lions. Animals that are normally upright, the anterior or most forward surface. e.g. Humans. Compare dorsal and see Fig. S11.

**ventricle** A cavity in an organ. e.g. **1** A muscular chamber of the vertebrate heart that pumps blood to the body. There is only one ventricle in the heart of fish and amphibians, while other vertebrates have two. In humans, the left ventricle (as far as the owner is concerned) pumps oxygenated blood to the tissues, while the right ventricle pumps blood to the lungs. See heart and Fig. H2. **2** Certain cavities in the vertebrate brain filled with cerebrospinal fluid.

**vermicular** Narrow and worm-like.

**vermin** Any undesired small animal which infests humans, their homes or agriculture. e.g. Rats, mice,cockroaches, lice, fleas.

**vernal** Of or relating to the season spring; appearing in spring.

**vernalisation** 1 The exposure of plants or their seeds to low temperatures to stimulate them to flower. e.g. Some *biennials need a cold winter after their first year of growth to induce flowering in the next spring. 2 Artificially, plants or their seeds are exposed to cold so that they will flower the next spring. e.g. The sowing of winter wheat, rye and oats in autumn so that they produce seed the next spring. In Russia, partially germinated seeds are exposed to cold before planting as the extreme cold of early spring would kill young plants that germinated in the soil. Exposing the young plants of biennials to cold can induce flowering the first year.

**vernation** The arrangement of unfolding of leaf buds.

**vermicose** Shiny and appearing as though polished; varnished.

**verrucose** Warty.

**vertebrae** Singular is **vertebra**. The individual bones of the vertebral or *spinal column. See Fig. V2.

Each vertebra consists of a solid central bone that connects to other vertebrae through cartilage, thus allowing some bending of the backbone. Outside of this is the *neural arch of bone through which the spinal cord runs. To the sides are the *transverse processes to which are attached body muscles, and ribs in certain cases. There are five groups of vertebrae, the number, shape and function varying for different animals. Thus with fish, all the vertebrae from head to tail are almost the same. In humans, there is much variation amongst the five groups which are made up as follows: 7 cervical vertebrae in the neck region (the top one is called the atlas and supports the head); 12 thoracic vertebrae carrying

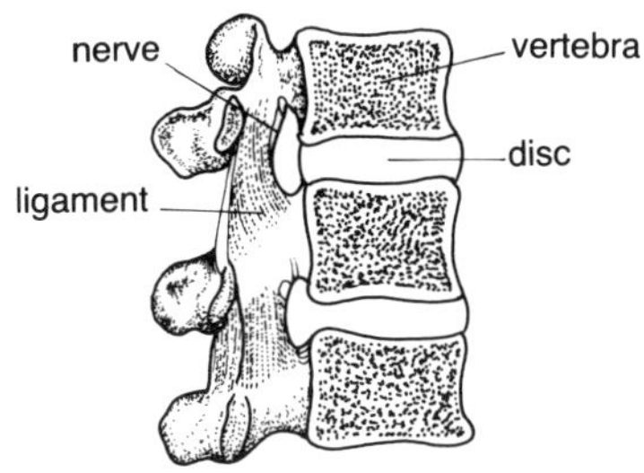


**Fig. V2 Vertebrae**

the ribs, 5 lumbar vertebrae which support the lower back, 5 fused sacral vertebrae (forming the *sacrum), and 5 fused caudal vertebrae (forming the *coccyx).

**Vertebrata** 1 Any animal possessing a backbone. 2 Any member of the phylum Chordata, subphylum Vertebrata, that has a vertebral column as distinct from a *notochord. e.g. Fish, amphibians, reptiles, birds, mammals.

**vesicle** A small sac, bladder, or cavity which contains a fluid. e.g. A small membrane-enclosed sac in the cytoplasm of cells, especially those produced by the *Golgi apparatus and carrying enzymes, hormones or neurotransmitters. See also synaptic vesicle, cisterna.

**vessel** 1 A tube or canal through which fluid passes. e.g. Blood or lymphatic vessel in animals. 2 In plants a tube within the *xylem of vascular tissue built up from *vessel elements joined end to end, their end walls having broken down to form *perforation plates. The main water-conducting cell system in the xylem of most angiosperms. See Fig. X1.

**vessel elements** A type of *xylem cell found in most angiosperms, and some ferns. Generally shorter and broader than *tracheids, vessel elements lack cytoplasm, have thickened cell walls (in various patterns) and wall *pits. The ends of the cells break down to form ★perforation plates allowing easier passage of water through the cell. Many vessel elements stacked end to end form a ★vessel. See Fig. X1.

**vestibular apparatus** That part of the inner ear concerned with balance and equilibrium. See inner ear.

**vestibule** A cavity which acts as an entryway or approach. e.g. **a** The start of the cochlea of the inner ear. **b** The space inside the nasal cavity, just inside the nostrils. **c** The cavity formed by the recessed ★vulva and ★labia giving entry to the vagina.

**vestigial organ** In evolution, any organ that is believed to have become reduced in size or function. e.g. Although the appendix in humans contains lymphatic tissue, it is believed to have been more important in the past, functioning like the ★caecum of present day mammals such as the rat.

**viable** 1 Capable of living and developing. e.g. Organisms of the same species are able to mate and produce viable offspring that can live and develop. 2 Used to describe seeds and spores that are able to germinate.

***Vibrio*** A genus of comma-shaped ★bacteria. Most live in the soil and are agents of decay, but some cause disease. e.g. *Vibrio cholera* causes the disease cholera. Compare bacillus, coccus, spirillum, spirochaete. See Fig. B1.

**vigour** Healthy both physically and mentally. e.g. ★Hybrid vigour.

**villi** Singular is **villus**. A finger-like structure that helps to increase the surface area of an organ. e.g. **1** The villi lining the ★small intestine help increase its surface area by 10 times for absorption of digested food. See Fig. V3. Inside the villi are blood capillaries where sugars and amino acids are absorbed and carried to the liver. Also present in each villus is a ★lacteal, part of the ★lymphatic system where small globules of fat (★chylomicron) from the walls of the intestine are absorbed. See also microvilli. **2** Villi also occur where the placenta meets the walls of the uterus, thus increasing the area for exchange of food and wastes between mother and embryo. See Fig. E4.

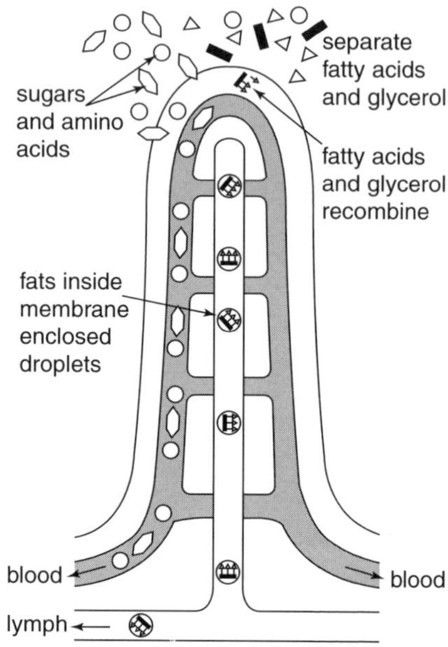


**Fig. V3 Villi**

**villous** Covered with long weak hairs.

**vine** A plant with a stem that rests and clings to other objects for support. A climbing or trailing plant. e.g. Grapes, passionfruit.

**virgate** Twiggy.

**virgin** 1 An unfertilised gamete. 2 An organism prior to taking part in sexual reproduction.

**virion** The inert phase in the lifecycle of a ★virus when outside a living cell. It consists of one or more strands of DNA or RNA inside a protein coat. During this phase, many viruses can be made to form crystals.

**viroid** Even smaller than a ⋆virus, a viroid is made of just a small, unprotected ring of RNA, which can replicate inside suitable cells. It is believed responsible for certain plant diseases. They are transmitted from plant to plant by insects. e.g. Potato spindle tuber viroid.

**virulence** A measure of the ability of a parasitic microorganism to cause disease. e.g. The virus that causes myxomatosis in rabbits has lost much of its effectiveness due to its reduced virulence (and increased resistance in the rabbits).

**virus** Any of a large group of organisms that are too small to be seen with a light microscope, and can only reproduce inside living cells. Viruses are all ⋆parasitic on their host cells, causing diseases in animals (smallpox, herpes, influenza), plants (tobacco mosaic virus), and bacteria (⋆bacteriophages). See Fig. V4.

Viruses can only just be seen with electron microscopes and when not inside a living cell they exist in an inert form (see virion above) where the one or more strands of DNA or RNA are enclosed in a protective protein coat. The coat, called a capsid, has a shape that can help in identification. When the virus enters a cell it leaves the coat behind. If DNA enters, it is incorporated into the chromosomes of the host cell. It may lay dormant there for some time, or may start to take control of the cell's genetic machinery almost immediately. If it is RNA that enters the cell, it must be first converted into DNA, before it begins to control the cell, directing its own replication. Eventually the host cell dies, splits open, and releases the multiplied viruses. Animal viruses can contain either DNA or RNA, while plant viruses contain RNA, and bacteriophages mainly contain DNA. Viruses which contain double strands of DNA include ⋆adenoviruses, ⋆herpes viruses, and ⋆poxviruses. Viruses which contain a single strand of RNA include ⋆arboviruses and ⋆myxoviruses.

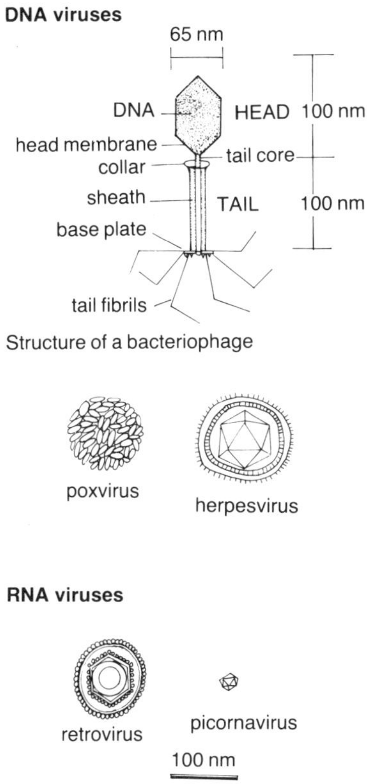


**Fig. V4 Virus**

**viscera** The internal organs of the body, especially those of the abdominal cavity.

**viscosity** A measure of the resistance to flow that a fluid offers, as when it flows through a tube, or when another body attempts to travel through the fluid. See viscous.

**viscous** Thick and sticky. The more viscous a liquid is, the harder it is to travel through it, and the harder it is to make the fluid travel through a tube. e.g. Honey is more viscous than water.

**vision** The sense which allows us to see. Sight. See eye.

**visual purple** See rhodopsin.

**vital staining** The technique of staining cells or tissues without killing them, so that they can be more easily viewed through a microscope. See also staining.

**vitalism** The theory that the functioning of living things is not entirely mechanical and due to chemical and physical processes, but is at least in part due to a non-physical 'vital force'.

**vitamin** An organic compound needed by living things in small amounts for their normal functioning. Plants can make their own vitamins, but animals must obtain most of them in their diet. Not all animals, however, need the same vitamins. Deficiency diseases can result from insufficient vitamins. e.g. In humans, scurvy and lack of vitamin C. For humans, the main vitamins exist in two groups: **1 Water-soluble vitamins** These include the ⋆vitamin B group, which function as ⋆coenzymes or are converted into coenzymes, and ⋆vitamin C which functions as a ⋆cofactor. **2 Fat-soluble vitamins** These include ⋆vitamin A, ⋆vitamin D, ⋆vitamin E, and ⋆vitamin K. Vitamin D can be made in the skin as long as there is sufficient sunlight, and vitamin K is partly supplied by microbes in the large intestine.

**vitamin A** Also called **retinol**. A fat-soluble vitamin needed for proper night vision (see rhodopsin), and body growth. Best sources include green vegetables, carrots, and liver.

**vitamin B** Nine water-soluble vitamins including ⋆biotin, ⋆cobalamin ($B_{12}$), ⋆folic acid, ⋆nicotinic acid, ⋆pantothenic acid, ⋆pyridoxine ($B_6$), ⋆riboflavin ($B_2$), and ⋆thiamine ($B_1$).

**vitamin C** See ascorbic acid and Fig. A8.

**vitamin D** A small group of fat-soluble vitamins that can be made in our bodies by action of sunlight, especially ⋆ultraviolet light, on the skin. Vitamin D has a role in calcium and phosphate absorption from the intestine. Lack of vitamin D results in weak and deformed bones. e.g. Rickets in those still growing.

**vitamin E** Also called **tocopherol**. A small group of fat-soluble vitamins common in wheat germ and green vegetables. It may be needed for healthy reproduction, and helps preserve cell membranes by preventing the fatty acids they contain from being oxidised.

**vitamin K** A fat-soluble vitamin needed for correct clotting of the blood. The two forms of this vitamin occur in the leaves of certain vegetables, and are made by certain bacteria living in our large intestine.

**vitreous humour** The clear, jelly-like substance which fills and gives shape to the eyeball behind the lens. Compare aqueous humour. See Fig. E9.

**viviparous** **1** In animals, giving birth to living young after they have developed and been nourished in a uterus. Compare oviparous and ovoviviparous. **2** In plants, referring to the germination of the seed while attached to the parent plant, as in ⋆mangroves.

**vivisection** The cutting into, or dissecting of, a living organism, especially for research purposes rather than as an operation (⋆surgery).

**vocal cord** More correctly **vocal folds**. A pair of elastic membranes which are stretched across the ⋆larynx (voice box) of air-breathing vertebrates. Sound is produced as air rushes past the folds, different pitches being produced by varying their tension.

**voice** The sounds produced by the vocal cords in the *larynx (voice box), especially those with meaning as produced by humans. The exact sound produced also depends upon resonance in the mouth, pharynx and nasal cavities, as well as the effects of lips, teeth, tongue and palate.

**voice box** See larynx.

**volatile** Easily vaporised and converted into a gas or vapour, especially when compared to water.

**voluntary** Under the control of will, or done by choice.

**voluntary muscle** Also called **skeletal**, **striped**, or ***striated muscle**.

**Vombatidae** The family of marsupial mammals that includes the *wombats. See Appendix 1.

**vomiting** The forcible ejection of the contents of the stomach through the mouth.

**vulva** The external opening of the vagina, which in women, is surrounded by two skin folds called *labia. See also vestibule.

# Ww

**wagtail** Also called **willie wagtail**. A member of the flycatcher family of birds (Muscicapidae), *Rhipidura leucophrys* is found throughout Australia, showing little fear of humans.

**wallaby** Any of the medium-sized kangaroo-like herbivorous marsupials in the family Macropodidae, especially genus *Macropus* which includes the Parma, Tammar, Whiptail, Agile, and many other wallabies. Compare kangaroo, pademelon. See Appendix 1.

**Wallace's line** An imaginary line across Indonesia, near the island of Bali, that divides the Oriental and Australasian biogeographical realms. See Fig. B6.

**wallaroo** Also called the **euro**. A small group of stocky ⋆kangaroos with often shaggy hair, and living in rocky areas. The most widespread is the common wallaroo, *Macropus robustus*.

**wambenger** See tuan.

**waratah** A spectacular red-flowering shrub of eastern Australia, Family ⋆Proteaceae, especially *Telopea speciosissima*, the floral emblem of New South Wales. The 'flower' is actually made up of 50 or more individual flowers.

**warm blooded** See **homoiothermic**.

**warning colouration** Often brightly coloured, easy to see markings on animals which are poisonous, bad tasting, etc. Predators usually only eat one or two such organisms before learning to leave them alone. Compare mimicry, protective colouration, camouflage.

**wart** Harmless but unsightly small growths from the skin. 1 Common warts are caused by a contagious virus. 2 Genital warts are a ⋆sexually transmissible disease (STD) and occur on the skin around the penis or vagina, or around the anus.

**wasp** Any of the numerous insects of the order Hymenoptera, with four membrane-like wings, a very thin waste and a sting at the end of the abdomen of the female. Some wasps are solitary (e.g. mud-daubers which build egg-chambers of mud into which they lay their egg onto a paralysed caterpillar), while others are social (e.g. paper-nest wasps). See Appendix 1.

**waste** Material eliminated from the body of an organism, including undigested parts of food, unwanted products of metabolism, etc. See faeces and urine.

**water** The compound $H_2O$, which makes up most of living things (e.g. 70% of humans). It is crucial to life because of its ability to form solutions, its high chemical stability, and its ability to absorb or take in much heat with relatively little change of temperature (see specific heat, latent heat).

**water balance** The need for living things to ensure correct concentrations of solutions in the body, and hence the correct amounts of water. In humans water intake is from drinking, from food, and some from metabolism. Water loss occurs in urine, faeces, and perspiration. Animals living in the oceans and in deserts have special adaptations to control the amount of water in their bodies. See osmosis, hypothalamus.

**water cycle** The circulation of

water between the atmosphere, land and oceans of the earth. See Fig. W1.

Water in the atmosphere, in the form of clouds, condenses to form rain, snow, etc. The water which falls on the ground may enter streams which eventually flow to the oceans. Other water re-enters the atmosphere by evaporating off the ground, streams and lakes. Water also re-enters the atmosphere via ★transpiration in plants. Some water soaks into the earth to form groundwater. Water that reaches the oceans also evaporates back into the atmosphere, to form clouds and eventually start the cycle again.

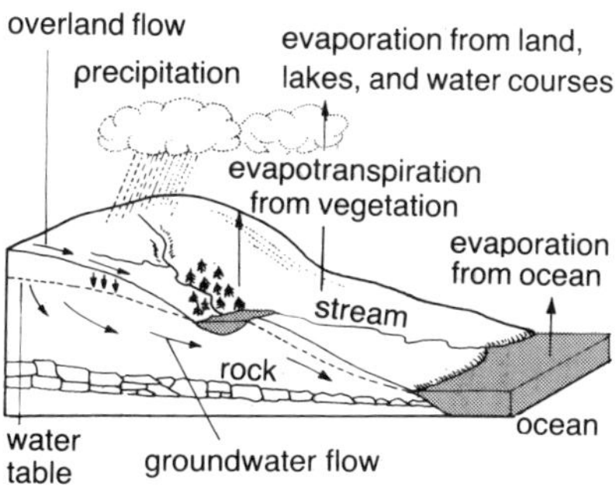


**Fig. W1 Water cycle**

**water strider** Insects in the order Hemiptera, family Gerridae, which are also called pond skaters. They are able to walk on the surface of water using surface tension to support their weight. See Appendix 1.

**water vascular system** In echinoderms, a system of canals containing water which is used to move the ★tubed feet. A system of small 'pumps' operates this hydraulic system.

**waterboatmen** Insects in the order Hemiptera. Two groups are often called waterboatmen. The backswimmers (family Notonectidae), swimming upside-down using their long hind feet, should not be called waterboatmen. True waterboatmen are in the family Corixidae, and swim upright. See Appendix 1.

**wattle** 1 Of plants, the genus *Acacia*, Family Mimosaceae, a group of 600 or more tree and shrub species common throughout Australia. Wattle is a ★legume, so can be important to the fertility of the soil. The golden wattle, *A. pyenantha*, is found on Australia's coat-of-arms. See Appendix 1. 2 A fleshy outgrowth on the head of a reptile or bird. e.g. Wattlebirds, a large honeyeater.

**wax** A group of ★lipids produced by both plants (carnauba wax, myrtle wax) and animals (bees wax, sperm whale oil, lanolin). In both plants and animals waxes often cover the body, thus helping to make them waterproof and cut water loss. e.g. Waxes on the leaves of ★xerophytes, and the ★cuticle of insects. Bees use wax to construct their honeycomb.

**weather** The state of conditions in the atmosphere, including ★humidity, precipitation (rain, snow, hail), temperature, cloud cover, and wind at any one place and time. Compare climate.

**weathering** The decay and breakdown of rocks by both mechanical means (e.g. abrasion by wind-blown or ice- or water-carried particles, extreme temperature changes) and chemical means (e.g. action of acidic water from the atmosphere). Weathered particles may remain in place to form ★soils or be ★eroded to other locations.

**web** A thin membrane-like structure. 1 The silken webs constructed by spiders and the larvae of some insects. Spider webs are classed as orbs, sheets, or tangle-webs. See also spinnerets. 2 A thin membrane between toes. e.g. Platypus, wading birds, tortoises.

**weed** Any plant growing in a place where it prevents proper growth of desired plants, or is otherwise undesired.

**weedkiller** See herbicide.

**weevil** Beetles belonging to the insect family Curculionidae, with an elongated 'snout', and which cause economic loss when they can reach grains, nuts, etc.

**weight** The force with which the *mass of an object is attracted by gravity. It is scientifically measured in units called Newtons, one Newton being the force with which gravity attracts one kilogram when at sea-level. However, in everyday use, the term weight is used when mass is meant.

**wetland** Land or areas such as tidal flats or swamps that contain much soil moisture. Vital for preserving *biodiversity.

**whale** Any of the air-breathing aquatic mammals in the order Cetacea, the largest living mammals. There are two suborders, the toothed whales (e.g. killer whales, sperm whales) and the baleen whales (e.g. humpbacks, and blue whales). The blue whale is the largest animal that has ever lived, often reaching a weight of 100 tonnes. See Appendix 1.

**whirligig beetle** Beetles (order Coleoptera) in family Gyrinidae, which usually swims in circles on the surface of lakes and streams.

**white ant** See **termite**.

**white blood cell** Also called a **leucocyte**. White blood cells move around in the bloodstream and are part of the *immune system, being involved in helping protect the body against toxins, and viral and bacterial infections, as well as the removal of worn out red blood cells (Fig. E6).

White blood cells have a nucleus, but carry no *haemoglobin. They move around the body in a manner similar to that of an *amoeba, using *pseudopodia. In humans there are normally 5000 to 10 000 white blood cells per cubic millimetre of blood. There are three main types of white blood cells: **a Granulocytes** or *polymorphs with a granular cytoplasm. The *neutrophils are most numerous, and are responsible for the defence against bacteria. Eosinophils and basophils have functions related to allergies that are not fully understood. **b Monocytes** do not have a granular cytoplasm. They form *macrophages which help remove invaders, as well as worn out red blood cells. **c *Lymphocytes** do not have a granular cytoplasm. B-lymphocytes (*B-cells) and T-lymphocytes (*T-cells) form part of the *immune system, helping to recognise invaders and producing *antibodies which coat the invader so that neutrophils and macrophages can remove them. Compare erythrocytes.

**white matter** That part of the brain and spinal cord made up mainly of *axon fibres covered with *myelin. Forms much of the interior of the brain, and the outside part of the spinal cord. The white matter is bundles of nerves that connect different parts of the nervous system. Compare grey matter.

**whooping cough** An infectious disease of the mucous membrane lining the respiratory system that results in coughing followed by a characteristic whooping sound. Caused by the bacterium *Bordetella pertussis*, which releases toxins. See also triple vaccine.

**whorl** **1** The circular arrangement of three or more parts, such as leaves, petals, sepals, etc., around a stem or axis. See perianth. **2** One of the turns of a spiral shell such as that of a periwinkle or snail. **3** One of the shapes of a human fingerprint due to the ridge pattern present.

**wild type** The characteristics of an organism that are most common in natural or wild populations of that organism. Especially applied to the genetic characteristics or genotype. The wild-type allele of a *gene is often represented by a + rather than a letter. e.g. *Drosophila* in the wild type has red eyes and normally shaped wings.

**wilderness** An area only inhabited by *wildlife. Especially applied to National Parks where wilderness areas are set aside and entry by humans is severely restricted.

**wildlife** Animals living in their natural environment.

**wilting** Drooping and limpness in the stems and leaves of plants. Occurs in fleshy stems and leaves after excessive loss of water causes the cell cytoplasm to shrivel, and the plant tissue loses rigidity. Compare turgid.

**wind pollination** The transfer of *pollen by the wind. Wind-pollinated flowers usually produce large amounts of smooth-walled pollen, and each grain is very small. Such flowers are usually inconspicuous and often green, but the *stigmas are large and feathery. e.g. Grasses. Gymnosperms are also wind pollinated.

**wing** 1 In animals, the organ of flight. Wings range from the thin membranes of insects, to the membrane-like skin of bats and the feather-covered appendages of birds. 2 In plants, the thin membranes attached to pollen, seeds and fruits to assist *dispersal by the wind.

**wisdom teeth** In humans, the last four molars to appear, one each at the rear of each side of each jaw.

**witchetty grub** The larvae of a number of different types of wood-boring moths and beetles, especially that of the wood moth (family Cossidae), *Xyleutes leucomochla*, whose larvae form a silk-lined tunnel in the soil where it feeds on the roots of *wattle trees.

**withdrawal symptom** Any nervousness, pain or otherwise distressing symptom which results from an addict ceasing to use the drug of their addiction. e.g. Nicotine of cigarettes, alcohol, heroin. See also dependence.

**wobbegong** A harmless shark that lives on the ocean bottom along the eastern coast of Australia, named *Orectolobus maculatus*.

**Wollemi pine** A new genus of *conifer, family Araucariaceae, found 200 km west of Sydney in 1995.

**woman** The adult female of genus *Homo*.

**womb** See **uterus**.

**wombat** Any of three types of large burrowing herbivorous and mainly nocturnal marsupials in family Vombatidae. The hairy nosed wombat, *Lasiorhinus latifrons*, is the animal emblem of South Australia. See Appendix 1.

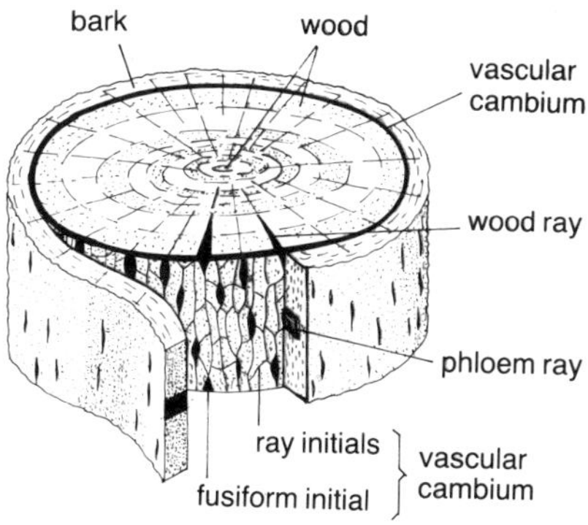


**Fig. W2 Wood**

**wood** 1 A term generally applied to the hard trunks of trees and shrubs. 2 The relatively hard *secondary xylem of dicotyledons and conifers. See Fig. W2. It is composed of two parts. *Sapwood makes up the outer layers that contain the living ★xylem tissue. The inactive innermost parts are called ★heartwood, which has been made harder by tannins, etc., and contains the annual ★growth rings. There are two main types of wood. ★Softwood is obtained from conifers such as the radiata pine, widely planted in Australia. Native eucalypts produce ★hardwoods due to a high proportion of strengthening fibres and vessels in their timber.

**woodchipping** The reduction of the wood of trees to small chips approximately 2 cm square so that it can be readily transported and used to produce paper.

**woodland** A habitat, common west of Australia's eastcoast mountains, with grassland and scattered trees. The trees usually have spreading crowns with a short ★bole, and are separated by at least a distance equal to the diameter of their crowns. Compare scrubland, grassland and mallee.

**woollybutt** Any of several species of eucalypt with thick fibrous bark only on the trunk. e.g. *Eucalyptus longifolia* in New South Wales. Alpine ash in Victoria is sometimes called woollybutt. Compare blackbutt.

**worm** Any long, slender, soft-bodied invertebrate in the phyla Platyhelminthes (flatworms), Nematoda (roundworms), Annelida (segmented worms), and a number of smaller groups.

**wound healing** The response of the body to damage of the skin including ★blood clotting to stop bleeding, ★inflammation to protect against invasion by ★pathogens and regrowth of ★tissues to replace the damaged skin. See also platelets.

# Xx

**X chromosome** One of the ★sex chromosomes. In humans, females have two X chromosomes, while males have only one. The X chromosome also contains many genes compared to almost none on the Y chromosome (see below). See also Barr body.

**X-ray** Part of the electromagnetic spectrum with wavelength shorter than ultraviolet light but longer that gamma rays. X-rays are potentially dangerous as they can cause cancer. Apart from their use in medical ★diagnosis (e.g. studying broken bones), X-rays are widely used to help determine the structure of complex macromolecules in a process called X-ray diffraction. See also CT scanning.

**xeromorphic** Of plants, containing features which help them survive in arid habitats. e.g. Features for reducing water loss from leaves (waxy surface, hidden stomata), to help leaves withstanding wilting (thickened cell walls), and for storing water (swollen stems or leaves). See also xerophyte, sclerophyll and structural adaptations.

**xerophyte** A plant which has adaptations that allow it to survive in places often short of water. e.g. Sunken stomata, thick waxy cuticle, highly reflective leaves, leaves rolled or reduced to spines or needles. Due to frequent droughts, most plants in the Australian interior are xerophytes. Plants living in sandy areas are also usually xerophytes. e.g. Sand dunes, sandy soils on sandstone country such as near Sydney. See also xeromorphic and sclerophyll. Compare halophytes, hydrophytes, mesophytes.

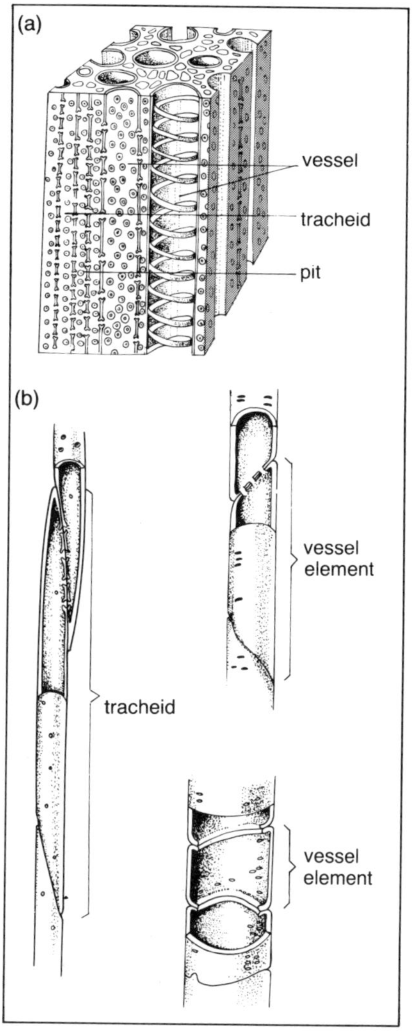


**Fig. X1 Xylem (a) Xylem tissue (b) Xylem cells**

**xylem** The tissue in ★vascular plants which carries water and dissolved minerals from the roots to the leaves. See Fig. X1.

Xylem is composed of ⋆vessels and ⋆tracheids (carrying water), ⋆fibres (for strengthening) and parenchyma packing cells. Together with the ⋆phloem it makes up the vascular tissue. *See* Fig. V1. Plants which grow in width by ⋆secondary thickening produce new xylem (wood) at the ⋆cambium (hence it is called secondary xylem). During periods of rapid growth (often spring) the new xylem cells are large with relatively thin walls, and thus light in colour, but they decrease in size and have relatively thicker walls as the season progresses and growth slows. This produces a pattern of ⋆growth rings. As the plant grows, the old xylem ceases to function as a conductor of fluids, and the ⋆wood hardens as tannins, resins, etc. are carried in by ⋆rays and deposited.

# Yy

**Y chromosome** One of the ⋆sex chromosomes. In humans, males have one Y chromosome and one ⋆X chromosome. In comparison to the X chromosome, the Y chromosome carries very few genes. Those present primarily help determine maleness. See also sex-linked.

**yabby** A freshwater ⋆crayfish, genus *Cherax* in class Crustacea, found throughout Australia. See Appendix 1.

**yallara** The lesser ⋆bilby or lesser rabbit-eared ⋆bandicoot, *Macrotis leucura*, found in the sand dune country of central Australia. See Appendix 1.

**yam** A climbing vine, genus *Dioscorea*, or more often its root which is starchy and widely used for food in the tropics.

**yeast** Any of a number of single-celled ⋆fungi, which reproduce by ⋆budding. They occur naturally on human skin (where they can cause the disease thrush), and on the surface of fruits such as grapes, but are also widely used in industry. In brewing and winemaking, the yeasts secrete enzymes which, during ⋆fermentation, convert sugars into alcohol and carbon dioxide. In baking, the carbon dioxide is used to make the dough 'rise' before baking.

**yolk** The food stored in an egg for the use of a growing embryo. The food stored mainly as protein and fat granules. The eggs of ⋆oviparous animals (e.g. birds) are usually relatively large. Compare albumin.

**yolk sac** A sac containing yolk that hangs from the under (ventral) surface of the embryos of various animals (e.g. sharks, bony fish, reptiles and birds). In certain animals, the yolk-sac is still visible after it hatches from its egg.

# Zz

**zero population growth** Called ★ZPG for short.

**zidovudine** Called AZT for short. A chemical used to treat ★AIDS. AZT interferes with the replication of the virus when it blocks the action of ★reverse transcriptase as it converts the ★HIV's RNA into DNA. It is not a cure but extends the life of sufferers.

**zonation** The existence of populations of organisms in ★zones. See Fig. Z1. e.g. **a** Zonation of a rock platform, determined by factors such as competition, the force of crashing waves, and the ability to withstand the drying heat of the Sun, and of freshwater during rain. **b** The zonation of Australia's vegetation is mainly due to rainfall, although patches of rainforest will grow with lesser rainfall if soil quality is high enough. See Fig. Z2.

**zone** An area that contains a particular set of organisms which is different from that on either side of it. Such zones of living things usually arise because of variations in environmental (physical and biological) conditions. e.g. Rock platforms can be divided into a number of zones such as kelp, cunjevoi, *Galeolaria*, barnacle zones, and the periwinkle zone. See also cline and succession. See Fig. Z1.

**zoo** An area where animals are kept for the public to look at. Compare botanical garden.

**zoogeography** The study of the geographical distribution of animals. Compare biogeographical realms.

**zoology** The scientific study of animals and animal life.

**zooplankton** The animal part of the ★plankton, including their eggs, larvae and adults. Many such animals are microscopic, but others are relatively large. e.g. ★Krill. Compare phytoplankton.

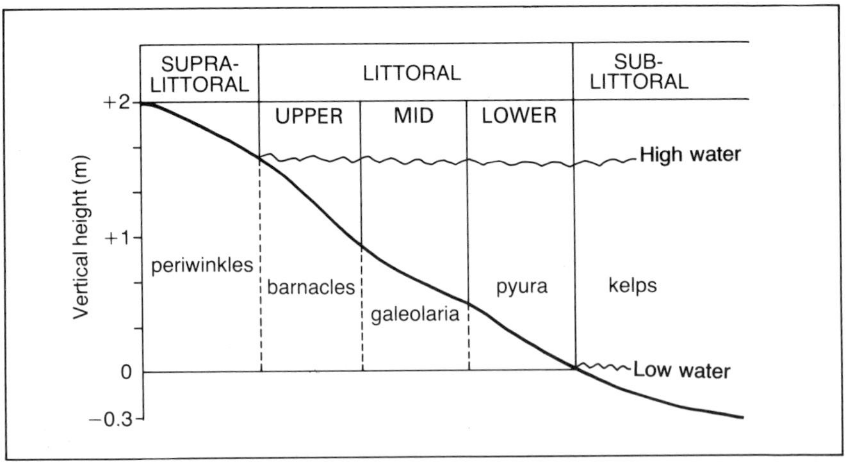


**Fig. Z1 Zonation of a rock platform**

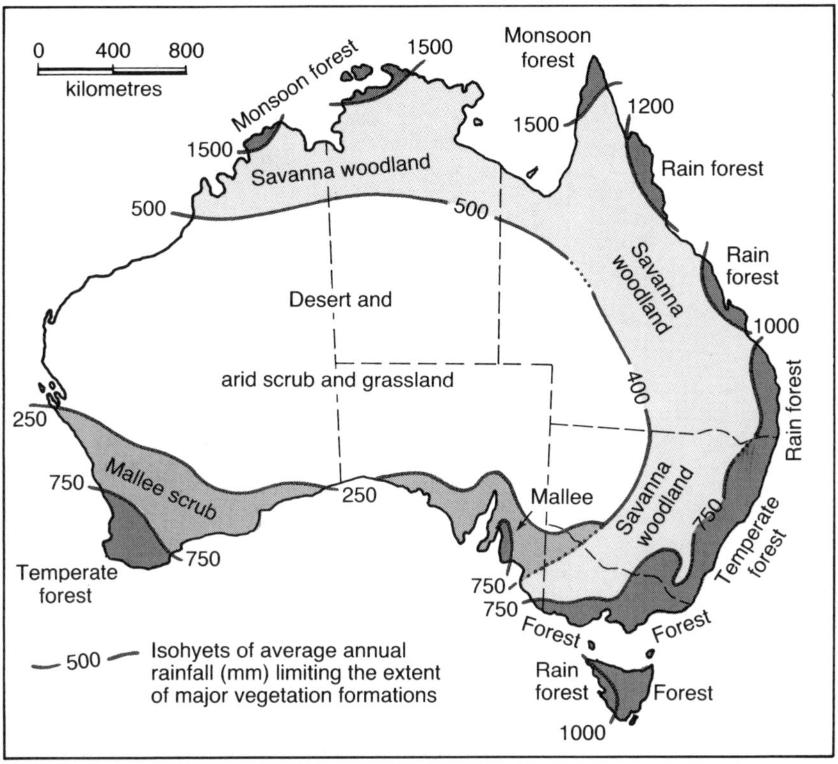


**Fig. Z2 Zonation of vegetation**

**ZPG** Short for **zero population growth**. The idea that humans should keep the population to present levels (zero growth) by only allowing the birth of enough children to replace their parents. i.e. Calculated at 2.11 children per family, on average.

**zygomorphic** Of flowers, bilaterally symmetrical or can be divided into two similar halves.

**zygospore** A thick-walled spore which is a *zygote, and which has the capacity to 'rest' until environmental conditions are suitable for growth to start again.

**zygote** The cell that results when a male and female gamete (sperm and ovum) join together, restoring the normal number of chromosomes. The fertilised egg or ovum.

**zygotene** A stage in *prophase I or meiosis where *homologous chromosomes come together forming pairs (see pairing).

**zymase** The group of enzymes present in yeasts which bring about conversion of sugar into alcohol during *fermentation.

# Appendix 1: A classification of living things

(This classification covers only the main groups, and is thus incomplete)

## Kingdom Plantae

| Phylum (Division) | Class | Subclass | Family | Comments |
|---|---|---|---|---|
| Tracheophyta | | | | Vascular plants: Contain vascular tissues; about 250 000 species. |
| | Angiospermae | | Only a few of the more than 300 families are listed below. | Flowering plants or angiosperms: Reproduce with flowers, seeds, fruits; about 200 000 species. |
| | | Monocotyledonae | | Monocots: One cotyledon in seed; usually parallel leaf veins; flower parts in multiples of 3; scattered vascular bundles in stem which are mainly soft (non-woody). |
| | | | Amaryllidacae | Amaryllis family e.g. Murray lily. |
| | | | Cyperaceae | Sedge family e.g. Rushes. |
| | | | Poaceae (Gramineae) | Grass family e.g. Wheat, rice, sugar cane, couch grass, bamboo. |
| | | | Iridaceae | Iris family e.g. Gladiolus, crocus. |

| Phylum (Division) | Class | Subclass | Family | Comments |
|---|---|---|---|---|
| | | | Liliaceae | Lily family e.g. Onion, garlic, tulip. |
| | | | Orchidaceae | Orchid family e.g. Greenhoods, slipper. |
| | | | Arecaceae (Palmae) | Palm family e.g.Cabbage palm, date palm. |
| | | | Xanthorrhoea | Grass tree family e.g. Grass trees. |
| | | Dicotyledonae | | Dicots: Two cotyledons in seed; usually network leaf vein pattern; flower parts in multiples of 4/5; vascular bundles around edge of stem which is usually woody. |
| | | | Caesalpiniaceae | Cassia family e.g. *Cassia*. |
| | | | Chenopodiaceae | Saltbush family e.g. Bluebush, saltbush. |
| | | | Asteraceae (Compositae) | Daisy family e.g. *Dahlia*, *Zinnia*, dandelion, lettuce, sunflower. |
| | | | Brassicaceae (Cruciferae) | Cabbage family e.g. Cauliflower, broccoli, raddish, stocks. |
| | | | Epacridaceae | Native heath family e.g. *Epacris*. |
| | | | Geraniaceae | Geranium family e.g. *Pelargonium*. |
| | | | Goodeniaceae | Goodenia family e.g. *Dampiera*. |

| Phylum (Division) | Class | Subclass | Family | Comments |
|---|---|---|---|---|
| | | | Lamiaceae (Labiatae) | Mint family e.g. Peppermint, sage, spearmint, thyme, sage. |
| | | | Malvaceae | Mallow family e.g. *Hibiscus*, cotton. |
| | | | Mimosaceae | Wattle family e.g. *Acacia*. |
| | | | Myoporaceae | Eremophila Family: e.g. *Eremophila*. |
| | | | Myrtaceae | Tea-tree family e.g. Bottlebrush, *Leptospermum*, *Eucalyptus*. |
| | | | Fabaceae (Papilionaceae) | Pea family e.g. Sturt's desert pea, 'eggs-and-bacon', clover, vetch. |
| | | | Proteaceae | Banksia family e.g. *Grevillea*, *Hakea*. |
| | | | Rutaceae | Citrus family e.g. *Boronia*, lemons. |
| | | | Solanaceae | Tomato family e.g. Potato, *Petunia*. |
| | | | Apiaceae (Umbelliferae) | Parsley family e.g. Flannel flower, carrots, celery, hemlock. |
| | Coniferopsida | | | Conifers: Woody plants with needle-like leaves; male cones produce much wind-carried pollen; seeds develop within female cones; about 700 species. e.g. Redwood, spruce, cypress pine, Bunya pine, Huon pine, Wollemi pine. |

| Phylum (Division) | Class | Subclass | Family | Comments |
|---|---|---|---|---|
| | Ginkgopsida | | | Ginkgo or maiden-hair tree: Fan shaped leaves; deciduous; two types of spore on separate plants; seeds. Only one species, in China. |
| | Gnetopsida | | | Rare plants, none native to Australia. 60 species. |
| | Cycadopsida | | | Cycads: Palm-like trees; two types of spore on separate plants; some cones large and fleshy. 100 species. e.g. *Macrozamia.* |
| | Filicopsida | | | Ferns: Sporophyte the main plant with usually the one type of spore being carried on the fronds; gamete plants small and inconspicuous; about 10 000 species. |
| | Sphenopsida | | | Horsetails: Sporophyte the main plant, spore being produced in cone like spore cases. Gametophyte small. About 30 species. |
| | Psilopsida | | | Fork-ferns: Very simple vascular system; no roots; sporophyte the main plant; gametophyte under ground. 2 genera. |
| | Lycopsida | | | Clubmosses and quillworts: Main plant the sporophyte with roots, stems and small leaves; two types of spore. About 1100 species. |

| Phylum (Division) | Class | Subclass | Family | Comments |
|---|---|---|---|---|
| Bryophyta | | | | Non-vascular terrestrial The gametophyte plant is the most conspicuous, the sporophyte plant being dependent on it. All live in mainly moist places. About 24 000 species. |
| | Hepaticopsida | | | Liverworts: Gametophytes a flat 'thallus'. About 8500 species. |
| | Bryopsida (Musci) | | | Mosses: Gametophytes usually erect. About 15 000 species. |
| | Anthocerotopsida | | | Hornworts: Like liverworts but the sporophyte plant is independent. About 50 species. |
| Rhodophyta | | | | Red algae or seaweeds: Multi-cellular, macroscopic, marine. Photosynthetic but chlorophyll masked. e.g. *Corallina*. About 2500 species. |
| Phaeophyta | | | | Brown algae or seaweeds: Marine, multicellular, macroscopic. Photosynthetic but chlorophyll masked e.g. Kelp, Neptune's necklace. About 1000 species. |
| Chlorophyta | | | | Green algae: Unicellular, colonial and multicellular forms. Distinct nucleus and cellulose cell walls. Photosynthetic. e.g. Sea-lettuce (*Ulva*) and other green seaweeds, *Volvox*. About 6000 species. |

## Kingdom Protista

| Phylum (Division) | Class | Order | Family | Comments |
|---|---|---|---|---|
| Protozoa | | | | Protozoans: Unicellular or colonial eukaryotic organisms, without chlorophyll or cell walls. Usually microscopic. About 20 000 species. |
| | Mastigophora (Flagellata) | | | Flagellates: Possess flagella. e.g. *Typanosoma*. About 2000 species. |
| | Sarcodina (Rhizopoda) | | | Amoeboids: Move by pseudopodia. Either naked cells (e.g. *Amoeba*), or intricate shell or skeletons (e.g. Foraminifera, Radiolaria). About 8000 species. |
| | Sporozoa (Apicomplexa) | | | Sporozoans: Microscopic with no means of locomotion. Complex life histories. e.g. Malaria. About 2000 species. |
| | Ciliata (Ciliophora) | | | Ciliates: Move and catch food using cilia. e.g. *Paramecium*, *Stentor*. About 5000 species. |
| Chrysophyta | | | | Golden algae and diatoms: Microscopic, unicellular, or as filaments or colonies. Silica in cell walls or as a skeleton. Photosynthetic but chlorophyll masked. About 6000 species. |
| Pyrrophyta | | | | Golden-green algae/dinoflagellates: Microscopic, unicellular or colonial with cellulose cell walls. |

| Phylum (Division) | Class | Order | Family | Comments |
|---|---|---|---|---|
| | | | | Photosynthetic but chlorophyll masked. Over 1000 species. |
| Euglenophyta | | | | Euglenoids: Microscopic, mostly unicellular with flagella but no cell wall. Green and photo synthetic. About 450 species. |

## Kingdom Monera

| Phylum (Division) | Class | Order | Family | Comments |
|---|---|---|---|---|
| Cyanobacteria | | | | Blue-green bacteria (or algae): Prokaryotic; cells very small as single cells or colonies; chlorophyll often masked. About 1500 species. |
| Eubacteria | | | | True bacteria: Very tiny just visible with light microscopes. Prokaryotic, usually unicellular. Some photosynthetic, with others chemo-synthetic. About 2000 species. |
| Archaeobacteria | | | | Ancient bacteria: Prokaryotic; very small; found in extreme environments; not well researched. About 100 species. |

## Kingdom Fungi

| Phylum (Division) | Class | Order | Family | Comments |
|---|---|---|---|---|
| Amastigomycota | | | | True Fungi: Plant-like with cell walls, but no chlorophyll (heterotrophic). No vascular tissues; hyphae either individual or in compact fruiting bodies. Reproduce by spores. About 250 000 species. |
| | Ascomycetes | | | Sac fungi: Varies from unicellular (e.g. yeasts), thread-like (e.g. mildews, wood-rot fungi), to large forms (cup fungi, edible truffles, morels). All produce spores in small sacs. About 200 000 species. See also lichen below. |
| | Basidiomycetes | | | Club fungi: Usually larger fungi such as mushrooms, puff balls, bracket fungi, but also some rusts and smuts. About 35 000 species. |
| | Deuteromyctes | | | Fungi imperfecti: Fungi that cannot be readily classified above as too little is known about their life cycle. About 25 000 species. |
| (Lichens) | | | | Not a 'true' classification group as they are a symbiotic association of fungi (often Ascomycetes) and green algae or blue-green bacteria. |
| Gymnomycota | | | | Slime moulds: Life cycle includes stages where they form masses of cells, a large mass of protoplasm with many nuclei inside, and spore producing structures. About 500 species. |

# Appendix 1



## Kingdom Animalia

| Phylum | Class | Subclass | Order | Comments |
|---|---|---|---|---|
| Porifera | | | | Sponges: Mostly marine with adult attached to a solid object. Body wall has two layers of cells, with spicules sometimes between them for support. Pores allow entry of water. About 1000 species. |
| Cnidaria (Coelenterata) | | | | Coelenterates: Bag-like body with tentacles around the single opening to the stomach; tentacles carry pneumatocysts or stinging cells; polyp and medusa forms. About 9000 species. |
| | Hydrozoa | | | Hydroids: Polyps either single or colonial; undivided stomach. About 3000 species. e.g.Freshwater *Hydra*, saltwater *Obelia*. |
| | Scyphozoa | | | Jellyfish: Floating medusa stage usually seen; divided digestive cavity. e.g. *Aurelia*, Box jellyfish (sea-wasp). About 200 species. |
| | Anthozoa | | | Anemones and corals: Attached polyps; corals secrete limy skeleton; Anemones generally larger; divided stomach. About 6000 species. |
| Ctenophora | | | | Comb-jellies: Looks a little like a jellyfish, but lacks the stinging cells in the tentacles. About 100 species. |

| Phylum | Class | Subclass | Order | Comments |
|---|---|---|---|---|
| Platyhelminthes | | | | Flatworms: Parasitic or free-living; flat shape with a single opening to the digestive cavity; no skeletal, circulatory or respiratory systems. About 15 000 species. |
| | Turbellaria | | | Planarians: Free-living mostly marine, but some freshwater; cilia on outside of body. About 3000 species. |
| | Trematoda | | | Flukes: Parasitic with no cilia on body; suckers to help attachment. About 8000 species. |
| | Cestoda | | | Tapeworms: Parasitic with body segments; no cilia or digestive system. About 4000 species. |
| Nemertea | | | | Ribbon-worms: Flat, unsegmented worms; digestive system has two openings. About 900 species. |
| Nematoda | | | | Roundworms: Parasitic or free-living; cylindrical, unsegmented body with two openings to the digestive system. About 80 000 species. |
| Rotifera | | | | Rotifers: Microscopic with mouth and anus separate; cilia around mouth; not worm-like. About 2000 species. |
| Bryozoa | | | | Bryozoans or moss animals: Colonial and marine; polyps with U-shaped digestive tract having tentacles around the mouth. About 4000 species. |

# Appendix 1



| Phylum | Class | Subclass | Order | Comments |
|---|---|---|---|---|
| Brachiopoda | | | | Lamp shells: Body attached to substrate by stalk, and enclosed by two calcareous shells. About 335 species. Many fossil species. |
| Chaetognatha | | | | Arrowworms: Marine, worm-like animals with dart-shaped body; usually part of plankton. About 70 species. |
| Mollusca | | | | Molluscs: Unsegmented, soft-bodied, usually a dorsal shell with ventral muscular foot or tentacles. Shell secreting mantle covers internal organs. Complex digestive, circulatory and nervous systems. About 90 000 species. |
| | Amphineura (Polyplacophora) | | | Chitons: Marine and covered with shell in 8 plates. Radula present. About 100 species. |
| | Gastropoda | | | Gastropods: Single shell twisted, if present. Body twisted inside. Head usually distinct and with eye- stalks. Radula present. e.g. Snails, slugs, periwinkles, cowries, limpets, sea-hares. About 70 000 species. |
| | Bivalvia (Pelecypoda) | | | Bivalves: Two shells hinged, flat-shaped foot protrudes to help in burrowing e.g. Clam, muscle, oyster. About 9000 species. |
| | Cephalopoda | | | Cephalopods: Tentacles with suckers replace muscular foot; teeth instead of radula; water jets for movement; excellent vision. e.g. Squid, octopus, nautilus. About 600 species. |

| Phylum | Class | Subclass | Order | Comments |
| --- | --- | --- | --- | --- |
| | Scaphopoda | | | Tusk or tooth shells: Long shell open at both ends; tentacles protrude from one end. About 400 species. |
| Annelida | | | | Segmented worms: Long, basically cylindrical worms segmented both internally and externally; Nervous, digestive and circulatory systems present; appendages, if present, not jointed. More than 9000 species. |
| | Polychaeta | | | Segmented marine worms: Paddle-like appendages on each body segment; head with tentacles; live in burrows or tubes. e.g. Beach worms, sand worms, bristle worms. More than 5000 species. |
| | Oligochaeta | | | Earthworms: Segmented body almost smooth; bristles assist movement. Some giant species. About 2500 species. |
| | Hirudinea | | | Leeches: Similar to earthworms but with suction caps each end; move end-over-end with these caps; feed on blood. About 500 species. |
| Onychophora | | | | Peripatuses: Caterpillar like with soft bodies and short legs with claws. Has annelid-like excretory system and internal segments. Has arthropod-like open circulation, and insect-like respiration. About 80 species. |
| Arthropoda | | | | Arthropods: Jointed legs; body segments (which may be fused). Body and legs covered with exoskeleton; open circulatory system. At least 800 000 species. |

| Phylum | Class | Subclass | Order | Comments |
|---|---|---|---|---|
| | Crustacea | | | Crustaceans: Two major body parts, further segmented; two pairs of antennae; at least 10 major appendages, with other smaller swimmerets; respiration by gills. About 40 000 species. |
| | | Branchiopoda | | e.g. Shrimps (e.g. brine, fairy), water fleas. |
| | | Ostracoda | | Minute freshwater and marine forms. |
| | | Copepoda | | Microscopic, free-living forms, often sold as 'fish food' for aquariums. e.g. Cylops. |
| | | Cirripedia | | e.g. Barnacles. |
| | | Malacostraca | | Larger crustaceans, usually with substantial carapace. |
| | | | Isopoda | e.g. Garden slaters, pill bugs, etc. |
| | | | Euphausiacea | e.g. Krill. |
| | | | Decapoda | e.g. Larger 10-legged crustaceans. |
| | | | SUBORDER Natantia | Swimming decapods e.g. Prawns and shrimp. |
| | | | SUBORDER Reptantia | Walking/crawling decapods e.g. Lobsters, crayfish, crabs. |
| | Arachnida | | | Arachnids: Two major body parts (cephalothorax and abdomen); few segments no antennae; 8 walking legs. About 30 000 species. |

| Phylum | Class | Subclass | Order | Comments |
|---|---|---|---|---|
| | | | Scorpionida (Scorpiones) | Scorpions: Long segmented body with poison stings on the end. |
| | | | Phalangida (Opiliones) | Harvestmen: Cephalothorax and abdomen joined for their full width, forming a small rounded body; very long legs. |
| | | | Acarina (Acari) | Mites and ticks: Small rounded bodies with fused abdomen and cephalothorax. Many feed on blood or are ectoparasites, while some are scavengers. |
| | | | Araneida (Araneae) | Spiders: Two major body parts without further segmentation; no antennae; 8 walking legs, small appendages near mouth assist in feeding; spinnerets; external digestion of food. |
| | | | Xiphosaura | King-crabs or horse-shoe crabs: Heavily armoured carapace over body; body ends in long spine. (Class Merostomata in some classifications.) |
| | Diplopoda | | | Millipedes: Many body segments; two pairs of legs on most segments and one pair of antennae on head. About 7500 species. |
| | Chilopoda | | | Centipedes: Many body segments; one pair of legs on most segments; one pair of long antennae; one pair of jaws on first segment. About 2000 species. |

| Phylum | Class | Subclass | Order | Comments |
| --- | --- | --- | --- | --- |
| | Insecta | | | Insects: One pair antennae; three body parts (head, thorax, abdomen) which may be divided into further segments; many species winged as adults. At least 750 000 species. |
| | | | Anoplura | Lice: Parasitic; sucking mouthparts. |
| | | | Blattodea | Cockroaches: Flat bodies; long antennae; rapid runners. |
| | | | Coleoptera | Beetles: Front pair of wings a hard covering with rear flight-wings folded beneath. e.g. Ladybirds, fire-flies; wood-borers, weevils. |
| | | | Collembola | Springtails: 'Spring' in tip of abdomen; found in leaf litter. |
| | | | Dermaptera | Earwigs: Large cerci abdomen tip. |
| | | | Diptera | Flies, mosquitoes, midges: One pair of wings; sucking mouthparts. |
| | | | Ephemeroptera | Mayflies: Very delicate, very long 'tails'; usually seen as aquatic larvae. |
| | | | Hemiptera | Bugs: Sucking mouthparts; two pairs of wings. e.g. Bugs, cicadas, water-striders; lerps; aphids. |
| | | | Hymenoptera | Wasps, ants, bees, sawflies: Front wings much larger than rear, and hooked together (if present); many social species; defensive sting. |

| Phylum | Class | Subclass | Order | Comments |
|---|---|---|---|---|
| | | | Isoptera | Termites or 'white ants': Soft-bodied social insects; feed on dry wood or grass. |
| | | | Lepidoptera | Butterflies and moths: Two pairs of wings covered with soft scales; coiled sucking mouthparts. |
| | | | Mantodea | Praying mantids: Large predatory insects with grasping forelegs and chewing mouthparts. |
| | | | Neuroptera | Lacewings (antlions): Slow flying with large lace-like wings; larvae construct conical sand traps. |
| | | | Odonata | Dragonflies and damselflies: Long bodied predatory insects with two pairs of transparent wings. |
| | | | Orthoptera | Grasshoppers, locusts, crickets: Large hindlimbs for jumping; winged; chewing mouthparts. |
| | | | Phasmatodea | Stick-insects, leaf-insects: Large, long and thin (stick-like) or long and flat (leaf-like). Chewing mouthparts. |
| | | | Siphonaptera | Fleas: Small; sucking mouthparts; wingless; external parasites. |
| | | | Thysanoptera | Thrips: Tiny insects with slender wings; sap feeders with piercing mouthparts. |
| | | | Thydanura | Silverfish: Small, always wingless with three long bristles on end of soft body. |

| Phylum | Class | Subclass | Order | Comments |
| --- | --- | --- | --- | --- |
| | | | Trichoptera | Caddis flies: Moth-like adults with aquatic larvae. |
| Echinodermata | | | | Echinoderms: All marine; internal limy skeleton often with spines; unsegmented; water vascular system. About 6000 species. |
| | Crinoidea | | | Sea lilies and feather stars: Cup-shaped body with mouth facing upwards between branched feathery arms used to catch food. About 550 species. |
| | Asteroidea | | | Sea stars: Flattened, star shaped body with mouth on underside of central disc (not always obvious). About 1500 species. |
| | Ophiuroidea | | | Brittle stars and basket stars: Central disc obvious with thin flexible arms extending from it. Mouth at base. About 2000 species. |
| | Echinoidea | | | Sea urchins and sand dollars: Spherical or disc-shaped with mouth at base; no arms but spines projecting from rigid skeleton of interlocking plates. About 950 species. |
| | Holothuroidea | | | Sea cucumbers: Cylindrical with tentacles around the mouth at one end. About 1100 species. |

## Animalia—Phylum Chordata

| Subphylum | Class | Order | Family | Comment |
|---|---|---|---|---|
| | | | | Chordates: Hollow dorsal nerve tube, a stiff notochord, and pharyngeal slits occur at least during some stage of development. About 46 000 species. |
| Hemichordata | | | | Acornworms: Marine worm-like mammals with three parts to their body. About 100 species. |
| Urochordata | | | | Sea-squirts or tunicates: Adults attached filter feeders, while larvae are free-swimming. e.g. cunjevoi. About 1250 species. |
| Cephalochordata | | | | Lancelets: Marine, free swimming, translucent fish-like animals; very small. About 20 species. |
| Vertebrata | | | | Vertebrates: Vertebrae of spinal column replace notochord during development; brain protected by cartilage or bone. 40 000 species. |
| | Agnatha | | | Jawless fishes: Cartilage skeleton; no paired fins; two-chambered heart. About 45 species. e.g. Lamprey, hagfish. |
| | Chondrichthyes | | | Cartilaginous fishes: Skeleton of cartilage; two chambered heart. About 600 species. e.g. Sharks, rays, skates. |

| Subphylum | Class | Order | Family | Comment |
|---|---|---|---|---|
| | Osteichthyes | | | Bony fishes: Skeleton atleast partly of bone; two chambered heart. Poikilothermic. About 20 000 species. |
| | | SUBCLASS Actinopterygii | | Ray-finned fishes: Most modern fishes including salmon, flounder, seahorse, puffer-fish, eels, bream. |
| | | SUBCLASS Sarcopterygii | | Lobe-finned fish: Lungfish and coelecanth. |
| | Amphibia | | | Amphibians: Restricted to fresh-water and moist habitats; skin is moist and slimy; tadpole stage; three-chambered heart. About 2500 species. Poikilothermic. |
| | | Urodela | | Salamanders, axolotls and newts: Lizard-like with long tailed body. None are native to Australia. |
| | | Anura | | Frogs and toads: Lack a tail and have large hind limbs for jumping. |
| | | Apoda | | Caecilians: Worm-like bodies lacking limbs; weak or no eyes. None are native to Australia. |
| | Reptilia | | | Reptiles: Dry scaly skin; most with three chambered heart; eggs have leathery shell allowing survival in dry environments; poikilothermic. About 6000 species. |
| | | Crocodilia | | Crocodiles and alligators: Large; with elongated skulls and large teeth; aquatic. |

| Subphylum | Class | Order | Family | Comment |
| --- | --- | --- | --- | --- |
| | | Chelonia | | Turtles and tortoises: teeth; enclosed in protective shell. Turtles have paddle-like limbs; tortoises have clawed and often webbed feet. Australian families listed below. |
| | | | Cheloniidae | e.g. Sea-turtles. |
| | | | Dermochelyidae | e.g. Leathery turtle. |
| | | | Carettochelyidae | e.g. Pitted-shell turtle. |
| | | | Chelidae | e.g. Side-necked tortoises. |
| | | Squamata | | Lizards and snakes: Teeth present; body covered with overlapping scales. |
| | | SUBORDER Sauria | | Lizards: Most have four limbs, or traces of limbs; ar opening; tail at least as long as distance from snout to vent. |
| | | | Gekkonidae | e.g. Geckos. |
| | | | Pygopodidae | e.g. Snake-lizards, legless-lizards. |
| | | | Agamidae | e.g. Dragon lizards, frill necked lizard. |
| | | | Varanidae | e.g. Goannas or monitor lizard. |
| | | | Scincidae | e.g. Skinks. |

| Subphylum | Class | Order | Family | Comment |
|---|---|---|---|---|
| | | SUBORDER Serpentes | | Snakes: Long, thin and legless; tongue forked; often poisonous bite, except pythons which suffocate their prey; flexible joint at jaw allows the mouth to open very wide. |
| | | | Typhlopidae | e.g. Blind snakes. |
| | | | Boidae | e.g. Pythons. |
| | | | Acrochordidae | e.g. File snakes. |
| | | | Colubridae | e.g. Colubrid snakes. |
| | | | Elapidae | e.g. Elapid snakes. |
| | | | Hydrophiidae | e.g. Sea-snakes. |
| | | SUBORDER Rhynchocephalia | | Tuatar: Teeth fused to jaw (not in sockets); front of skull beak-like. One species in New Zealand. |
| | Aves | | | Birds: Homiothermic animals covered with feathers; toothless; limbs modified for flight in most types; four chambered heart. About 9000 species.e.g. Robins, kookaburra, eagle, seagull, owl, penguin, emu, cormorant, wren. |

| Subphylum | Class | Order | Family | Comment |
|---|---|---|---|---|
| | Mammalia | | | Mammals: Homoiothermic animals with at least a little hair on their bodies; females produce milk from mammary glands; four chambered heart; teeth. About 4500 species. |
| | | Monotremata (or subclass Prototheria) | | Monotremes: Egg laying; mammary glands lack nipples. e.g. Platypus, echidna. |
| | | Marsupialia (or subclass Metatheria) | | Marsupials: Young undeveloped when born; development completed in a pouch; nipples in pouch. |
| | | | Macropodidae | e.g. Kangaroos, wallabies, quokka, tree kangaroo, rat kangaroo. |
| | | | Phalangeridae | e.g. Large possums: brush-tailed and scaly tailed possums, cuscus. |
| | | | Petauridae | e.g. Ring-tailed possums, gliders, Leadbeaters possum, triped possum. |
| | | | Burramyidae | e.g. Pygmy possums, feather-tailed gliders. |
| | | | Tarsipedidae | e.g. Honey possum. |
| | | | Paramelidae | e.g. Bandicoots. |

| Subphylum | Class | Order | Family | Comment |
|---|---|---|---|---|
| | | | Vombatidae | e.g. Wombats. |
| | | | Phascolarctidae | e.g. Koala. |
| | | | Dasyuridae | e.g. Quolls, tiger-cat, Tasmanian devil, numbat, phascogales, antechinus, dunnart. |
| | | | Thylacinidae | e.g. Tasmanian wolf. |
| | | | Notoryctidae | e.g. Marsupial mole. |
| | | | Didelphidae | e.g. American opossums. |
| | | Mammals below sometimes placed in subclass Eutheria | | Placentals: Young attached to parent by a placenta when in the womb; better developed when born; mammary glands and hair. |
| | | Chiroptera | | Bats: Webbings of skin between limbs allows flying. |
| | | Insectivora | | Insectivores: Many teeth. e.g. Moles, hedgehogs, shrews. |
| | | Primates | | Primates: Eyes usually directed towards the front; nails instead of claws. |
| | | SUBORDER Prosimii | | Prosimians: Arboreal; furred tails; pointed faces with nostrils at tip. e.g. Tarsier, lemur, bush baby. |

| Subphylum | Class | Order | Family | Comment |
|---|---|---|---|---|
| | | SUBORDER Anthropoidea | | Monkeys, apes and humans: Tree- and ground-living; often no tail; snout less pointed; eyes directed more fowards. |
| | | | Celibae | New World Monkeys: Usually long prehensile tail; nostrils open partly to side; thumb and big toe opposable. e.g. Squirrel monkey. |
| | | | Callithricidae | Marmosets: Long tail is not prehensile; nostrils open foward; only big toe opposable. |
| | | | Cercopithecidae | Old World Monkeys: Tail not prehensile; nostrils open foward; thumb and big toe opposable. e.g. Baboon. |
| | | | Pongidae | Apes: No tail; nostrils close and slanted into a V shape; thumb and big toe opposable; generally larger. e.g. Gibbon, orang-utan, gorilla, chimps. |
| | | | Hominidae | Humans: No tail; face almost flat with nostrils close and facing downwards; thumb opposable but not big toe; walks upright; enlarged brain. |
| | | Edentata | | Edentates: No front teeth, although molars present in some species. e.g. Giant anteater, sloth, armadillos. |
| | | Pholidota | | Pangolins: No teeth; body enclosed in scales. |

| Subphylum | Class | Order | Family | Comment |
|---|---|---|---|---|
| | | Tubulidentata | | Aardvark: No teeth in adults. |
| | | Rodentia | | Rodents: Chisel like incisors that are constantly growing; no canines but broad molars. e.g. Rats, mice, squirrels, porcupines, gophers, guinea-pigs, capybaras, jerboas. |
| | | Lagomorpha | | Lagomorphs: Teeth similar to rodents but with four incisors in upper jaw; short tail. e.g. Rabbits, hares, pika. |
| | | Cetacea | | Cetaceans: Marine mammals with front limbs being flippers and no hind limbs. e.g. Whales, dolphins, porpoises. |
| | | Carnivora | | Carnivores: Small incisors but large canine teeth; premolars and molars sharp for shearing flesh; claws usually sharp. e.g. Wolves, foxes, jackals, dogs, hyaenas, bears, racoons, pandas, weasels, ferrets, badgers, skunks, otters, mongooses, lynxes, cats. |
| | | Pinnipedia | | Seals: All four limbs are flippers; carnivore-like teeth; marine. e.g. Sea lions, walruses, seals. |
| | | Proboscidea | | Elephants: Upper incisors very large tusks; molars replaced as they wear; upper lip is trunk. |
| | | Hydracoidea | | Coneys and hyraxes: Chisel-like incisors; no canines; four toes on front feet, three on rear. |

| Subphylum | Class | Order | Family | Comment |
|---|---|---|---|---|
| | | Sirenia | | Sea-cows: Aquatic herbivores; no hindlimbs but tail flattened, forelimbs as lippers. e.g. Manatee, dugong. |
| | | Perissodactyla | | Odd-toed ungulates: Herbivores; odd number of toes (1,3,5) in form of hooves. Horses, donkeys, tapirs, rhinoceroses. |
| | | Artiodactyla | | Even-toed ungulates: Herbivores; even number of toes (2,4) in form of hooves; often have horns and antlers; complex stomachs. e.g. Sheep, goats, gazelles, bison, cattle, buffaloes, antelopes, deer, giraffes, llamas, camels, pigs, hippopotamuses. |

Based on Australian Academy of Science, Canberra, 1990.

# Appendix 2: Geological time scale

| Era | Period | Epoch | Age from start (millions of years ago) | Life forms |
|---|---|---|---|---|
| Cainozoic | Quaternary | Recent | 0.01 | Recorded history. |
| | | Pleistocene | 2.5 | Evolution of humans. Many large mammals become extinct. |
| | Tertiary | Pliocene | 7 | Large carnivores. First human-like primates. |
| | | Miocene | 26 | Whales, apes, grazing animals. Grasslands spread as forests contract. |
| | | Oligocene | 38 | Large browsing animals. Apes appear. |
| | | Eocene | 54 | Early horses, tiny camels, giant birds. |
| | | Paleocene | 65 | First fossil primates and carnivores. |
| Mesozoic | Cretaceous | | 136 | Dinosaurs become extinct. Modern flowering plants become common. |
| | Jurassic | | 190 | Dinosaurs common. Flying reptiles. Many gymnosperms, ferns and cycads. |
| | Triassic | | 225 | First dinosaurs in forests of gymnosperms and ferns. |
| Palaeozoic | Permian | | 280 | Many reptiles and insects evolve. First seed plants. Much sea life becomes extinct. |

# Appendix 2

| Era | Period | Epoch | Age from start (millions of years ago) | Life forms |
|---|---|---|---|---|
| | Carboniferous | | 345 | Age of amphibians. First reptiles and many modern insects. Many sharks. Huge forests buried and form coal. |
| | Devonian | | 395 | Age of fish. First amphibians appear. Many shellfish and lunged fish. Coral reefs common. Land plants. |
| | Silurian | | 430 | First land plants and land invertebrates (insects). First coral reefs. |
| | Ordovician | | 500 | First vertebrates (fish). Life only in the sea and mostly marine algae and invertebrate. |
| | Cambrian | | 570 | First marine (shelled) invertebrates. Marine algae (seaweeds) the only plants. |
| Proterozoic | Pre-Cambrian | | 700 | Most living things have soft bodies. |
| Archean | | | 3500 | Earliest known fossils of bacteria and stromatolites. |

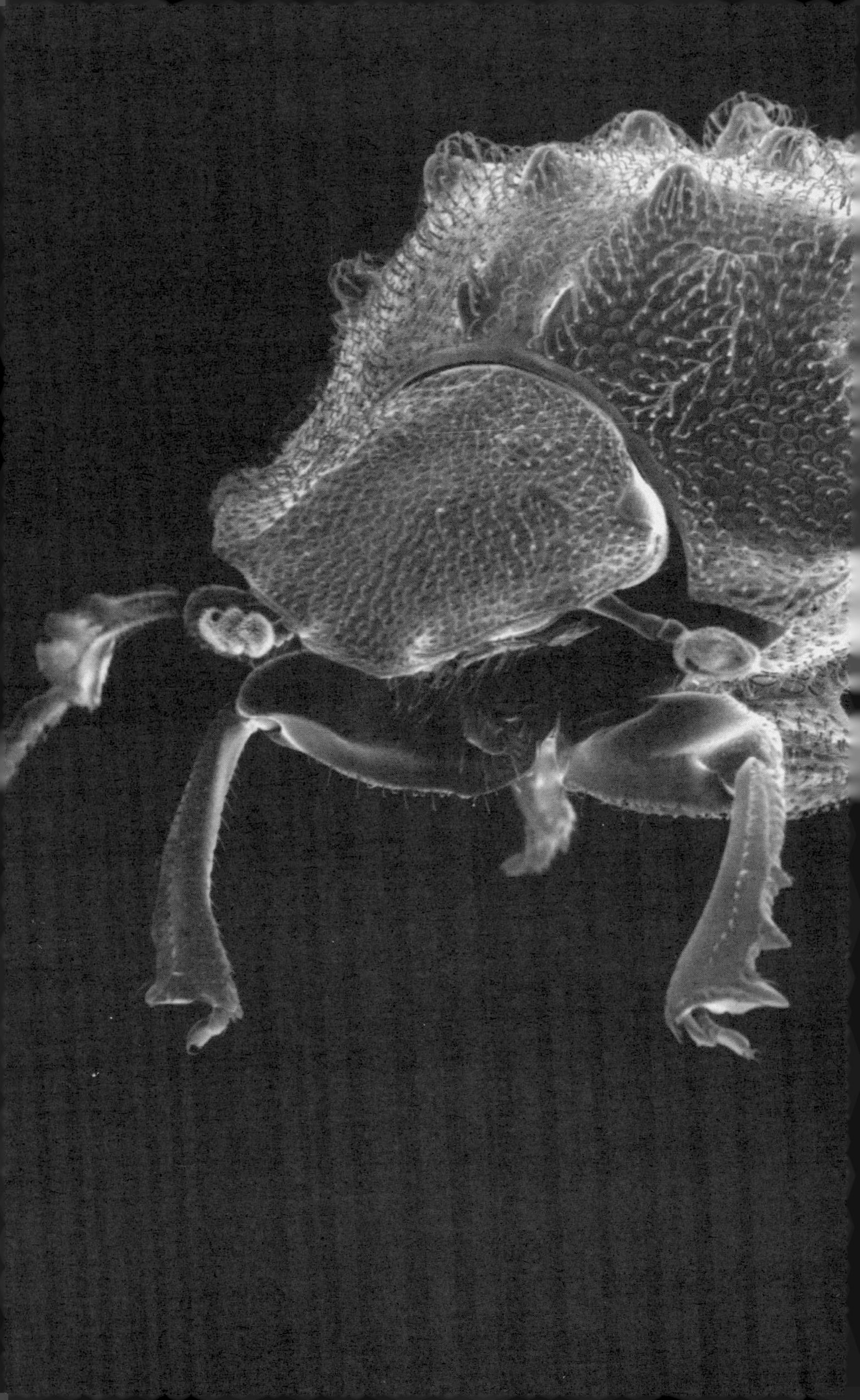